VIDEO

Digital Communication & Production

Third Edition

by

Jim Stinson
Portland, Oregon

Publisher
The Goodheart-Willcox Company, Inc.
Tinley Park, IL
www.g-w.com

The Goodheart-Willcox Company, Inc. Brand Disclaimer: Brand names, company names, and illustrations for products
and services included in this text are provided for educational purposes only and do not represent or imply endorsement or
recommendation by the author or the publisher.

The Goodheart-Willcox Company, Inc. Safety Notice: The reader is expressly advised to carefully read, understand, and
apply all safety precautions and warnings described in this book or that might also be indicated in undertaking the activities
and exercises described herein to minimize risk of personal injury or injury to others. Common sense and good judgment
should also be exercised and applied to help avoid all potential hazards. The reader should always refer to the appropriate
manufacturer's technical information, directions, and recommendations; then proceed with care to follow specific equipment
operating instructions. The reader should understand these notices and cautions are not exhaustive.

The publisher makes no warranty or representation whatsoever, either expressed or implied, including but not limited to
equipment, procedures, and applications described or referred to herein, their quality, performance, merchantability, or
fitness for a particular purpose. The publisher assumes no responsibility for any changes, errors, or omissions in this book.
The publisher specifically disclaims any liability whatsoever, including any direct, indirect, incidental, consequential, special,
or exemplary damages resulting, in whole or in part, from the reader's use or reliance upon the information, instructions,
procedures, warnings, cautions, applications, or other matter contained in this book. The publisher assumes no responsibility
for the activities of the reader.

The Goodheart-Willcox Company, Inc. Internet Disclaimer: The Internet resources and listings in this Goodheart-Willcox
Publisher product are provided solely as a convenience to you. These resources and listings were reviewed at the time of
publication to provide you with accurate, safe, and appropriate information. Goodheart-Willcox Publisher has no control over
the referenced websites and, due to the dynamic nature of the Internet, is not responsible or liable for the content, products, or
performance of links to other websites or resources. Goodheart-Willcox Publisher makes no representation, either expressed
or implied, regarding the content of these websites, and such references do not constitute an endorsement or recommendation
of the information or content presented. It is your responsibility to take all protective measures to guard against inappropriate
content, viruses, or other destructive elements.

Library of Congress Cataloging-in-Publication Data

Stinson, Jim
    Video : digital communication and production / by Jim
Stinson. -- 3rd ed.
        p. cm.

    Includes index.
    ISBN 978-1-60525-817-1
    1. Video recording. 2. Video recordings--Production and
        direction. I. Title.

T850.S7375   2012
777--dc23   2011043571

About the Author

After graduating from Harvard, Jim Stinson studied theater history at the Yale Graduate School and directing at the Yale School of Drama before transferring to the UCLA film school, where he earned the degree of Master of Fine Arts. Although he has worked on filmed commercials, TV series, and feature films, he has spent most of his career as a writer, producer, director, videographer, and/or editor of educational and corporate video programs.

In the classroom, Mr. Stinson has taught film production at Art Center College of Design, film studies at California State University, Los Angeles, and video production at La Cañada High School, La Cañada, CA.

For twelve years, he was a columnist and contributing editor at *Videomaker* magazine. In addition to *Video: Digital Communication & Production*, his published works include five novels: *Double Exposure, Low Angles, Truck Shot, TV Safe,* and *Tassy Morgan's Bluff,* as well as the study *Reconstructions of Elizabethan Public Playhouses.*

About This Book

Video: Digital Communication & Production fulfills the promise of its title by covering both the ways in which video communicates with viewers, and the methods by which it does so. Communication is featured because production by itself has no purpose. If communication were excluded, this book would be like a carpentry manual that covered sawing, drilling, and nailing, without ever explaining how to build anything.

This book treats video as a mature and independent medium, rather than merely a variant of television or a recording alternative to film. Video has become fully empowered by the digital revolution that is transforming so many aspects of twenty-first century life.

The topics in this book have been selected and organized with two groups of readers in mind: students preparing for careers in communications media and creators of personal programs who expect to make videos of professional caliber. Though the text does not pretend to include all there is to know about video, it does cover all you need to get started.

To organize this sprawling subject, *Video: Digital Communication & Production* is presented in six major sections:

- Chapters 1 and 2 help you start making videos immediately.
- Chapters 3 through 8 cover video communication—the concepts and principles behind the hardware and production techniques.
- Chapters 9 through 11 present the crucial process of preproduction—preparing to make successful programs.
- Chapters 12 through 17 introduce all major aspects of videography, lighting, and audio.
- Chapters 18 and 19 survey the art of directing—both the camera and the people it records.
- Chapters 20 through 23 explain the basics of postproduction. This edition provides greatly expanded coverage of digital editing processes and techniques.

This organization may be termed "semi-random access" because on one hand, it is possible to read only the chapters desired in any order. On the other hand, individual chapters are generally more useful in conjunction with the other chapters in the section. In most cases, larger subjects have been distributed among multiple chapters for simplicity of presentation. Since the same concepts and techniques may apply to procedures covered in different chapters, expect to find occasional duplication of material.

About This Edition

Since the previous edition of this book, the pace of innovation in video has increased to a breathtaking speed. Today…

- …tape is being supplanted as the primary video recording medium.
- …both professional and amateur still cameras have added high-definition audio/video capabilities.
- …shirt pocket cameras, mobile phones, and other personal electronic devices have put high-definition video recording in the hands of millions of casual users. These people are also using their personal devices to distribute and watch video, as well as shoot it.
- …digital cameras have grown so popular in Hollywood that cinematography journals and websites devote as much space to video as to film. And whether a program is recorded on film or video, all postproduction is now, essentially, digital.
- …digital broadcasting and widescreen high-definition TV sets have revolutionized the way people look at television.
- …many viewers now consume video on the desktop rather than in the den, with movies and TV programs available for downloading on demand, and thousands of short videos uploaded to Internet sites every day.
- …commercials inspired (or actually produced) by amateurs have been shown on network and cable TV.

Meanwhile, video hardware and software have kept pace with these developments. File-based recording systems are changing the production workflow, from preproduction through postproduction. Ultra low-power LED units are revolutionizing lighting. Editing programs are increasing their capabilities to keep pace with computer technology, while streamlining their interfaces to flatten the user learning curves.

To address these innovations and developments, we have revised every chapter in the text (substantially rewriting half of them) and have provided over 400 new or revised graphics. Most of these revisions are intended to update topics treated in earlier editions of the book. However, one completely new chapter covers script writing for several key video genres, and the chapter on analog postproduction has been sent into honorable retirement.

Once again, the author and publisher hope we have improved the book in the process of updating it. We cordially invite readers to e-mail corrections and suggestions to www.g-w.com.

As always, we hope that *Video: Digital Communication & Production* will enhance your pleasure in creating videos as much as it increases your skills.

Jim Stinson
Portland, Oregon
www.jimstinson.com

Brief Contents

Contents

CHAPTER 4 Video Space

CHAPTER 5 Video Time

CHAPTER 6 Video Composition

CHAPTER 7 Video Language

CHAPTER 11 Production Planning

CHAPTER 12 Camera Systems

CHAPTER 13 Camera Operation

CHAPTER 20 **Editing Operations**

CHAPTER 21 **Editing Principles**

CHAPTER 22 Digital Editing

CHAPTER 23 Mastering Digital Software

Acknowledgments

Dedication: For Sue, still more than ever.

The following organizations supplied photographs for the book:

ACD Systems
Adobe
Anton Bauer
Apple Inc.
Avid
Azden
Barco, Inc.
Bogen Manfrotto
Canon USA
Cinelerra.org
Corel
Edirol
Equipment Emporium, Inc.
Final Draft
Focus Enhancements, Inc.
Footage Firm
FrameForge
Gene Lester/Archive Photos
Gitzo/Bogen
Hollywood Lite

Johnny Chung Lee, http://steadycam.org
Jungle Software
JVC Corporation
Litepanels Inc.
Lowel Light Inc.
Michael Melgar
Microdolly Hollywood
Minnesota Public Radio
Movie Magic Budgeting, Creative Planet
NASA
Panasonic Corporation of North America
Photoflex
Pretec
Sekonic Corporation
Sennheiser
Sound Forge
The Tiffen Company
U.S. Fish and Wildlife Service
Ulead
Western Recreational Vehicles

Special thanks to Sue Stinson for the many photos she contributed to the book.

Many photos feature the stars of The McKinley Dramatic Arts and Vintage Tractor Society, especially, Jake Bennion, Ileana Butu, Claudia Conroy, James Conroy, Mark Conroy, Dan Dover, Dylan Kidd, Jessica Kidd, John Kidd, Garrett Nasalroad, Mauro V. Escobar, Jerry Ferree, Louanne Ferree, Tommie L. McAuley, Sam Millar, Dalan Musson, Annegret Ogden, Dunbar Ogden, Ian Panoscha, Juergen Panoscha, Mairi Panoscha, Nora Panoscha, Jennifer Schalf, J. D. Schall, Louanne Silveira, Marge Simas, Norm Simas, Alex Stinson, Anji Stinson, Lillie Stinson, Sue Stinson, Vincent Vaughn, and Alicia Wilkinson.

The fictional town of McKinley was created from shots of people, places, and events in the Humboldt County, California, cities of Arcata, Eureka, Ferndale, Fortuna, McKinleyville, and Trinidad. The annual three-day Kinetic Sculpture Race begins in Arcata and ends, more or less, in Ferndale.

After studying this chapter, you will be able to:

- Compare the advantages and drawbacks of both film and video media.
- Recognize the impact of video communication.
- Understand the nature of the video world.
- Summarize the tasks and responsibilities involved in each of the three major phases of video production.

About Video

Today, everybody is making video recordings, and they are shooting them with webcams, still cameras, and mobile phones, as well as conventional camcorders (**Figure 1-1**). We are sharing them by tape, disc, portable drive, e-mail, telephone, and Internet website. Every day we find more ways to capture and distribute video recordings.

But a simple recording is not the same as a *video*. It is only the raw material from which you can create a true video program, if you know how. This book explains how to do this.

So much video is intended for the Internet that we now say it is "published," rather than "distributed."

If you already know how to make basic video recordings, you may want to pick up a camera right now and get started. But, if you first take the time to review this chapter and the next, you will have a better idea of what true video is all about, and you will be able to make better video programs from the start.

How Video Developed

Exactly, what is "video"? The answer is not as obvious as it seems. Until a few years ago, video (as we treat it in this book) did not exist. Instead, there were just two audiovisual media: *film* and *television*.

Words in **bold and italicized type** are defined in the Technical Terms list at the end of each chapter.

Film was the medium used for creating most audiovisual programs—from movies to TV commercials. Film was (and still is) an excellent production medium for several reasons:

- Film equipment is relatively portable, so location filming is practical.
- Film's ability to reproduce quality images in black and white or color is highly refined.
- Film picture and sound tracks are usually recorded separately (on audio tape or digital storage), so sophisticated editing is possible.
- Properly archived (stored permanently) film can last a century or longer.

Television was the medium used for broadcasting studio programs "*live*" (as they happened) and other programs that were previously produced on (or copied onto) film, **Figure 1-2**. Originally, television was not an ideal production medium for several reasons:

- The equipment was heavy, complex, and tied down by cables to its control systems.
- TV's image quality was markedly lower than that of film, and its ability to render shades of gray from black to white was limited. Color could not be reproduced at all for television, except in experimental setups.

Figure 1-1 People often shoot video footage with mobile devices.

Figure 1-2 Comedians Dean Martin and Jerry Lewis in a 1950s TV studio.

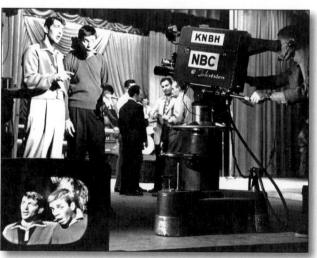

(Gene Lester/Archive Photos)

- Television could not be recorded for later editing, except by copying a TV screen display of it to film and then treating it as if it were a filmed program. These "kinescope" films degraded picture quality even further.

As the popularity of television grew over the years, equipment manufacturers gradually solved many of its problems. They miniaturized hardware until a broadcast-quality camera and its recorder could be combined in a package smaller than the size of a film camera, plus its attendant sound audio recorder. This combination *camcorder* also considerably reduced the tangle of studio cables, **Figure 1-3**. At the same time, engineers greatly improved picture sharpness and gray scale range, and developed high-quality color. They perfected videotape recording, so that the television signal could be electronically copied and edited.

Today, you can obtain professional quality high-definition video and audio from camcorders that weigh less than three pounds.

In all these ways, *digital* recording of high-definition images improved picture and sound quality of video until most viewers could not tell video from film. Today, you can produce professional-quality programs on either film or video with comparable ease and practicality. Though film has continued to evolve over the years, it is still essentially the same medium. By contrast, the electronic production technology that began as "television," has changed so

| **Figure 1-3** A compact professional camcorder. |

(Panasonic Corporation of North America)

drastically that it needs a name of its own. That is why we now call it *video*.

Video Versus Film

Some film makers still think that video is less desirable than their own medium in some ways:
- Even high-definition video, as displayed on digital TV sets, is not quite as sharp and clear as film shown on a theater screen, **Figure 1-4**. (However, when shown in theaters, digital release prints from film originals can be as high-resolution as film prints; and some digital cameras can now equal the resolution of film.)

Figure 1-4 At close range, the low resolution of traditional video is obvious compared to the visible detail in a high-definition video image.

Traditional video

High-definition video

- Video color looks subtly different in a way that is hard to describe, but easy to see, when compared to film. That is why some TV commercials are recorded on film, even though they will be seen only on video.
- For technical reasons, most video cameras are not capable of the highly selective focus techniques essential to theatrical films.
- Long-term video storage is more expensive, and truly archival storage (100-year) has not yet been achieved, except by converting the finished program to film.
- Before the digital revolution some years ago, video offered less flexibility in sound editing. It has always been common for even simple films to mix at least eight or more audio tracks. Old-fashioned "analog" sound editing was more cumbersome in video.

In other ways, however, film has been less desirable than video:

- Film is far more expensive to shoot and process into a final composite positive print. While small format video can cost as little as a dollar an hour, finished film costs hundreds of times as much. When stored on hard drives or solid-state media, video recording costs essentially nothing, after the initial investment in hardware.
- Film is somewhat less tolerant of different light levels, so it has to be supplied in several grades of light sensitivity to suit different conditions.
- Film sound recording is more cumbersome, since it is almost always captured on a separate audio recorder.
- In the pre-digital era, film color balancing was time-consuming and expensive.
- In traditional postproduction, film titles and effects, such as dissolves and double exposures, could not be added in real time and evaluated on the edited working copy. Instead, they had to be created separately in a film laboratory.
- Traditional film editing requires negative cutting, a tedious and expensive extra step. Since the original camera film is almost never used to create an edited program, a negative cutter must match it, frame-for-frame to the program's completed "work print," before viewable "release prints" can be made.

Converging Technologies

Today, however, the arguments for and against film or video are outdated. Whatever the historical weaknesses of film or video, they are now less important as the two media grow ever closer together.

Hybrid Forms

At the same time, visual hybrids are being created. For example, many commercials are shot on film, which is then immediately converted to a digital recording. The rest of the production process is pure video. In other cases, digital work prints are made from the film original, editing is completed, and then the film negative is cut to match.

Today, the original negative may not be cut at all, if the takes selected for editing have been digitized at full theatrical quality.

In many theatrical films, the special effects are created electronically. Production "workflow," as it is called, now varies considerably from one project to the next. A great many productions intended for both theaters and broadcast are now shot digitally to begin with.

The Digital Intermediate

The advantages of computer-based postproduction are so great that even movies recorded on film are usually transferred entirely to ultra-high-definition video before editing begins. The process produces a *digital intermediate (DI)* of a quality as high as that of the original film. Since color management, titles, transitions, and special effects are visible immediately, most of the drawbacks of film postproduction are eliminated. When postproduction is complete, the programs are scanned back onto film and/or encoded for digital projection, **Figure 1-5.**

Improvements in Video Sound Editing

The complete conversion to digital postproduction helps improve video recordings as well as film, especially in sound editing. Since computerized editing software can handle an almost limitless number of audio channels, multilayer video sound tracks are now the rule in professional production.

Figure 1-5 In digital postproduction, the original film is converted to video, all the postproduction operations are performed, and the finished video is converted back to film.

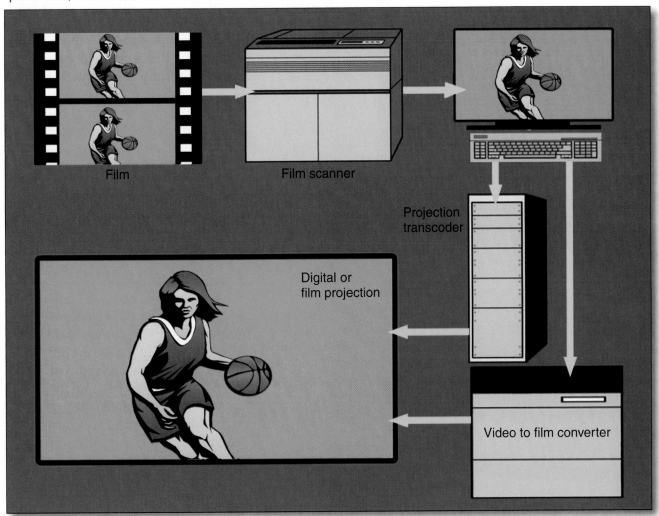

Film

Film scanner

Projection transcoder

Digital or film projection

Video to film converter

Converging Job Skills

Historically, the professional skills needed for film and television production and postproduction differed in so many ways that most people worked exclusively in one medium or the other (**Figure 1-6**). Today, however, the two media share so many procedures—most of them digital—that many people can (and do) move back and forth between them.

Of course, the procedures for actually recording on film and video remain quite different.

Most importantly, film and video communicate using the same audiovisual "language." If you master this language, you can use it fluently in either medium or in any hybrid of both.

Types of Video Production

Today, people make a wide variety of video programs, ranging from five-second commercials to thirteen-hour miniseries. These are distributed by broadcast, cable, and satellite TV, and by discs played in schools, businesses, and homes.

As noted, many feature films are recorded digitally, and video recording and editing has opened movie making to independents who could not afford the high costs of film.

The Internet is a major medium of video distribution. TV programs and feature films are available on the desktop, and seemingly numberless websites offer videos in every length and format. Using the Internet, movie rentals can be delivered "on-demand" directly to home video systems.

Figure 1-6 Video display technology has come a long way.

(Source: robococquyt)

(Panasonic Broadcast)

Some videos are produced to entertain, to persuade, or to teach. Others capture a family holiday, inventory a stamp collection, or document a vacation. Increasingly, videos are made and shown purely for self-expression.

Many computer games use video for everything from live action segments to motion control for modeling action—and new applications are constantly being developed.

Beyond the kinds of videos we think of as "programs," lie still other types—such as the specialized cinematography employed in areas as diverse as medicine, industry, science, and law enforcement. In fact, professional uses for video are expanding in much the same way as computer applications. It is probably safe to say that all but a few 21st century careers will involve video in some way.

Video Talents and Jobs

If you are considering a career in video, you have a wide range of specialties to choose from.

If you like the story side of production, try writing, directing, or editing. Though these crafts involve different skills and techniques, they are really all part of the same process. If you have graphic talents, art direction, set and costume design, and makeup are vital to sophisticated video production. You can also create postproduction video graphics and titles.

If you are intrigued by the nuts and bolts of production, then camera operation, lighting, and audio recording are skills that command respect.

If you have technical aptitude, audio and video engineering are challenging occupations. If management is your aim, video producing and production management require well-developed skills in organization, personnel, and finance.

Every type of business expertise is employed in running a video production unit or company. To function effectively, corporate managers need to know the fundamentals of video as thoroughly as their colleagues on the production side. In particular, they need to understand how the video medium communicates its message.

Above all, video is a digital medium—from the earliest production planning and story-boarding, to the finished program ready for showing. If you look toward working in computer hardware, software development, or applications, video is an exciting place to do it.

Not for Professionals Only

This book treats the art and craft of video on a professional level. Making video programs is usually a collaborative process carried on by people working in organizations—that is, by professionals.

However, you may not anticipate a career in video production. Instead, you may want to master this medium purely for personal expression. If so, you need not be concerned that video is discussed here in a professional context.

Like music, painting, photography, or cabinet making, video is an art in which "amateur" practitioners can and should develop exactly the same skills that full-time professionals use to make their living in this medium, **Figure 1-7**.

Video Communication

You may become the best camera person on your block, the finest sound recordist, or the most talented lighting director; but, you still will not know how to make a video unless you master the art of video communication.

This book includes several chapters on video communication, which cover three broad topics: the nature of the video world, the language of video expression, and the construction of video programs. (These broad topics are surveyed in Chapter 3, *Video Communication*.)

Figure 1-7 Consumer camcorders offer professional features.

(Panasonic)

The Nature of the Video World

When you watch TV, you may think you are looking at a picture of the actual world, but you are not. The TV screen is a window that looks out on a completely different universe—a strange cosmos in which the normal laws of space, time, and gravity do not work at all. In the video universe:

- An actor can open a door in London and walk through it to Los Angeles.
- A car can turn a corner and jump forward a week in time.
- An actor can fall ten stories onto concrete and walk away unhurt.

Except during the most obvious special effects, most of the strange behavior of the video universe is quite invisible to the audience. Writers, directors, designers, and editors understand how to use the laws of the video world to fool the viewer, **Figure 1-8**. These laws are so powerful that unless you understand them, you cannot effectively record a single scene or competently edit two shots together.

Chapter 4, *Video Space* and Chapter 5, *Video Time* explain the rules and regulations of the video universe and show how to take advantage of them to make effective programs.

The Language of Video Expression

Video communication uses a visual language—a language with rules much like those of a written language, such as English or Chinese. For example:

- An image is much like a single word.
- A shot is like a complete sentence.
- A scene is like a paragraph.
- A sequence is like a chapter.

Unlike their written languages, the video languages of England and China (and almost all other nations) are the same. Video is a powerful social force, partly because almost everyone on the planet speaks the universal language of film and video.

Visual Literacy

Even if your career never involves producing video, you spend considerable time *consuming* it—several hours a day, if you are a typical viewer. Video is the most persuasive and powerful system ever invented for delivering facts, ideas, and opinions. You may not think video affects *you* all that much, but it does:

- It persuades you about what to buy and how to vote.
- It explains who is important in the world and why.
- It teaches you how your community, your nation, and your planet are working, and how they *should* work, if the world were an ideal place.
- It shows you what is fashionable, what is desirable, and what you are supposed to want in life.
- It models ways to love, fight, worship, work, and dream.

Video does all this with a vividness and immediacy that make you feel that you are experiencing the actual sights and sounds of the real thing.

In fact, you are not seeing the real thing, of course. Instead, you are presented with a sophisticated and carefully contrived imitation of reality that the video makers want you to accept. They may present the "reality" that soap A is better than soap B. They may "prove" that political candidate X is superior to candidate Y. They may "demonstrate" that wolves are essential to sound ecology (or, alternatively, that wolves are livestock killers that must be exterminated).

Even when they are not trying to sell you a product, a person, or an idea, makers of video programs do not present reality. Rather, they offer their own versions of it. Video makers cannot help doing this, because even the most objective programs simply cannot be made without selecting and condensing reality. This essential process imposes a certain point-of-view on every video program, because the selection and presentation of material reflect a set of standards adopted by the program's makers.

A video documentary can show only what the camera person chooses to record.

(JVC)

What does that mean to you as a viewer? If you understand the techniques of video communication, you can separate the information you are watching from the methods used to organize and present it. At best, you will be able to get past the artificial reality on your screen to search for the actual reality beyond it. At least, you will not be deceived by a medium that looks real *always*, but is real *never*.

Understanding how video affects viewers is widely called ***visual literacy***; and several of the following chapters are devoted to it.

At the basic level, video has a grammar, with the equivalents of subjects and verbs and tenses. On a more sophisticated level, video has its own rhetoric—a wealth of techniques for creating distinctive styles of expression. These techniques include the management of composition and camera movement, the creation of visual continuity, and the control of program rhythm and pace.

Like literary styles, powerful visual styles are recognizable. Just as a knowledgeable reader can tell the work of Dickens from Hemingway, an informed viewer can identify a film by Fellini, Bergman, or Scorcese.

Figure 1-8 The laws of the video world can fool the viewer.

One side of a door is at a location.

The other side is on a set.

The Construction of Video Programs

Video communication is like written communication in yet another way—it is not enough to write a grammatically correct sentence or compose a short paragraph. You must also be able to organize and develop a coherent story or essay, or even a whole book. To make a professional video program of any length, you have to design and construct it as carefully as you would a piece of professional writing.

Effective nonfiction programs, such as training films and documentaries, depend on logical subject organization, clear presentation, and energetic pacing. Fiction, music, and performance videos also demand the skills of the artist and storyteller.

The shortest videos can be as hard to construct as the longest. A 30-, 10-, or even 5-second commercial must organize and present its material with rigorous economy, so that it builds to its final effect.

Phases of Video Production

Video production falls into three broad topics: preproduction, production, and postproduction.

Preproduction

The preproduction phase includes everything you do before actual shooting begins. Preproduction includes scripting (or storyboarding) the video (**Figure 1-9**), as well as scouting locations, gathering cast and crew, and planning production equipment and other requirements.

Preproduction may not seem as exciting as shooting or as creative as editing, but experienced professionals know that it is a mistake to short-change the preproduction phase. Video programs are generally made in small, disconnected pieces. Advance planning is essential to determine how these pieces should fit together. Also, in professional video production, time really is money. So, the more thoroughly you have planned your *shoot*, the faster you can complete it. In creating professional programs, the preproduction phase is often longer and more complicated than the production phase.

Production

The production phase covers the actual shooting of the material that will become the video program. In most projects, responsibility for the visual character, or "look," of the program is shared by the director, cinematographer, and production designer. The director also guides the performers through their roles and ensures that the program content is recorded from appropriate points-of-view.

The director is usually supported by a production management staff, while the cinematographer is assisted by a gaffer (lighting director),

Figure 1-9 Storyboarding a program is done in the preproduction phase.

A storyboard is a common alternative to a script.

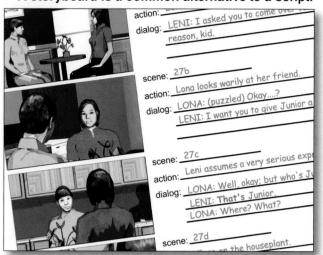

Storyboarding is often performed on computers.

(FrameForge)

Figure 1-10 Shooting a documentary program.

(Sue Stinson)

sound recordist, and key grip. *Grips* are staff members who practice many of the technical crafts associated with production.

Working in close collaboration, the production people, technicians, and performers stage and record the video footage that will become the basis of the finished program, **Figure 1-10**.

Postproduction

At the end of the production phase, many video newcomers are essentially finished. "Shoot it and show it" might be their motto. But for more experienced video makers, the postproduction phase is just as important as production, and just as enjoyable.

When you finish shooting, you do not yet have a video program, but only a collection of footage. In postproduction, you select the shots you want to include, assemble them in order, add music and sound effects to the audio, and create titles and visual effects. Collectively, this process is called *editing*, **Figure 1-11**. When practiced skillfully, editing is so creatively satisfying that some video makers think of the production phase as primarily a source of raw material for postproduction.

The director Alfred Hitchcock said he did not find the production phase very interesting because by the time it started, he had already finished making the entire movie in his head.

The Tip of the Iceberg

This introductory chapter has provided a very brief overview of video communication and production. The rest of this book is devoted to the details of this rich and fascinating subject.

You do not have to master every chapter, however, before you begin to shoot videos. Chapter 2, *Getting Started* contains the essentials you need to make short, simple programs. When you have completed that chapter, you should be able to achieve satisfying results in beginning video production.

Figure 1-11 Most of the postproduction phase is spent editing the program.

(JVC)

Workflow: The Bridge across Production Phases

In today's digital world, the components of video exist mainly in computers. In preproduction, scripts are created on specialized word processor software. Storyboards are often computer-generated. Planning tools are spreadsheets.

During production, the video and audio are identified with file names, loaded with "metadata," and recorded on digital media. Images are checked against digital profiles called "look-up tables" (L.U.T.s) to ensure consistency through postproduction.

Postproduction now takes place entirely on the desktop. Some productions may even start postproduction in the field, using laptops.

With the complete transition to digital production, old methods of identification no longer work well enough. Once, scripts had scene numbers and shots were identified by slates (which are still used, though slightly differently). The film itself had numbers pre-printed along one edge, so that original negatives could be cut to match completed work prints.

Today, the three phases of production are, increasingly, stitched together by identification systems that begin with initial scripting and continue through the completion of postproduction. These innovations are covered in Chapter 9, *Project Development*.

Summary

- The predominant technologies involved in film and video production are converging as advancements in digital processes continue. However, both film and video media have unique advantages.
- The video world presents a version of reality that is selected and condensed by video makers.
- A professional video program of any length is carefully designed and constructed.
- Video preproduction includes all the tasks and planning that must be completed before shooting begins.
- During video production, the material that will become a video program is recorded.
- In postproduction, shots are selected and assembled, music and sound effects are added to the audio, and titles and visual effects are created for the program.

Technical Terms

Camcorder: An appliance intended solely for capturing sound and motion pictures (camera), and stores them on tape, disc, or other media (recorder).

Digital intermediate (DI): Original camera film that is converted into ultra-high-definition video for use in postproduction.

Digital: To record images and sounds as numerical data, either directly in a camera or during the process of importing them to a computer.

Film: An audiovisual medium that records images on transparent plastic strips by means of photosensitive chemicals.

Grip: Production staff member who practices many of the technical crafts associated with program production.

Live: A program that is recorded and, sometimes, transmitted for display continuously, in real-time.

Shoot: To record film or video. Also, "a shoot" is an informal term for the production phase of a film or video project.

Television: Studio-based, multi-camera video that is often produced and transmitted "live."

Video: An audiovisual medium that records on a magnetic tape or digital storage media by electronic means. Also, single-camera program creation in the manner of film production, rather than studio television.

Visual literacy: The ability to evaluate the content of visual media through an understanding of the way in which it is recorded and presented.

Review Questions

Answer the following questions on a separate piece of paper. Do not write in this book.

1. List some benefits of using film as a production medium.
2. Explain how visual hybrid media is used.
3. What is a *digital intermediate*? How is it used?
4. *True or False?* In creating professional programs, the preproduction phase is often longer and more complicated than the production phase.
5. Responsibility for the "look" of a program is most often shared by which production members?
6. Identify the tasks involved in the editing process.

STEM and Academic Activities

STEM

1. Technology. Research the evolution of the video camera. How has the video camera changed in size, quality, and cost over time?
2. **Engineering.** The three phases of video production are preproduction (planning), production, and postproduction (completed product). Relate these phases to another major task. What are the steps in planning for the task? What are the steps involved in performing the task? What is done with the finished product?
3. Social Science. What types of video programs are available via the Internet? How has access to videos on the Internet changed our expectations of video quality, subject matter, and the production process?

CHAPTER 2 Getting Started

Objectives

After studying this chapter, you will be able to:

- Recall how to operate basic video equipment.
- Explain the process of recording a simple program.
- Identify steps to avoid common shooting mistakes.
- Understand how to conduct a safe and courteous shoot.

About Video Production

This chapter is designed to get you ready for successful video making as quickly as possible. Of course, you could skip this chapter and begin right away. Many new camera owners unpack their hardware and start shooting immediately. But, the results are often unsatisfying, because even the simplest video projects demand that you know at least a few fundamental ideas and techniques.

To create a good program, you need to know how to operate the equipment, how to record good quality video and audio, and how to record footage that can be edited effectively. By the time you have finished this chapter, you should be able to do these things comfortably. We will begin with a quick run-through of your camera controls and other equipment.

Equipment Basics

Cameras are unusual because they are both simple and complicated at the same time. On one hand, cameras are so simple that small children can use them without instruction. On the other hand, they are so complicated that many owners never figure out how to use all the buttons on their hardware.

This section assumes that you will begin shooting with small format equipment. Mobile phones, still cameras, and personal electronic appliances are generally simpler to use, **Figure 2-1**. On the other hand, if you are using industrial or broadcast cameras, you will need extra time to cover their more advanced features.

Basic Camera Controls

To operate a camera successfully, you need to understand four essential controls: *power*, *record*, *zoom*, and *white balance*.

Power Switch

The ***power switch*** turns the camera on and off. Many power switches include a small light to remind you when the unit is running. This information is important because a running camera uses battery power even when it is not recording.

Figure 2-1 Mobile phones are easy to shoot with.

Technically, a video *camcorder* includes both a camera and a video recorder in a single unit, while a studio video *camera* is connected by a cable to a separate recorder. In practice, however, camcorders are often referred to as "cameras," and we use that term throughout this book. ("Camera" also includes still cameras, mobile phones, and other appliances that record video.)

Some cameras include power-saving circuits that put the unit into "standby" or "sleep" mode when it has not been recording for a certain period of time. This will happen often when you are shooting, so you need to learn how to return the camera to full power. Different units are reactivated in different ways.

Record Switch

The ***record switch*** starts and stops the actual recording process. Some cameras have two record switches—one in a convenient location on the hand grip, and another on the camera body.

Many cameras also have a record switch on a wireless remote control. Also, note that you do not have to hold down the record switch continuously while shooting. Simply press it once to begin and then press it again to stop.

Zoom Control

The ***zoom control*** may be a pair of buttons or a flat bar that rocks back and forth, **Figure 2-2**. In either system, you press the forward button (or bar end) to zoom in and fill the screen with a narrower part of the scene and the rear button

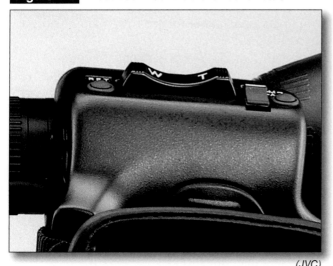

Figure 2-2 Most zoom controls are rocker bars.

(JVC)

(or bar end) to zoom out and fill the screen with a wider part of the scene. On some cameras you can *zoom* the lens at different speeds, depending on how firmly you press the zoom controls.

White Balance

The *white balance* switch matches your camera to the overall color quality of the light in which you are shooting. All cameras have an "automatic" setting for white balance. For now, use this setting to let the camera read the incoming light color and adjust itself automatically.

In some cameras, the white balance and many of the other controls are controlled via menus displayed in the viewfinder, the external viewing screen, or on a special LCD panel on the camera.

Automatic Camera Controls

White balance is not the only control that can be set to work automatically. The camera can also automate lens focus and exposure control.

Auto Focus

The *auto focus* control focuses the incoming light to keep the picture sharp and clear. (Like your eyes, the camera lens must be set one way to record distant objects and another way to record close ones.) On the automatic setting, your camera estimates the distance to the central subject in the viewfinder. Then the camera sets the lens for that estimated distance. For now, leave the auto focus enabled, even though this

may present occasional focus problems. Later chapters discuss how and when to use auto focus to best advantage.

Throughout this book, *enabled* means that a function is turned on (activated) and *disabled* means that it is turned off (deactivated).

Auto Exposure

The *auto exposure* control regulates the amount of light admitted through the camera lens. Whether shooting a candle-lit birthday or a day on bright ski slopes, the camera must have exactly the same amount of light to form a good picture. The auto exposure system delivers that precise amount. Like auto focus, the auto exposure system can be fooled sometimes. Leave it enabled for now, however.

Default Settings

A *default* is an action or condition that is automatically chosen by the equipment, unless you actively select a different one. In most cameras, the auto focus and auto exposure functions are default settings, enabled automatically when the power is turned on.

In some cameras, the white balance does not default to the automatic setting. Instead, it remains on the setting selected before the unit was last turned off. If the light color is different when you turn it on again, you may record footage with an unpleasant color cast (**Figure 2-3**). For this reason, always check the white balance before you shoot. For now, set it to automatic.

Recording Media

Videotape was once the only medium on which the camera recorded picture and sound. Today, cameras more often record on discs, hard drives, flash memory cards, or other media, though tape is still used in some consumer and a few entry-level professional cameras.

Cards and similar flash media act like computer hard drives, recording until they are full. Make sure that any material previously recorded on a card has been transferred to a computer, not because you might record over it (you cannot do that) but because you want to clear the entire storage capacity of the card for your new shooting session.

Managing Videotape

All older cameras and a few current models record on videotape. If your camera uses tape, you will need to prepare it for shooting. Here are some tips for doing so.

All camera videotape is supplied in closed plastic cassettes. Handling tape is simple because it is well-protected in its plastic armor. Store a tape in its sleeve or case when not in use, and keep it out of direct sunlight and hot car interiors. Never open the cassette's protective front flap and expose the tape inside. There is no reason to open this flap, except, on rare occasions, to inspect a damaged tape. Touching the delicate tape can degrade its recording ability.

Loading videotape can be easy or difficult, depending on the design of the camera. Some units will accept a tape, however carelessly you insert it. Others demand that you position the cassette precisely and install it with care. No camera, however, will permit you to insert a tape upside-down or backward. Most cameras accept the tape with the cassette face toward you and the tape flap on the lower (or inner) edge. One caution: unlike full-size VCRs, miniaturized cameras are somewhat delicate. You can break them by forcing a tape. If the tape does not slide in at once, make sure that it is properly oriented.

Once the tape is inside the camera, it must be prepared for use. If the tape is brand-new, roll it forward about 30 seconds. Do this by pressing the record button with the cap still covering the lens. This is done because the first few inches of tape will gradually stretch as the cassette is repeatedly rewound to its beginning, eventually ruining any material recorded there. By starting recording 30 seconds in, you can avoid this problem.

If you have used some of the tape before, you must prepare it by ***rolling down to raw stock***. This means previewing the tape to find the end of your previous footage, recording a few seconds of black (again, with the cap on the lens) and then leaving the tape at that position.

To save precious battery power, it is best to do this with the camera's AC/charger cord plugged into an outlet.

Tape preparation is more important than it may seem, because video makers usually review their footage after returning from a shoot, and they sometimes forget to roll past their final shot in preparation for their next session. Too often, precious footage is taped over and erased by the next day's work.

All-in-all, these precautions suggest why tape is being supplanted by other storage media.

An image indicates previously recorded tape.

"Snow" or solid blue indicates a blank section.

Black color indicates recording with the lens cap on or black color from computer.

Image

Snow

Black

Batteries

Cameras run on battery power, though they can use household current when available. Batteries are covered in detail in Chapter 12, *Camera Systems*. For now, you need to know how to charge them, handle them, and ensure an adequate supply of camera power.

Though batteries appear simple and tough, they do require care. Dropping them can split their cases. Leaving them outdoors in cold weather

Figure 2-3 Used in daylight, the indoor white balance makes everything look blue. Enabling the outdoor white balance records colors accurately.

Indoor white balance

Outdoor white balance

can reduce the amount of power they deliver. Most importantly, be careful not to short-circuit a battery by bridging its contact terminals. This will produce high heat levels and eventually ruin the battery.

A camera battery can usually be charged in two ways: by leaving it in the camera and plugging in the camera's power/charging cord, or by docking it directly in the camera's charging unit. Whichever method you choose, always charge a battery immediately after use, then remove it for storage. Just prior to your next use, recharge the battery again.

Finally, never go out shooting without at least one fully-charged spare battery. Though camera batteries have advertised capacities of up to an hour or more, they seldom last as long as their advertised rating. Older batteries, in particular, will stop working relatively quickly, so a spare is indispensable.

Tripod

Another essential accessory is a camera tripod. Although a skilled camera operator can shoot good quality footage by hand-holding, it is difficult to obtain steady images without the support of a tripod. Shaky pictures are the most obvious signs of amateurish production.

Many beginning video makers avoid tripods because they can be clumsy and a nuisance to manage. But if you routinely use a tripod from

the start, you will come to find it a natural part of the shooting process.

You can adjust the stiffness with which tripods move sideways and up and down. Locate these controls (**Figure 2-4**) on your particular unit, and set them so you can move the camera smoothly.

Be careful to set the vertical tension tightly enough to hold the camera level even when it is unattended.

Some tripods include quick-release mechanisms that make it easier to attach and remove the camera. Before using one of these tripods, make sure you understand how the release mechanism works.

Now that you have located the camera controls and learned about recording media, batteries, and tripods, you are ready to start

Figure 2-4 A typical tripod head.

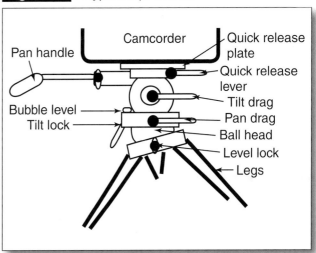

shooting. But before you do so, use the accompanying checklist to ensure that you are ready to go.

Pre-Shoot Checklist

- Battery charged and inserted in camera.
- Spare battery charged and ready for use.
- Blank recording media provided.
- White balance control set to automatic.
- Tripod included in shooting kit.

Some cameras can imprint a visible shooting date on the footage. Unless you want this special feature, make sure that the date is not visible in the viewfinder. If it is, disable the "date" function on the camera.

Camera Operation

Cameras are so simple to operate that anybody can obtain images of some sort. But if you follow just a few professional practices, you can improve your footage dramatically. The following suggestions assume that you will edit your first projects in the camera—that is, you will record each shot in order and make each one last as long as it will in the completed program. (Some of these tips do not apply when you are shooting material to edit later.)

Checking the Viewfinder

All camera viewfinders display several types of information, **Figure 2-5**. The problem is that many people look right through this information and pay no attention to it. By checking the viewfinder data before making each shot, you can avoid many common shooting problems.

Different cameras display different types of information, but most of them will show you the following, in one form or another:

- *Battery charge:* Shows roughly how much power is left in the camera battery. When it gets too low, change batteries, so that you do not lose power in the middle of a shot.
- *Time code or counter:* Shows how much footage has been shot.

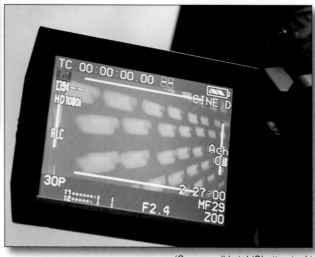

Figure 2-5 The viewfinder on different camera models display different types of information, including a time counter, battery power level, and aperture setting.

(Source: nikkytok/Shutterstock)

The term "footage" actually means the number of minutes, seconds, and frames recorded. Strictly speaking, it should apply only to tape; but no equivalent (such as "timeage," perhaps?) has been adopted for other recording media.

- *Record symbol:* Shows whether you are actually recording or only previewing the image in the viewfinder.
- *White balance:* Shows whether the camera is set for outdoor or indoor color, or whether the control is set to the automatic position.
- *Date:* Shows whether or not a date stamp will be printed over the shot.
- *Zoom:* Shows the image size, compared to the extreme wide angle setting. 2X, 3X, and 4X indicate two, three, and four times the magnification of the widest setting. Some viewfinders indicate zoom position on a graphic scale.

In addition to these common displays, some cameras show other kinds of information in the viewfinder. To interpret them, consult the instruction manual for your particular model.

Checking Camera Settings

No matter how carefully you set up your camera at your base of operations, check it again when you are ready to shoot. In particular:

- Make sure the white balance switch is set to automatic.
- Verify that the auto focus and auto exposure controls are enabled. Though they generally default to automatic, they can be accidentally changed after the camera is turned on.
- Check the other camera controls and review the information in the viewfinder to ensure that unwanted special features, like the date stamp, have not been enabled.
- Just before making each shot, ensure that the camera has not gone into standby (sleep) mode.

Using the Tripod

Here are some tips for mounting the camera on the tripod and using the tripod effectively.

- Make sure the tripod head is level by adjusting the lengths of the legs.
- Screw the camera onto the head firmly enough so it does not wobble, but not tight enough to make it difficult to remove.
- If the tripod has a center column, do not raise it unless needed for a high angle. Doing so will make the unit less stable.
- Point one tripod leg at the subject to be recorded. Doing this will let you stand close behind the camera, in the space between the other two legs.
- When you *pan* the camera (pivot it from side to side), stand facing the center of the move (**Figure 2-6**). To make the shot, twist your upper body to frame the start of the shot, then follow the camera, twisting your body the opposite way until you frame the end of the shot. By doing this, you will avoid tangling yourself in the tripod.

Hand-Holding the Camera

Because a tripod will sometimes be impractical, here are some tips for making steady hand-held shots:

- Whenever possible, brace yourself on something. Lean your elbows on a wall or table. Prop yourself up with a tree, light pole, or wall.
- To shoot a low angle, sit with the camera firmly in your lap, swing the viewfinder or viewing screen up and look down into it to make the shot.
- Do not walk while shooting if you can avoid it.
- Unless the shot is quite long, take a deep breath and let half of it out before starting, then hold your breath as you shoot.
- Use the widest angle lens setting. The wider the angle, the less obvious any camera shake will be.

Most cameras are equipped with "image stabilization," optical or electronic systems that compensate for camera shake. If your camera has image stabilization, use it by all means; but remember that shake reduction cannot compensate for negligent hand holding technique.

Figure 2-6 Start a pan with your feet aimed at the center of the pan, and twist your body as you rotate the camera.

Avoiding Camera Problems

The dos and don'ts of good camera work are covered more completely in later chapters. At this stage, just remember four tips:

- Avoid swinging the camera around to center one subject, and then another, and then another. Instead, get a good-looking picture of each subject and shoot it as a separate shot.
- Do not make shots too brief for the viewer to look at. Three to five seconds is generally a good *minimum* length for this kind of project.
- Do not pose subjects against the sky, white walls, or other backgrounds that are lighter than their faces. This will deceive the auto exposure system and result in poorly-lit pictures.
- Avoid zooming while shooting. If you want to change from a wider to a closer view, for example, shoot each angle as a separate shot.

As you will see later on, professional video makers rarely use the zoom except in studios or when covering live activities like sports.

If you follow these four tips, your footage will look technically competent, but it still may not be very pleasing. To record good-looking video and clearly audible sound, you need to remember just a few more basic ideas.

Quality Video and Audio

Your program will be judged on how it looks and sounds, so here are some tips for improving both video and audio quality.

Good-Quality Video

You can make better-looking images by using four key ideas: *head room, look room, lead room,* and *the rule of thirds.*

Head Room

Allowing proper *head room* means positioning subjects at a pleasing distance from the top of the picture. When recording people, most beginners center the subjects' heads in the frame because that is how we look at people with our own eyes. The problem is that the results look awkward on a video screen, **Figure 2-7**. For better composition, place the subject's eyes in the top third of the frame, except in wide shots.

Look Room

Look room is similar to head room. When watching people we tend to center them left-to-right as well as top-to-bottom, **Figure 2-8**. As you can see, this appears pleasing only when the subject is looking straight at the camera. If the subject is looking to one side, the composition seems constricted on that side. The remedy is to shift the subject away from the direction of the look.

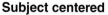

Figure 2-7   **Head room.** Centering the subject's eyes in the frame results in too much head room. Try to keep the subject's eyes in the upper third of the frame.

Subject centered

Proper head room framing

Lead Room

Lead room is look room with the subject moving, **Figure 2-9**. If you center moving subjects left-to-right, they seem about to run into the edge of the picture. By positioning subjects away from the frame edge toward which they are moving, you achieve more satisfying results.

The Rule of Thirds

The so-called *"rule of thirds"* offers the simplest way to achieve good pictorial composition. People tend to center subjects in their pictures. A tree is photographed dividing the frame vertically. The horizon is placed so it divides the image horizontally. The resulting picture looks balanced and rather dull. You might call this kind of composition "the rule of halves" because the frame is divided in half on both axes.

If you imagine a tic-tac-toe grid in front of your picture, you can divide the image into thirds instead of halves, **Figure 2-10**. The resulting compositions will be much more interesting.

An "axis" is the same in video as in graphed algebra equations. The X axis is horizontal (left-to-right) and the Y axis is vertical (top-to-bottom).

Good-Quality Audio

For your first video projects, you will probably use the camera's built-in microphone. Here are three simple tips for recording better-sounding audio.

Stay Close to the Subject

Place the camera, with its microphone (mike), as close as possible to the subject. The farther away the microphone is, the more it picks up interfering background noise. Instead of setting the lens at full telephoto and shooting from, say, 20 feet away, set it at wide angle to obtain a similar composition from a five foot distance instead, **Figure 2-11**.

Minimize Background Noise

Set up your shoot so that the camera is aimed away from major noise sources. Do not place subjects in front of a busy street because the mike will be pointed directly at the traffic noise behind them. Instead, position your subjects so that the camera mike points away from the traffic.

Figure 2-8 **Look room.** Looking forward, the subject may be centered in the frame. When looking to one side, a centered subject crowds the frame edge. Allow some space on the side toward which the subject is looking.

Looking forward/centered

Looking to one side/centered

Looking to one side/shifted

Figure 2-9 **Lead room.** Do not allow a moving subject to crowd the frame edge. Allow extra room ahead of a moving subject.

Too close to the frame edge

Proper lead room

Figure 2-10 **Rule of thirds.** A centered composition may look stiff. A composition organized on a grid of thirds is more visually interesting.

Centered

Grid of thirds

Figure 2-11 Moving the camera/microphone closer captures better quality sound.

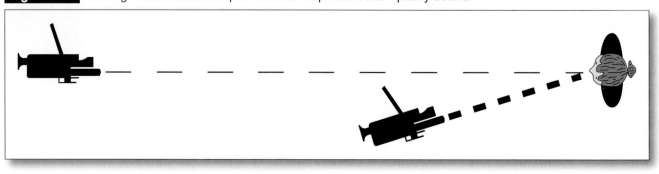

Camera Angles

When people begin making videos, they often set up shots that look very similar. These figures simply document a person walking. Because the person is nearly the same size in every shot and the camera is in nearly the same position, the resulting shots are not very interesting.

As you can see, the sequence is more interesting when the shots are more varied.

Varying subject size and camera position makes the shots more interesting.

Shots that are too similar make uninteresting video.

Notice that the shots vary in both size and camera position. Changing only the image size makes the camera appear to jump toward or away from the subject.

On the other hand, changing only the camera position produces shots that are too similar.

Only subject size changes.

Changing only the horizontal camera angle makes a jump cut.

Direct Silently

Do not give verbal instructions from the camera position while shooting. Camera mikes pick up sounds to the sides as well as in front. You can ruin sound takes because your own voice is mixed with the production sound.

The Shooting Session

Here are some simple guidelines for guiding both the performers in front of the camera and the crew behind it.

Directing the Shoot

On more advanced video projects you will be shooting to obtain raw material for later editing. On beginning projects, however, you may be editing in the camera to create a finished video in the process of shooting it.

To edit in the camera successfully, you need to observe a few rules. First, record every shot in the order in which it will appear in the finished program. If you are recording continuous action, such as somebody walking, try to leave the screen empty between shots. For example, let people walk off-screen at the end of one shot and then show them walking again at the start of the next shot, **Figure 2-12**.

Figure 2-12 The first two shots (A and B) make a jump cut. Allowing the subject to leave the frame (C and D) hides the mismatched action.

A

B

C

D

Correcting Common Mistakes

Symptom	Cause	Cure
Viewfinder blank	Power not on	Turn on power
	Battery dead	Replace battery
	Camera on standby	Reactivate camera
Camera won't record	No recording media installed	Install media
	VCR mode enabled	Enable camera mode
Colors too blue or too orange	White balance set incorrectly	Switch white balance to automatic setting
Date appears on screen	Date function enabled	Disable date function
Shaky pictures	Hand-held camera	Use tripod
Too much head room	Subject's head centered in frame	Place subject's eyes in top third of frame

The rule works as well in reverse: show people walking at the end of the outgoing shot; then have them walk into an empty screen at the start of the incoming shot. By doing this, you will not have to match the actions in the two shots.

Try to make each shot look different from the preceding one. Varied images will make your video much more interesting to watch. (See the sidebar *Camera Angles* for tips on varying image look.)

Finally, it is worth repeating that the crew should keep quiet during each shot, because the camera mike will pick up all the sounds of the crew as well as those of the cast.

Managing the Shooting Session

Shooting video can sometimes become such an intense activity that otherwise responsible people get carried away. A few reminders about video security, safety, and courtesy may help avoid this problem.

"Security" refers to the care of your expensive and somewhat delicate video equipment. To keep your hardware safe:

- Designate one person to take responsibility for the camera and tripod at all times. That person is never to leave the immediate area of the equipment.
- If you are hand-holding the camera, never set it down on anything except its tripod. It is easy to grab a camera from a table or a sidewalk and run. It is much harder and more conspicuous to snatch both camera and tripod. Also, the tripod makes the camera more visible, so it is less likely to be bumped or knocked to the ground.
- Detach the camera from the tripod when transporting them. If the camera has a handle, never use it to lift both camera and tripod together. The handle is not strong enough to carry the weight of both pieces.

- Always protect the camera from the weather. Keep it out of hot sun except when actually shooting. Do not allow it to get wet, particularly the delicate glass of the zoom lens.

Security for cast and crew is as important as it is for the equipment. Many video projects involve action sequences such as fights and car chases. To keep everyone safe:

- Do not ask people to perform feats for the camera that they would not normally attempt. Professional stunt people are highly trained, very experienced, and well paid for taking risks. Your cast is not.
- Do not put the crew at risk either (say, by hanging out over balconies or climbing on roofs) in pursuit of interesting camera angles.
- Remember that the cinematographer is concentrating intently on the viewfinder. It is easy for that person to run into objects or stumble on stairs. An assistant should guide the cinematographer during moving shots.

Finally, practice simple courtesy. You have a right to shoot outdoors in public places as long as you observe the rules of both law and good manners. Do not shoot inside (in stores or offices, for example) without permission. Though passers-by may walk through the background of your shots, do not use them as actual subjects without their permission. Some people feel uncomfortable or irritated when suddenly confronted with a camera. Respect their right to privacy.

Getting Started

By following the suggestions in this chapter you can make simple but satisfying videos. Soon, however, you will want to undertake more ambitious projects and create programs of fully professional caliber. To do this, you will use the more sophisticated production techniques covered in the rest of this book. All of these techniques have essentially the same purpose: to communicate effectively with viewers in the medium of video. For this reason, Chapters 3 through 8 cover the principles of video communication.

Summary

- To operate a camera successfully, you need to understand four essential controls: power, record, zoom, and white balance.
- In most cases, automatic camera controls (such as auto focus and auto exposure) are enabled automatically when the power is turned on.
- No matter how carefully you set up your camera at your base of operations, check it again when you are ready to shoot.
- Better-looking images can be composed by keeping four key ideas in mind: head room, look room, lead room, and the rule of thirds.
- For good quality audio, stay close to the subject, minimize background noise, and direct silently.
- Being mindful of video security, safety, and courtesy are key in successfully managing a shooting session.

Technical Terms

Auto exposure: An automatic camera control that regulates the amount of light admitted through the camera lens; may be disabled for manual operation.

Auto focus: An automatic camera control that focuses the incoming light to keep the picture sharp and clear; may be disabled for manual operation.

Default: An action or condition that is selected automatically by the equipment, but the user may change it manually.

Head room: The distance between the top of a subject's head and the upper edge of the frame.

Lead room: The distance between the subject and the edge of the frame toward which the subject is moving.

Look room: The distance between the subject and the edge of the frame toward which the subject is looking.

Pan: To pivot the camera horizontally (from side to side) on its support.

Power switch: A camera control that turns the camera on and off.

Record switch: A camera control that starts and stops the actual camera recording.

Roll to raw stock: To advance a videotape through previously recorded sections to a portion of blank tape in preparation for additional recording.

Rule of thirds: An aid to pictorial composition in the form of an imaginary tic-tac-toe grid superimposed on the image. Important picture components should be aligned with the lines and intersections of the grid.

White balance: A camera control system that neutralizes the color tints of different light sources, such as sunshine and halogen lamps, and matches the camera to the overall color quality of light in the shooting environment.

Zoom: Change the focal length of the camera lens while recording to magnify or reduce the size of an element in or a portion of a shot.

Zoom control: A camera control that zooms the camera lens in to fill the screen with a narrow portion of a scene, or zooms the camera lens out to fill the screen with a wider portion of the scene.

Review Questions

Answer the following questions on a separate piece of paper. Do not write in this book.

1. Identify the four basic camera controls and explain the function of each.
2. What is the function of auto exposure?
3. An action or condition that is automatically set by the equipment is called a(n) _____.
4. Identify some of the cautions presented related to the use and handling of camera batteries.
5. What information is typically displayed in a camera's viewfinder?
6. Explain how to pan a camera that is mounted on a tripod.
7. *True or False?* Creating proper lead room means positioning subjects at a pleasing distance from the top of the picture.
8. How is the rule of thirds used to achieve good pictorial composition?
9. Why should the camera's built-in microphone be placed as close as possible to the subject?
10. Identify the directing guidelines that should be followed to successfully edit in the camera.

STEM and Academic Activities

STEM

1. **Science.** When an image passes through a zoom lens, it is turned upside down (inverted). The human eye perceives images the same way—images formed on the retina are upside down. Research human vision and explain why we do not see everything upside down.
2. **Technology.** Create a timeline that illustrates the evolution of video recording media. Explain the technological advancement that led to the development of each different recording media type.
3. **Social Science.** Review several photographs and make a note of the portion of the image your eye is drawn to when first looking at the photo. Draw a grid representing the rule of thirds on a sheet of vellum or transparency film, and place the grid over each photo. Which quadrant of each photo do you look at first? What do your findings tell you about placement of the most important information contained in a video image?

CHAPTER 3 Video Communication

Objectives

After studying this chapter, you will be able to:

- Summarize the process of video communication.
- Explain how the video world differs from the real world.
- Recognize how video alters "truth."

The Moon Is Not Green Cheese—It Is a Tortilla

The shot on the opening page of this chapter depicts a NASA probe approaching the third moon of Tralfamador. This image is a multilayer digital *composite*.

First, a tortilla was shot against a green screen. The sun and stars were created in a paint program. Then, the sun and stars were composited with the tortilla.

Tortilla against green screen.

Sun and stars painted as background.

Tortilla matted onto background.

The space probe is a NASA photo placed on a large green background in the paint program. In video editing, digital panning "moved" the probe across the green background and zooming made it fly "farther away." Finally, the moon, sun, and stars background was composited with the moving probe.

Probe photographed on large green background "moves" by digital panning.

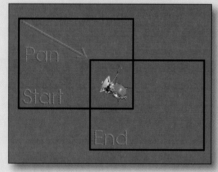

Probe moves across moon face.

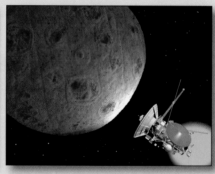

The entire process was created on a desktop without specialized professional animation software.

About Video Communication

Defined simply, *video communication is the process of recording and presenting information through moving images and related sounds.* Some video can actually be as simple as that definition suggests. For example, if you select the video function on your mobile phone, aim the unit at something, and activate it, you will record moving images and sounds. If you then send the recording to an Internet site or to another phone, you will communicate what you have recorded. Many home movies are little more than collections of such shots.

Although you can record video with many different appliances, this book calls all of them *cameras*, for simplicity.

Though these simple shots or shot collections are videos recordings, they are not *video*, as the medium is presented in this book. To make a true video, you must shape the raw recordings to communicate coherent information, ideas, and emotions. To do this effectively, you have to understand how video works; and to do *that*, in turn, you need to master the concepts that underlie everything else in the medium of video.

Chapters 4–8 examine these concepts one at a time. But since they all work together, it is helpful to introduce them that way. That is the purpose of this chapter.

Video communication is simply film communication transferred to the video medium.

A Different Universe

First, and most importantly, the screen does not show the real world. It shows the video world—a whole different universe in which space and time do not work the way they do in the actual universe. A good video presents an *illusion* of the real world by manipulating the laws of its own, strange universe. To explain this universe, Chapter 4 is about video *space* and Chapter 5 covers video *time*.

Video Construction

A video is not merely a recording of "everything" that has been shortened for convenience. It is not, as someone has said, "life with the boring parts cut out." A video is a planned, complex construction, assembled with skill and purpose from hundreds or even thousands of individual pieces. The skills required for video construction include knowledge of video *composition*, video *language*, and video *sound*. These are the subjects of Chapters 6, 7, and 8.

Video Truth

Since the video world is not actually the real world and videos are artificial constructions, video programs do not necessarily communicate the truth or even the facts that make up the truth. (In fiction programs, of course, they are not required to.) How, then, does an ethical video maker deal with the slippery concepts of "facts" and "truth?" This chapter concludes with a discussion of "truth" in the video world and honesty in presenting it.

The Video World

Because the components of the video world operate together, affecting and reinforcing one another (**Figure 3-1**), it is useful to review them as a single system of *space* and *time* constructed through *composition*, *language*, and *sound*.

Video Space

The laws of video space are aspects of two basic concepts:
- The power of the frame.
- The illusion of size and distance.

If you understand these ideas, you can manage video space with confidence.

The Laws of Video Space

For quick reference, here are the laws of video space (as they are detailed in Chapter 4):
- What is outside the frame does not exist, unless that existence is implied.
- Height and breadth are determined solely by the frame, and depth is only an illusion.
- Size, position, distance, relationship, and movement are not fixed.
- Direction is determined solely by the frame.

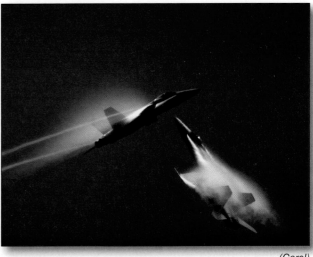

Figure 3-1 A video shot of aircraft acrobatics involves the use of space, time, composition, and sound.

(Corel)

The Power of the Frame

The *frame* around the image is the most powerful shaper of the video world because it determines what exists in that world and what does not. To use it, you:

- *Frame* (include) everything you want to exist.
- *Frame off* (exclude) everything you do not want to exist.

This means, for example, that all your production crew and equipment "do not exist" in the video world because you always keep them outside the frame. It also means that your shooting locations do not have to match the places you wish to show the audience, as long as you include the parts that do fit and exclude the parts that do not.

Suppose, for example, you need to record a sequence in the courtyard of a Spanish royal palace. By framing the parts you want and framing off the parts you do not want, you can conceal the fact that you are really in a similar courtyard of an American municipal building, **Figure 3-2**. So, when hunting locations, do not look for the places you need, but for the places in which to *create* what you need by selective framing.

Figure 3-2 A modern city hall built in the Spanish style is the shoot location. Inside, the frame excludes the modern offices, signs, and other features in the courtyard to "create" the courtyard of a Spanish royal palace.

Modern city hall building

Courtyard of a Spanish royal palace

Figure 3-3 Suggesting things exist outside the borders of the frame.

The climber could be thousands of feet up...

...instead of a few feet from the ground.

(Sue Stinson)

The frame also allows you to suggest that some things exist outside the borders of the image. It can do this in several ways. If something in the frame, such as a rock face, continues past the frame border, then viewers tend to think that it goes on for some distance. In fact, the rock may be very small (**Figure 3-3**).

Things outside the frame can also be suggested by actions. In **Figure 3-4**, the runner is poised on the starting line. When the starter's gun goes off, he sprints out of frame, suggesting that he is in a stadium, rather than merely on a small pad painted blue with white lines. (Like the rock face, the lines and blue paint that continue past the frame border also suggest that the track continues outside the frame.)

The most powerful suggestive tool is sound. Anything can be made to exist outside the frame if its characteristic sounds are laid on the audio track. The unseen starter's gun sound is typical example, **Figure 3-5**.

You can combine these techniques to reinforce their suggestive power. For instance, you could enhance your city hall courtyard sequence by starting with stock footage of a real Spanish palace, **Figure 3-6**. In the same way, a stock establishing shot can make your climbers appear to ascending the Rock of Gibraltar,

Figure 3-4 Suggesting things exist outside the frame by actions. The rest of the "track" is suggested by the action of running out of frame.

(Adobe)

Figure 3-5 Suggesting things exist outside the frame by sound. A non-existent starter gun can be suggested by sound.

(Adobe)

Figure 3-6 A stock shot of an actual Spanish palace establishes the courtyard "inside."

Stock shot

Courtyard scene

(Sue Stinson)

Figure 3-7. (You might reinforce the illusion with the dramatic sound of blowing wind.)

A *stock shot* is a piece of video obtained from a commercial library of commonly used material. For example, stock shots of New York exteriors are often inserted into situation comedies shot in Los Angeles studios, **Figure 3-8**.

In addition to deciding what exists, the frame also determines what is up and down. "Vertical" (height) is always parallel to the sides of the frame, while "horizontal" (width) is aligned with the top and bottom. By rotating the camera onto its side, you can reverse the horizontal and vertical dimensions, **Figure 3-9**. For example, if you turn the camera completely upside down, you can create human fly or zero gravity effects.

By convention, the frame can determine the compass direction, with left-pointing subjects moving "west" and right-pointing ones moving "east" (**Figure 3-10**). In the real world, camera position can create direction. For example, two sets of planes could be flying in the same direction, but if shot from opposite sides, their direction appears opposite. See **Figure 3-11**.

The frame can also affect apparent subject motion. Allowing the subject to cross the frame and leave it implies strong movement, while a subject that remains stationary with respect to the frame can seem motionless, **Figure 3-12**.

The Control of Distance

In the real world, distance is a function of the third dimension: *depth*. But, the video

---

Figure 3-7 The Rock of Gibraltar establishes the location of the rock climber.

Establishing shot of the Rock of Gibraltar

Rock climber ascending

Figure 3-8 Stock shot of a New York street.

Figure 3-9 The "wall climber."

Figure 3-10 Left-pointing planes appear to be heading west and the right-pointing planes appear to be heading east.

Left-pointing

Right-pointing

world has no *true* depth because everything is displayed on the flat surface of the screen.

In the real world, people and objects remain the same size no matter how far away they are. They only *seem* to grow smaller as they move farther from us. The video world is just the opposite: people and objects cannot move nearer or farther away because they always remain on the surface of the screen. However, they *seem* to move nearer or farther because they shrink and grow in size on that surface.

In the real world, we judge distances by the apparent sizes of people and things because we know how large they actually are. In using this ability when we watch videos, we can easily be fooled because the real-world rules of size and distance do not apply. Subjects that appear smaller are not always farther away.

Figure 3-11 From camera A's point of view, the first set of planes is flying "west" (right to left). Camera B sees the second set as flying "east" (left to right).

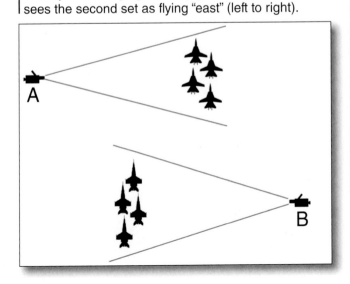

Figure 3-12 The racing boat flashes across the frame, but the space craft appears motionless.

Strong movement

Motionless

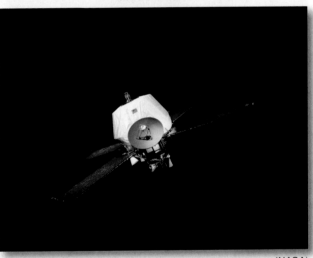

(NASA)

For instance, on movie sets, children or little people extras are sometimes used in the background to make rooms, and especially hallways, appear deeper than they are. Because we interpret them as full-height people, we judge that they are farther away.

The Management of Scale

The same characteristics that govern apparent distance in the screen world also control *scale*—the apparent size of subjects and things. On the screen, scale is provided by comparing things to other things whose size is known by the viewer. In **Figure 3-13**, the man in the foreground provides our sense of scale. In reality, the other man, who is 50 feet away, is the same size, but because we compare him the nearer man's hand, he looks like a leprechaun. If the distant man were elsewhere

in the background, he would be considered normal size. The reverse is also true. Without comparison to other objects, small models can appear full-size. Properly photographed, a model steam engine can look like the real thing. But, the illusion is destroyed when a human hand is added to create a sense of scale, **Figure 3-14**.

Notice the "worm's-eye" camera angle used in **Figure 3-14**; as low as it can get, to simulate the five-foot high eye level of a standing person.

By providing different clues about scale, you can make people and things seem to be any size you choose, from much smaller to much larger than life.

Notice that in the leprechaun shot, the hand *creates* the illusion, while in the locomotive shot, the hand *destroys* it.

Figure 3-13 The "leprechaun" effect is created by the foreground hand. Standing independently, the distant man looks normal size.

Figure 3-14 The steam engine looks real until compared to a human hand.

Confusing Scale

Sometimes, screen world scale can be confusing. Figure A shows a worker on a rock face with the sun behind him. Figure B reveals that the "worker" is actually a bronze sculpture. In the second figure, the standing human provides a sense of scale, revealing that the sculpture is more than twice life-size.

The "worker"…

...is really a sculpture.

A

B

(Sue Stinson)

But is it that big? The human provides a true sense of scale only if standing directly beneath the sculpture, at the same distance from the camera. If the image was taken with a long telephoto lens, the man may actually be many feet farther away than the statue. In that case, the sculpture would be much smaller than it appears.

To discover the truth about the sculpture image, compare the apparent size of the pickup truck in the background to the apparent size of the man, who is indeed, standing directly in line with sculpture.

Video Time

In making videos, you are constantly dealing with time, as well as space. Unlike real-world time, which never changes its characteristics, video time can be shaped in any way you choose.

Speed

You can change the speed of time within a shot by slowing it with slow motion, or speeding it up with fast motion.

Flow

You can control the way video time flows. In addition to the customary single stream of events, you can present multiple streams, either by placing one after another or by cutting back and forth among them. These streams may run parallel in time to the main story, or they may represent the past, the future, or both.

Direction

Usually, videos move forward in time—from the beginning of events to their ending. Time can also move backward when you reverse a shot in editing.

Coherence

In the video world, you can separate the video and audio, moving them around in time independently. The most common way to do this is the split edit, in which either the video moves forward in time before the audio does, or vice versa. Another common use is *voice-over* narration, **Figure 3-15**, where the off-camera speaker is commenting on visuals that happened in the past (or will happen in the future).

Figure 3-15 NARRATOR: Some tombs have remained standing for thousands of years.

(Sue Stinson)

A Christmas Story is a well-known film in which the narrator is in the present, while his story (on the screen) takes place entirely in the past.

Space, Time, and Video Language

Though space and time can be managed in individual shots—space through the control of framing and perspective, and time through slow, fast, or reverse motion—the most powerful tool for shaping the video world is *editing*.

Editing is the application of video language: the visual grammar that guides the selection and sequencing of the separate shots. In placing one shot after another, you literally create video worlds as you build sequences.

A **sequence** is an assembly of shots organized to present a single action, process, or idea.

Using the special language of video, you can compress or expand both space and time—usually both at once. By switching camera angles, you can cut out unwanted parts of actions (such as walking long distances). In the courtyard example, (**Figure 3-16**) notice that editing eliminates both the unneeded *space* in the courtyard and the unwanted *time* required to cross it.

Editing can create single locations out of two or more places that may be far apart in the real world. This happens each time a subject starts through a door at one location and finishes their entrance in a different location, or even a studio set. For example, suppose you need a scene in an historic mansion, but are not permitted to shoot inside it. Instead, you can record establishing shots outside the building, and then use a matching studio set for the interior, **Figure 3-17**.

Editing is the most powerful way to control time flow, direction, and coherence, **Figure 3-18**. Notice that the reversed shot in this jumping sequence, **Figure 3-19**, would be meaningless without the shots placed before and after it by the editor.

Parallel time streams are dependent on editing to show first a piece of one, then the other, then the first again. Sometimes, two streams are shown simultaneously, either in different parts of a multiple image or superimposed for a "double exposure" effect. See **Figure 3-20**.

Figure 3-16 A—The full, wide courtyard (not shown in the video). B—The subject takes the first few steps across it. C—A cut to the last few steps eliminates the entire walk through the center of the courtyard.

A

B

C

Figure 3-17 Creating a single location.

A location shot of this Victorian mansion...

...is matched with an interior shot in a studio.

Video Construction: The Role of Composition

Chapter 6, *Video Composition* explains that **composition** in video is the purposeful arrangement of the components of a visual image, using the principles of simplicity, order, balance, and emphasis. Good composition enhances the overall quality of the image and leads viewers' attention where the director wants it to go.

Directing the Eye

Composition directs viewers' attention to certain elements in the image by managing their *position* in the frame, *relationship* to other elements, *significance* (importance for their own sake), *contrast* (difference from other elements in the image), and *movement* of subjects. Movement is the most powerful tool for emphasizing a pictorial element. Doing this involves controlling composition continuously throughout a shot.

Figure 3-18 In the actual world, the subjects jumped backward off the wall.

Figure 3-19 The subjects jumping upward precedes the shot of the backward jump down, which is run in reverse, and ends with the subjects "landing" on the wall top.

Figure 3-20 Parallel time streams may be shown sequentially or simultaneously on a split screen.

Time stream A

Time stream B

Split screen

Managing Depth

Most importantly for creating the video world, composition controls the illusion of depth—the false third dimension on the flat screen. The techniques for suggesting depth in an image are collectively termed *perspective*.

The most common tool for managing perspective is lens focal length, **Figure 3-21**. A *wide angle* setting increases apparent depth, rendering distant subjects smaller than "normal" and making backgrounds seem more distant than they really are. A *telephoto* setting does the opposite: enlarges distant subjects, making the background seem closer. As a result, wide angle images have more than average "depth" and telephoto images have less.

In making videos, you use these lens characteristics constantly. You may select a wide angle lens setting to make a small space look bigger, to de-emphasize a distant background, or to make movement along the "Z axis" (apparent depth) more dynamic, **Figure 3-22**.

You may pick a telephoto lens setting to make subjects appear closer together than they are, to make an environment feel close and congested, or to bring a distant background closer (**Figure 3-23**).

You can also use perspective to fool the eye by changing the apparent scale of pictorial elements, like the leprechaun and the locomotive.

Perspective is discussed here because it is an aspect of composition. Perspective is also a major tool used to manage video space.

Figure 3-21 Managing perspective using lens focal length.

Wide angle lens

Telephoto lens

The Principal Tools of Perspective

Size

Overlap

Convergence

Vertical position

Sharpness

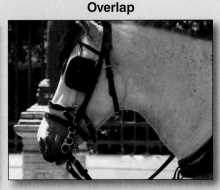

Color intensity

Video Construction: The Power of Sound

Because all the parts of the video world work together, we have already noted the role of sound. Most often, audio adds to the information presented by the visuals, but it also plays important roles in creating the video world.

Implication

As mentioned previously, sound can imply the presence of something that is outside the frame.

Reinforcement

Sound can also make a created environment feel more realistic. For example, you can make

Figure 3-22 Wide angle lens setting.

Figure 3-23 Telephoto lens setting.

(Corel)

a single table against a blank wall appear to be in a restaurant by adding running sound effects of silverware, glassware, and conversations.

Blending

Finally, sound can pull together visuals recorded at different times and/or places by running continuously under the edited shots.

In summarizing the roles of video space, time, composition, sound and editing, we see how they create each video world.

It is "each" video world because the world of every program is different—custom-made for each program during production and postproduction.

Video Organization and Video Language

Like a masonry wall, a video world is constructed of many individual pieces, each one obeying the rules of video space, time, composition, and sound. Despite these hundreds and often thousands of separate parts, the resulting programs *appear* to be seamless and unified. That is because each and every separate element has been individually *created, selected, modified,* and *sequenced* to give the illusion of continuity.

Why not simply record and present the real world unmodified—the way videos such as reality programs, news reports, and documentaries appear to do it? The fact is that these programs do *not* present the real world at all. Even the most "neutral" news report (**Figure 3-24**) or "unbiased" documentary is not a recording of reality, but someone's *version* of it.

No matter how hard people may try, it is impossible to present a truly complete or unbiased "reality" in a video because the techniques required to make it cannot avoid creating a video world, rather than capturing the real one. This is unavoidable because,

- No video can show everything, but only a few pieces.
- To make sense to viewers, those pieces must be created, selected, and sequenced according to a plan (usually in the form of a script).

Figure 3-24 A "standup" news report.

- A plan is a way of organizing material, and any organization requires a point of view—a particular way of seeing the subject matter of the program.

By the time the raw material has been fully organized for presentation, it has been shaped by *recording, selecting, modifying, **sequencing,*** and *reinforcing.* For a simple example of how video manipulates the real world, study the sidebar, *The Lighthouse.*

The following topics are fully covered in later chapters on directing and editing.

Recording

During production, every single frame of video footage is shaped by two essential decisions: what subjects to frame/what to frame off, and when to start taping/when to stop.

Framing the Subject

In the lighthouse example, the fairground light is framed by the camera, while its base and background are framed off. As a result, the fairgrounds "do not exist" in the video world, while the rest of the "lighthouse" below the frame is implied. (Review the images in the sidebar *The Lighthouse.*)

Timing the Recording

The sequence is shot at sunset, the only hour at which both the lighthouse lamp and the subjects are easily seen. The rest of the day, with its unsuitable lighting conditions, is omitted.

Selecting

As postproduction begins, the material selected for inclusion in the program is identified and processed.

Choosing Shots

Usually, the action is covered in more camera angles than needed, and each angle may have two or more takes. Everything that is *not* selected is excluded from the program, **Figure 3-25**.

A **shot** is a continuous recording, from camera start to stop. A **take** is one attempt to record a shot.

Figure 3-25 All the shots recorded are evaluated for use in the final program.

Discarded establishing shot

Selected establishing shot

Trimming Shots

Each shot is *trimmed* to length, so that all the material before the start point and after the end point is discarded.

Modifying

Digital postproduction offers powerful tools for changing the color, speed, screen direction, and framing of each shot.

Altering Shots

In our lighthouse example, both shots are darkened and tinted to simulate night.

Compositing and Computer Graphics (CG)

Elements can be added to the original footage. To intensify the seaside effect in our lighthouse example, flying gulls might be superimposed on the shots of the fairgrounds light. Elements that are created using computer graphics programs to be added to original program footage are called *CG* effects or elements.

Sequencing

Changing and shaping the real world continues as individual shots are edited together.

Condensing

Different angles are cut together to delete unnecessary space and time. In **Figure 3-26**, for example, cutting from the base of the lighthouse to the light itself eliminates a long walk up the lighthouse staircase.

Repositioning

In a finished program, shots are rarely placed in the order in which they were recorded. Sometimes, they can be used in ways different from the original intention. For example, a performer's reaction shot intended for one point in a conversation can often be used (or even reused) at another point.

Relating

Editing can bring together elements that have no real connection. "Glance-object pairs" are edits in which the subject in the first shot

The Lighthouse

In a feature film, a scene supposedly set in a working lighthouse was actually staged elsewhere. Here is how it could have been done, using just the very top of an old lighthouse that had been moved miles inland to decorate the entrance of a county fairgrounds.

The "lighthouse," as it would appear in the movie.

The light in its actual location—atop a county fair entrance.

(Sue Stinson)

To return the "lighthouse" to the "seashore," you would start with a slightly low angle that excludes most of the inappropriate background…

The light from a camera setup inside the gate.

(Sue Stinson)

A slight zoom in excludes the entrance roof from the frame.

…then frame off the fairground gate supporting the light.

To make both the performers and the lighthouse's own light visible, you would shoot the sequence near sunset. Then in postproduction, change the warm sunset light to a bluish tint that suggests twilight…

The shot changed to "twilight" in postproduction.

Ocean sound effects added (suggested here by drawings of gulls).

…lay in an audio background of waves and gulls to enhance the effect…

…and blend it with an establishing shot taken at an actual lighthouse.

The actual lighthouse makes the fairground light look real. The combined shots appear to have been made at the same location.

In this example, the real lighthouse is on the coast of Maine, 3,000 miles away from the fairgrounds. Edited together, the establishing shot made at the actual lighthouse and the following shots of the fairground light seem to exist at the same place.

Why not shoot the whole sequence at the real lighthouse? Camera placement is much easier at the fairground light, which is only 20 feet off the ground, instead of 100 feet.

The fairground lighthouse offers easier access… …than the real one.

looks off screen, and then another subject is shown in the second shot. The result makes it appear that the first subject is looking at the second. In **Figure 3-27**, the man and the sundial are actually in two completely different locations. The man's actions are parts of a single shot, with the sundial edited in.

The sundial shot is called a "cutaway" because the editor literally cuts the main shot, inserts the different one, and then cuts back again.

Reinforcing with Audio

Sounds added in postproduction both enhance individual shots and tie together whole sequences.

Sound Effects

Effects from a sound library can be laid in to synchronize with on-screen subjects or suggest things, such as gulls at the fairground lighthouse, that are not seen.

Background Tracks

Background tracks are audio tracks of continuous noise, such as city traffic, that can tie shots together and smooth out differences in their separate production sound tracks (**Figure 3-28**).

Narration

Narration relates different shots to one another by, in effect, explaining why they belong together (**Figure 3-15**).

Figure 3-26 Cutting from the base of the lighthouse (Angle A) to the light itself (Angle B) eliminates a long walk up the lighthouse staircase.

Angle A

Angle B

Figure 3-27 Glance-object pairs.

The man consults his watch.

To check it, he *glances* off screen…

…at an *object* (the sundial)…

…then looks back at his watch.

Figure 3-28 Because background noises in different camera angles do not match, they are often replaced by background tracks that run under all the edited shots.

Techniques for Different Programs

The lighthouse example was suggested by a sequence in a feature movie—an obvious work of fiction. Conscientious documentary makers, on the other hand, might insist that their programs are not manipulations of reality, but simply condensed versions of it. As we have seen, however, the moment they start selecting the material to show and the order in which to show it, they begin building a video world that is both less than and different from the real one.

Video and "Truth"

Most viewers of "realistic" programs think that the video world is the same as the real one. This is partly because they do not know or care how video worlds are constructed, and partly because the resulting "reality" is so convincing. The ability of video to present a convincing "truth" is a powerful tool and, like any tool, it can be misused. To prevent this, responsible program producers need to know how far they can go in changing reality.

Types of Programs

Is altering reality unethical? That depends on the type of program. Fiction programs, news and sports, documentaries, educational programs, and commercials all demand different standards of truthfulness.

Story Programs

Feature movies, TV series programs, and music videos are all made purely for entertainment. Viewers do not expect them to deliver facts or tell the "truth," but to present stories that are obviously invented. In programs like these, literally anything goes because viewers completely understand that they are not watching reality. In these programs, you can ethically exploit all the tricks available in creating video worlds.

Sports

Sports programs and reports are, perhaps, the closest to reality, especially when they are presented "live"—that is, in real time.

Most "live" programs, often including sports, are actually "live-on-tape." This means that the programs were recorded and assembled in real time, but presented later.

Even so, most productions have many cameras capturing images at once, and each camera is framing some subjects and framing off others (**Figure 3-29**). In the control booth, the director, at any given moment, is using one of those shots and excluding all the others. So, even the most straightforward sports program alters reality by making decisions about what and when to shoot, and what to select from the footage. The conscientious producer tries to present viewers with the most important action, as seen from the most informative angles.

However, sports footage can be easily manipulated by selecting shots that fit the producer's agenda and supporting them with editorially-weighted narration. For example, a nation's performance at the Olympic Games can be "improved" by presenting all the wins and omitting the losses, or at least by excluding their worst moments.

News

Breaking (live) news is very much like sports, but many news reports concern events to which cameras were invited—interviews (**Figure 3-30**), press conferences, and *media opportunities* set up specifically to be recorded by the cameras. Here, the people and organizations staging the events are also working to shape the "reality" presented.

Figure 3-30 Many news features are interviews.

This practice is so common that it has a name: *spin*. And, changing reality to reverse damage is "spin control."

Whether staged or spontaneous, news reports are selected and edited by channel news departments. Even the most honest ones tend to favor more sensational stories.

A newsroom phrase for this practice is, "If it bleeds, it leads." This means that the most violent or otherwise sensational events get the most prominent coverage.

In addition, TV channels, and even whole media corporations, have been known to distort reality in support of their social and political agendas.

Figure 3-29 Sports program directors have several cameras shooting at once.

(Corel)

Documentary Programs

Documentary programs aim to *document* reality, **Figure 3-31**. In their most rigorous form (sometimes called *cinema verité*), documentaries present the essence of their subjects, painstakingly culled from hundreds of hours of unstaged footage, often without the editorial comment of narration or the emotional enhancement of music. But, as noted previously, the selection process must follow a plan, and a plan represents a point of view. As a result, even the strictest documentary presents a video reality rather than the real one.

Many programs could be called *editorial documentaries* because they attempt to persuade viewers, as well as inform them. For example, one program might urge the value of forest conservation, while a rival program "proves" the importance and value of logging, with both videos using much of the same raw footage (**Figure 3-32**). Another term for these documentaries with hidden agendas is *propaganda*.

The first great editorial documentary was Leni Riefenstall's *Triumph of the Will*, made in Germany in the 1930s. Unfortunately, its editorial purpose was to celebrate Hitler and Nazism.

Figure 3-31 Animals and travel are popular documentary subjects.

Figure 3-32 Logging and anti-logging documentaries could use many of the same shots, but for opposite purposes.

(Sue Stinson)

Whether or not propaganda is ethical is a complex (and important) debate that has never been satisfactorily resolved.

Commercials

Commercials are advertising videos that use every tool in the video maker's kit to sell their products. Typically, they announce that they are *not* realistic through lavish use of computer graphics, extreme situations, and humor (**Figure 3-33**).

Commercials generally do not distort the obvious facts about their products because their manufacturers could be open to lawsuits or penalties from regulatory agencies. Nonetheless, they characteristically use video tricks, such as wide angle lenses to make cars look longer and sleeker, with roomier interiors. Colors may be intensified in postproduction, and lengthy processes (such as mowing a lawn or painting a house) may be condensed by editing or fast-motion recording until they seem to take no time at all.

Political commercials, however, often distort facts, or at least misinterpret some of them and omit others that may be inconvenient for the candidate. Even when the facts are unreliable, political commercials use every video trick available to sell their products: the candidates, **Figure 3-34**.

Figure 3-33 Commercials make extensive use of computer graphics.

(Adobe)

Figure 3-34 Political commercials work to "sell" candidates to voters.

Commercials can distort reality by hinting that you will be more stylish in a certain brand of clothing, or more important in a certain make of car. But, these distortions do not always rely on the tricks of the video world. 30- or 60-minute-long commercials disguised as programs are called **infomercials**. They are often flagged in TV schedules as "paid programming."

Infotainment

Cable channels are filled with "factual" videos that present information by frankly and openly shooting to a script and editing the footage to entertain, as well as inform. Often labeled *infotainment*, these hybrid programs combine the characteristics of documentaries with the techniques of fiction movies. At their best, they do so in order to present information more clearly and vividly; but the makers of infotainment programs never forget that the programs are designed first of all to entertain.

Most of the programs on cable channels devoted to history, science, health, and nature are infotainment.

Since they are imitating, or at least evoking, reality rather than distilling it from actual footage, these hybrid programs use many video techniques, such as:

- Shooting action that has been staged for the camera.

Figure 3-35 History programs rely heavily on period graphics, creating movement by panning and zooming the camera.

- Re-enacting action that happened in the past.
- Combining stock shots with actual footage.
- Using still photos and graphics where live action is not available, **Figure 3-35**.
- Most importantly, editing together elements that occurred at different times and places.

These techniques are covered thoroughly in Chapter 10, *Program Creation*.

The Ethical Producer

The question remains: How close to the real world should an honest producer try to make the video world? There is no simple answer to this complex question, but some guidelines can be suggested.

- In a documentary, the edited material should represent the actual subject as closely as possible.
- In an infotainment program, the essentials of the subject should be accurate and truthful, even though the material has been manipulated for presentation.
- Commercials should be honest, where facts are concerned. However, it is generally accepted that they may exaggerate the good qualities of the product being sold.
- Sports and news should be as accurate and unbiased as possible. However, it is traditional to present news features with a "slant" or "hook" to increase viewer interest.

- Fiction programs make no claim to show reality at all, so they can employ every technique in the video toolbox.

Video Communication Techniques

This introduction to the video world has summarized the topics of video space, time, composition, language, and sound. The next four chapters treat each subject in depth, beginning with video space. In the chapters that follow, expect to revisit some of the images introduced here, along with detailed discussions of the techniques that they illustrate.

The Great Persuader

What is it about video that makes it so compelling, so good at selling false realities? There are several answers, but perhaps the central one was articulated by the French critic Andre Bazin in an essay called "The Ontology of the Photographic Image." (*Ontology* is the formal study of existence, of being.) The essay was translated and published in *What is Cinema?* University of California Press, 1967.

Though Bazin's essay is concerned with the medium of film, his ideas are equally applicable to video. And, though his reasoning is extensive and subtle, it is based on a simple idea. Bazin suggests that photographs compel belief because they appear to prove the existence of their subjects.

His argument is something like this: the original photos that most people see are snapshots made with simple cameras and without manipulation by the photographer. Therefore, if something appears in a snapshot, it *must* be real or, at least, it must have been real when that photo was taken. After all, you cannot take a snapshot of something that does not exist. So, when you look at a photo of someone on a ferry boat with the Statue of Liberty behind her, you see apparent evidence of three things. At the instant when that photograph was snapped:

1. That person existed.
2. The statue existed.
3. The person stood in front of the statue.

A snapshot seems to prove that the person was at the Statue of Liberty. But, the person was composited into the original image.

Even the background image is not quite "real." Taken before 9/11/01, it originally included the twin towers of the New York World Trade Center. To update it, the towers were digitally removed.

The background as it originally appeared.

This belief in the literal truth of photographs has been reinforced by over 150 years of photography, most of it news pictures, portraits, and snapshots—all records of actual people and places. Consequently, an unconscious assumption has developed that moving photographs too (cinema and video) do not lie, just as mathematical figures do not lie.

But remember the old cliché: "Figures don't lie, but liars figure." You could say with equal truth, "Photographs don't lie, but liars photograph."

Summary

- Video space is created by use of the frame and by the illusion of size and distance.
- Video time, unlike real world time, can be shaped in any way you choose.
- Editing is the most powerful way to control time flow, direction, and coherence.
- Composition controls the illusion of depth in the video world using tools of perspective.
- Audio adds to the information presented by the visuals, but also plays important roles in creating the video world.
- Video programs do not present the real world. After raw material has been fully organized for presentation, it has been shaped by recording, selecting, modifying, sequencing, and reinforcing.
- Fiction programs, news and sports, documentaries, educational programs, and commercials all demand different standards of truthfulness.

Technical Terms

Background track: An audio track of the characteristic sounds of an environment, such as ocean, city traffic, or restaurant noises.

CG: Short for computer graphics, which are visuals created (in whole or in part) with a computer rather than recorded by a camera.

Cinema verité: A style of documentary that presents material with as little intervention by the program makers as possible.

Commercial: A very short program intended to sell a product, a person, or an idea.

Composite: A multi-layer image that digitally combines foreground subjects with different backgrounds. Creating these images is called "compositing."

Composition: The purposeful arrangement of visual elements in a frame.

Documentary: A type of nonfiction program intended to communicate information about a real-world topic.

Editorial documentary: A documentary with a specific position or point of view that attempts to persuade viewers, as well as inform them.

Frame: (1) The border around the image. (2) To *frame* something is to include it in the image by placing it inside the frame.

Frame off: To exclude something from an image by placing it outside the frame.

Infomercial: A program-length commercial masquerading as a regular program.

Infotainment: A form of documentary whose primary purpose is to entertain viewers.

Media opportunity: An event created specifically for the purpose of being covered by news organizations.

Narration: Spoken commentary on the sound track provided by a person not seen in the frame.

Parallel time streams: Two or more lines of action presented together, either by alternating pieces of the various streams or by splitting the screen for simultaneous presentation.

Perspective: The simulation of depth in two dimensional visual media.

Scale: The perception of the size of something by comparison to another object.

Sequence: A segment of a program, usually a few minutes long, consisting of related, organized material.

Sequencing: Determining the order in which individual shots are placed.

Shot: An uninterrupted recording by a video camera.

Spin: The management of information for somebody's benefit.

Stock shot: A shot purchased from a library of pre-recorded footage for use in a program; collectively called "stock footage."

Take: A single attempt to record a shot.

Telephoto: A lens setting that magnifies distant subjects and reduces apparent depth.

Trimming: Removing unwanted material from the beginning and/or end of a shot.

Voice-over: Spoken commentary on the sound track from someone who is not in the image.

Wide angle: A lens setting that reduces subjects in size and exaggerates apparent depth.

Review Questions

Answer the following questions on a separate piece of paper. Do not write in this book.

1. The _____ shapes the video world because it determines what exists in that world and what does not.
2. Explain how the frame can affect apparent subject motion.
3. How is scale depicted on a flat video screen?
4. How is a split edit used to achieve coherence?
5. *True or False?* Fiction programs, documentaries, and commercials are all held to the same standard of truthfulness.
6. What is a *sequence*?
7. Explain the role of composition in video construction.
8. A(n) _____ lens setting increases apparent depth, which makes distant subjects appear smaller than normal.
9. Explain why programs, even news reports, do not present the real world unmodified.
10. How are glance-object pairs used in editing?
11. Identify the ways in which audio is used to reinforce video.
12. What is the difference between a "live" program and a "live-on-tape" program?

STEM and Academic Activities

STEM

1. **Engineering.** Digital video effects have changed the way videos are made. What do you think the next advancement in digital video technology will be? How would this advancement affect the video industry?
2. **Language Arts.** Record a few video clips of action only (no audio). Add voice-over narration to describe the action and events in the video clips.
3. **Social Science.** Because computer-generated video footage can be manipulated to look convincingly real, courts are now suspicious of "eye witness" video. Discuss other impacts, both positive and negative, that digital video effects have had.

CHAPTER 4 Video Space

Objectives

After studying this chapter, you will be able to:

- ● Understand how to use the video frame to control what viewers see.
- ● Explain how to compose images and action for the two-dimensional screen.
- ● Recall how to create the illusion of depth in the image.
- ● Explain the impacts of exploiting scale, distance, position, and relationship in the video world.

About Video Space

The previous chapter introduced you to the concept of video space. This chapter explains the subject in greater detail, beginning with the very different laws that govern space in the video universe. You will discover how to turn these odd-seeming laws to your advantage in communicating information and feelings to your audience.

When viewers watch video, they unconsciously think they are looking through a sort of window at the events on the other side. In real life, both sides of a window belong to the same world and observe the same laws of space and time. For this reason, viewers assume the same about the world on the other side of the screen "window." However, the screen is not a window and it has no "other side." Video images are not *behind* the screen, but *on* it. That creates two fundamental differences from the real world.

First, video information is limited by the borders of the image, **Figure 4-1**. By opening a real-world window and leaning through it, you can see more of the scene outside. But no matter how close you get to a video screen, you can never see anything beyond its borders and, of course, you cannot thrust your head through it.

Second, the screen itself is flat, which means it has only two dimensions instead of three. There is no actual depth in a video or movie image. No matter how dramatically that spaceship seems to zoom away from you, it is not actually moving farther back from the screen surface, but merely growing smaller on it.

In short, space (and time, as well) work very differently in the special world on the screen. It is these crucial differences that make video possible. If you master the laws of video space, you will be able to use some of the major tools of video communication.

In Chapter 3, we introduced video space in terms of the power of the frame and the illusion of size and distance. We begin here by returning to the frame.

Figure 4-1 Screen proportions (aspect ratios).

Traditional video frame (4 to 3 / 1.33 to 1)

Wide screen video frame (16 to 9 / 1.78 to 1)

(Sue Stinson)

The Laws of Video Space

This chapter discusses the four laws of video space that were introduced briefly in the previous chapter. To help keep track of these laws as we discuss them, here they are again in summary form:

- What is outside the frame does not exist, unless that existence is implied.
- Height and breadth are determined solely by the frame, and depth is only an illusion.
- Size, position, distance, relationship, and movement are not fixed.
- Direction of movement is determined solely by the frame.

The Frame

The frame is not just a passive border around the picture. It is a powerful video communication tool. First, the frame defines and controls what the viewer sees. As we have seen, even the most honest videos show only selections from reality, and those selections are made by including them within the frame. Second, screen composition and movement are defined by reference to the frame around the image.

In video, the word **frame** has two completely different meanings: 1) the border around the screen, and 2) a single film or video picture, as in "Video is displayed at 30 *frames* per second." See **Figure 4-2**.

The Frame Controls Content

The first law of video space is perhaps the most important: *what is outside the frame does not exist* in the video world, no matter how real it is in the actual world. This is the law that makes media production possible. The soundstage or location may be packed with people, cameras, lights, and sound equipment, but because the camera operators exclude these extraneous elements from the frame, they are completely invisible to the audience. And because the audience cannot see it, it does not exist for them.

Video makers use the frame to alter reality in countless ways. Here are a few examples:

- By having a small number of spectators sit close together and framing off the many empty seats around them, you can make an audience seem bigger than it really is. Since all the empty seats are outside the frame, they do not exist.

Note that to *frame off* something is to exclude it completely from the image, while to *frame* something is not only to include it but to feature it. The command, "Frame off that tree" means keep it out of the picture, while "Frame that tree" means make it the center of interest in the picture.

Figure 4-2 A "frame" is both the border around the screen, and a single video image in a series.

- By framing off the many run-down homes in a poor neighborhood and framing only the few well-kept houses, you can make a slum look like a prosperous area. In the video world, the dilapidated homes outside the frame do not exist.
- By showing the skyscraper tops behind your actor, but framing off her feet and the flat roof on which she is standing, you can make her seem dangerously close to the edge of the building. On the screen, the roof outside the frame does not exist.

Using the frame to include and exclude information from a shot is perhaps the most universally exploited technique in video communication.

Although real-world elements outside the frame do not exist in the video world, the converse is also true: elements that do *not* exist in the real world can exist in the video world, if the video maker implies that they lie somewhere outside the frame.

For example, suppose viewers see three shots: (1) a rodeo from a spectator's point of view, (2) the spectator herself, and (3) another shot of the rodeo (**Figure 4-3**). Crowd noises and the announcer on the public address system fill the sound track across all three shots. Imagine, too, that the director has sent other people walking in front of the spectator.

In fact, the spectator never was at a rodeo, as you can see from the wider view of her in **Figure 4-4**. The passing extras and the sound effects laid under her closeup *imply* that the rest of the racetrack extends far beyond what viewers can see on the screen. Because the frame prevents viewers from seeing the reality of the desert background, they believe that the spectator is at a rodeo.

Rodeo sounds help stitch the shots together by continuing through her closeup.

To include this ability to suggest things that are not really there, we must amend our law of video space to include this idea: what is outside the frame does not exist, *unless that existence is implied.* By exploiting this law, video makers use the frame to determine what information viewers receive and how they interpret it.

The Power of the Frame

The frame that bounds the video world is a versatile tool.

The frame can conceal production. The frame isolates the subjects from the production around them.

(Photoflex)

The frame can alter reality. Though the subject seems to be high in the air…
…he is actually standing on firm ground.

The frame can create a visual composition. The frame isolates a composition within the random elements of the scene.

(Corel)

The frame can define movement. Against a featureless background…
…the car's movement is only in relation to the frame.

Figure 4-3 Rodeo background sounds fill the sound track in all three shots.

Figure 4-3 Rodeo background sounds fill the sound track in all three shots.

Spectator's perspective
of the rodeo

Spectator

Shot of the rodeo

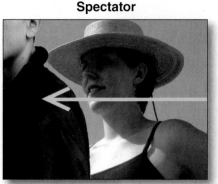

The Frame Determines Composition and Movement

Video makers also use the frame to control how information is presented. The frame is viewers' only fixed reference (though they are almost never consciously aware of it). Viewers depend on the frame to locate and organize the parts of the pictures they are watching, and to judge the speed with which things move. That is, viewers use the frame to perceive *composition* and *movement*.

Composition

Composition is the organization of the elements (parts) of a picture. These elements are arranged in relation to one another. More importantly, they are also arranged by their relationship to the frame that surrounds them. Without the defining border supplied by the

Figure 4-4 The "spectator," as actually recorded.

frame, composition would be difficult or even impossible. (Composition is covered fully in Chapter 6, *Video Composition*.)

Movement

Movement refers to motion on the screen, and viewers perceive that motion by using two sets of clues:

- How rapidly the background (such as scenery flying past a car window) changes in relation to the subject in the foreground.
- How quickly the subject moves in relation to the frame.

Since this discussion concerns the frame, we are more interested in that second clue to movement: motion in relation to the frame.

Think of the video images you have seen of satellites in orbit, as recorded from a space lab. The satellites do not move on the screen at all, but seem to float in space as quietly as a boat on a pond, **Figure 4-5**. In truth, of course, those satellites are whirling along their orbits at thousands of miles per hour. But since the cameras recording them are moving at exactly the same speed, the images of the satellites do not move *in relation to the frame*.

Now, think of the spaceships that zoom across movie and TV screens at warp speeds, spanning whole galaxies in minutes. In reality, these spaceships are often miniature models that do not move at all. They are fixed in place and the illusion of rapid motion is created by moving the camera past them, **Figure 4-6**. Making the camera fly past the stationary ship models causes them to move violently *in relation to the frame*.

Figure 4-5 A fast-moving satellite does not move in the frame.

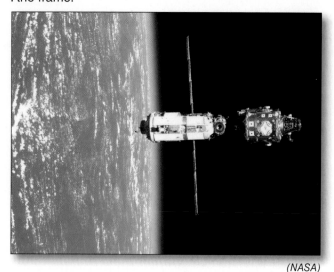

(NASA)

Figure 4-6 Plan of shooting setup: the spaceship remains fixed, while the camera moves right to left. On screen, the spaceship seems to enter the frame from the left and leaves the frame on the right.

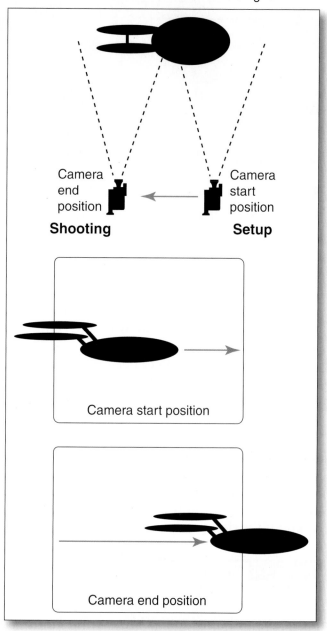

Today's spaceships are usually computer graphics, especially in programs made for television. In theatrical films, however, large models are often used as well because they can have a more realistic "feel" than even the best CG effects.

In the actual world, the satellite in space is traveling at thousands of miles per hour, while the model spaceship is not moving at all. But in the video world, the exact opposite is true: the satellite does not change position with respect to the frame, while the model is.

It is fair to say that the frame around the screen world is the video maker's most powerful single tool. The frame determines what does or does not exist in that world, how objects in that world relate to one another, and how things move in that world. Every decision made by the director, the cinematographer, the lighting director, the production designer, and even the sound recordist, is determined, first and foremost, by the characteristics of the frame.

Video Dimensions

While the frame around the screen forms the boundaries of the video world, it is the screen itself that displays that world—a strange, flat universe in which space acts differently from real-world space. In the video world, the dimension called "height" is not always vertical and the dimension called "breadth" is not always horizontal. The dimension called "depth" does not even exist. What viewers perceive as that third dimension is merely an illusion.

This expresses the second law of the video world: *height and breadth are determined solely by the frame, and depth is only an illusion.* If you understand how this second law operates, you can make it do all kinds of tricks.

Naming Screen Dimensions

In technical discussions of screen-world space, the dimensions (height, breadth, and depth) are sometimes given the names of the axes of a mathematical graph, because such a graph, like a video screen, occupies a two-dimensional space.

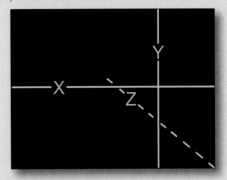

The X, Y, and Z axes of the video world.

The horizontal dimension is called the "X axis" and the vertical dimension is called the "Y axis." The apparent third dimension, at right angles to both the others, is called the "Z axis." It is indicated here by a dashed line because in video, depth is only an illusion.

Height, breadth, and depth.

Artists working in traditional two-dimensional animation have long used a different system. In calling for the animation camera to move up, down, left, or right on a drawing, they often use compass directions: north, south, east, and west.

Though all three sets of names have advantages, this book uses the familiar terms—height, breadth, and depth.

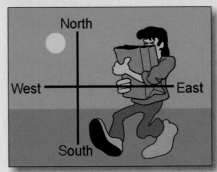

In classic animation, the screen axes are called north, south, east, and west.

Horizontal and Vertical

In the real universe (at least on Earth), height and breadth are defined by gravity. Height is the dimension parallel to the direction in which things fall, and breadth is the dimension at right angles to that direction. Newton's apple always falls in a vertical path to the horizontal ground because the orientations of height and breadth are unchangeable, **Figure 4-7**.

In the video world, however, height and breadth are not defined by gravity, but by the frame around the image. Height is the dimension parallel to the sides of the frame. Breadth is the dimension parallel to the top and bottom.

The difference is important because it means video height and breadth are determined entirely by the video maker. By tilting the camera sideways, you can make "vertical" and "horizontal" whatever directions you choose.

Playing Tricks with Video Space

Since height and breadth are not fixed in the video world, you can play all kinds of tricks with them. For example, by turning the camera 90 degrees, you can reverse the two dimensions. In the example introduced in the previous chapter, an actor standing on a walkway appears to be scaling a vertical concrete wall. The second shot shows how the trick is done.

In the second example, the subject appears to be able to do push-ups with just one finger. (In a moment, he will remove his finger from the "floor" and float in midair!) Note how the cup affixed to the wall helps sell the gag that the wall is actually a floor.

Because the camera is turned 90 degrees, the subject seems to be floating above a concrete floor.

The subject appears to be scaling a wall, but the camera has actually been turned sideways.

In movie jargon, a *gag* is a trick or stunt, and to *sell* an illusion is to include information that makes it more believable.

The Illusion of Depth

Though screen height and breadth are different from their real-world counterparts, they do exist in some form, at least, because the screen itself has height and breadth. However, the flat screen surface has no depth, and so the image on it cannot have depth either. In video, depth is purely an illusion.

In both the real and video worlds, the farther away objects are the smaller they appear. But though the appearance is the same in both worlds, the realities are exact opposites. In the actual world, receding objects really do get farther away, but they only seem to shrink in size. In the video world, receding objects only seem to get farther away (they remain on the surface of the screen), but they really do shrink in size, **Figure 4-8**.

The lack of true depth in the video world is both an advantage and a disadvantage. On the one hand, since depth is just a trick, video makers can manipulate it like any other illusion. On the other hand, since the screen is really flat, they must use a variety of compositional tricks to create the illusion of depth upon it.

Figure 4-7 In the real world, horizontal and vertical are defined by gravity.

Figure 4-7 In the real world, horizontal and vertical are defined by gravity.

So-called "3-D" movies do not present a truly three-dimensional world within the frame. If they did, viewers at opposite sides of the audience would see somewhat different views from their differing viewpoints. In fact, exactly the same perspective is seen from every seat in the house. (Holography images *can* show shifting perspective, but they are not yet practical for media.)

Spatial Relations

In the previous chapter, spatial relations in the video world were introduced as matters of size and scale. Translating these ideas into the terms of our video laws, the second law of the video world (height and breadth are determined solely by the frame, and depth is only an illusion) leads directly to the third law: *size, position, distance, relationship, and movement are not fixed*. Video space is completely fluid and is controlled by the video maker. In the actual world, objects keep the same size and position; but in video space, nothing is fixed. You can use this fact to your advantage.

Size, Position, and Distance

In the actual world, an inch is always an inch and a meter is always a meter. That means that the size of an object is always the same. The object's position also remains the same (at least until it actually moves). The distance between objects stays the same, too. None of this is true in the video world. For example:

- Two actors may escape from a full-size automobile just before it explodes. But, the car that actually blows up is only a foot-long model. An object's size can be altered as long as it looks unchanged to the viewer.
- A reporter sitting close to an interviewee in a wide shot may be **cheated** several feet farther away for a closeup (to allow more room for camera and lights). Though the reporter's position has changed, the shift is not visible to the audience (**Figure 4-9**).
- Two aircraft that actually pass each other a safe distance apart seem, on screen, to be a hair's breadth from colliding. The trick is done with a telephoto lens, which greatly reduces the apparent distance between the two objects (**Figure 4-10**).

As these examples suggest, movie makers routinely change the space in the screen world to fit the needs of the moment. Changes in size, position, and distance are all forms of changes in scale. In the video world, the apparent length of an inch or a meter or a mile can vary tremendously from one shot to another.

Figure 4-8 Though the spaceship seems to have traveled away from the viewer, its image simply shrinks on the screen surface.

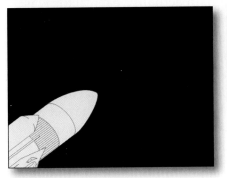
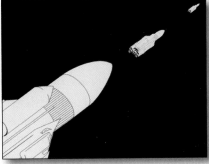
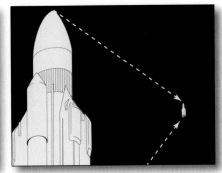

Figure 4-9 In the wide shot, the reporter is close to the interviewee. Preparing for her closeup, the reporter is moved back to allow room for lights and camera. The move is not apparent in her closeup.

Interview positions

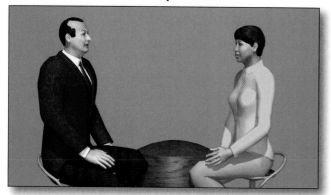

Reporter moved

Closeup

Figure 4-10 From the camera's point of view, the planes nearly collide. In reality, they are a safe distance apart.

(Corel)

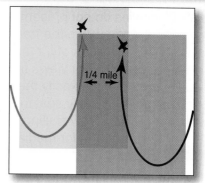

Relationships among Objects

You can control more than just the scale of screen world space. You can also control the relationships among different objects. Controlling spatial relationships usually means putting things together in the screen world that are far apart in the real world. This is so common in movie making that professionals take the process for granted. Typically, spatial relationships are

altered by editing two or more separate shots together, into what seems like a continuous action. For example:

- On screen, a character walks down a street, turns a corner, and continues up a side street. In the actual world, the main street is a location and the side street is a construction on the studio lot.

- On screen, a character climbs six flights of stairs. In actual shooting, a single flight was used repeatedly, with different floor number signs and camera angles.

- On screen, two characters carry on a conversation. In actuality, each actor was photographed separately on similar-looking sets, playing the scene with a crew member. In postproduction, the editor replaced the crew members with the absent actors.

- On screen, a character crosses a courtyard in fifteen steps (as we saw in Chapter 3). The real-world location is 50 steps wide and obstructed by a large fountain. But, the director and editor have condensed the space by means of clever camera *setups* (individual camera positions) and editing, **Figure 4-11**.

In short, screen-world relationships between unconnected places and people can be created by recording them separately and editing the different shots together.

Direction of Movement

In addition to controlling the relationships between screen-world places, video makers can control the way in which on-screen people get to those places. That is, they also control direction

in the video world. To state this fact in the form of the fourth law of video space, *the direction of movement is determined solely by the frame.*

In the real world, we express directions in compass terms (north, south, east, west), and those directions are fixed. If the street on which you live runs north and south, it *always* runs north and south. In the screen world, by contrast, compass directions are quite meaningless. Instead, direction is perceived with reference to the borders of the screen. For instance, successive shots may show a subject walking south, west, and then north. But as long as she is oriented in approximately the same way *on the screen*, the audience perceives her to be continually moving in the same direction (**Figure 4-12**).

Figure 4-11 Controlling spatial relations.

The subject starts to cross the courtyard.

The audience sees the actor complete the cross.

The central portion of the courtyard was omitted from the two shots.

Figure 4-12 Viewers perceive direction of movement in relation to the frame. A—In the real world, the subject walks south, turns east, and then turns north. B—On the screen, however, she always walks left to right.

A

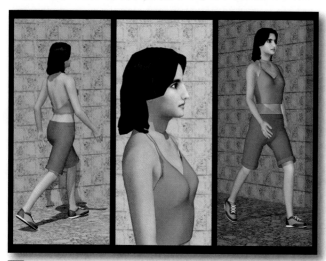

B

Figure 4-13 A plane flying from Europe to the U.S. is conventionally shown heading screen-left.

Conventional map

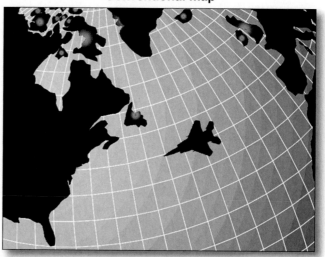

Plane headed west

(Corel)

Directions in the video world are subject to certain conventions. For instance, a plane traveling from England to the U.S. is usually shown moving from screen-right to screen-left, **Figure 4-13**. That is because conventional maps show North America on the left side of the Atlantic Ocean and Europe on the right.

If the plane flying west were photographed from the north side, it would move from screen-left to screen-right, instead. Though the resulting video would be a perfectly accurate record of the plane's movement in the actual world, it might confuse audiences watching it on the screen.

Video Communication

Today, video makers are often less concerned about screen direction than formerly, as modern audiences have learned to follow quicker, jumpier editing. This does not lessen the importance of screen direction or any of the other aspects of video space. All these "rules" and principles are communication tools, and the more completely you master them, the more effectively your videos will communicate to your audiences. As noted previously, video space and time are really aspects of the same thing. So, the next chapter takes a more thorough look at time in the video world.

Summary

* The frame defines and controls what the viewer sees.
* In the video world, height is the dimension parallel to the sides of the frame, and breadth is the dimension parallel to the top and bottom.
* The flat screen surface has no depth, so the image on it cannot have depth either—depth is purely an illusion.
* Video space is completely fluid and is controlled by the video maker.
* Direction in the video world is perceived with reference to the borders of the screen.

Technical Terms

Cheat: To move a subject from its original place (to facilitate another shot) in a way that is undetectable to the viewer.
Frame: A single film or video image.
Gag: Any effect, trick, or stunt in a movie.
Sell: To add details in order to increase the believability of a screen illusion.
Setup: A single camera position, usually including lights and microphone placements, as well.

Review Questions

Answer the following questions on a separate piece of paper. Do not write in this book.

1. Identify the four laws of video space.
2. How is the frame used as a video communication tool?
3. *True or False?* The dimension of depth is an illusion in the video world.
4. Explain the relationship between the frame and composition.
5. How do viewers perceive motion on the screen?
6. *True or False?* In the real world, objects grow smaller as they move farther away.
7. Changes in size, position, and distance are all forms of changes in _____.
8. Explain how a video maker can control the relationship between objects on screen.
9. What is the difference between screen direction and real-world direction?

STEM and Academic Activities

STEM

1. **Technology.** Research holography recording. How does this recording technique produce images that truly appear three-dimensional?
2. **Mathematics.** On a piece of graph paper, draw a graph that illustrates the X, Y, and Z axes. Label each axis with the corresponding screen dimension: height, breadth, and depth.

The model engine looks convincing in full shot.

In closeup, dust and poor detailing give it away.

CHAPTER 5

Video Time

Objectives

After studying this chapter, you will be able to:

- Explain the differences between actual time and video time.
- Understand the impact of changing video time speed.
- Identify the characteristics of the three forms of video time progression.
- Recognize the uses of changing video time direction.

About Video Time

This chapter examines the nature of video time, following Chapter 4, *Video Space*. Like video space, video time is quite different from real-world time, and operates by its own set of rules. If you master these rules, you can alter video time in your programs—cutting, pasting, stretching, squeezing, and otherwise manipulating it to suit your own purposes. You can time-travel backward, forward, and sideways, or operate in multiple time flows at once. In short, you can control video time with an ease and power that any science fiction writer might envy.

Because space and time are as closely connected in the video world as they are in the real one, they were introduced together in Chapter 3. Examining each in greater depth, however, we are considering them separately in Chapters 4 and 5 to simplify the discussion.

Not only *can* you control video time, you *must* control it. Otherwise, you could not possibly fit the actions of your program into the brief time allotted to it. And, if you did present a program in real-world time—say a two-hour hike that actually lasted two hours on screen—your audience would be profoundly bored long before your literal video ended.

The film *2001: A Space Odyssey* begins in the pre-human past and concludes in the 21st century—a real-world time period of as much as a million years.

To manage video time, you have to understand the laws by which it operates. A straightforward way to do that is by comparing the nature of video time with the laws of actual time, as we perceive it in the real world (**Figure 5-1**).

Real-World Time vs. Video-World Time

The most important characteristic of real-world time is its constancy. Time always behaves in exactly the same way with respect to its *speed, flow, direction,* and *coherence.*

We are talking about time as humans commonly experience it. How time may be viewed by philosophers or scientists lies outside our subject area.

Figure 5-1   Real-world time always flows at the same rate.

Speed

Time always passes at exactly the same speed. A second is always one second long, and a year (as the calendar counts it) lasts always one year—no more, no less. Despite our subjective feelings that time sometimes drags and sometimes flies, it actually proceeds at a rigidly fixed pace.

Flow

The pace of time is never interrupted because real-world time is always continuous. Though we may sometimes feel that time seems to stop, it actually rolls along without even the briefest pause.

Direction

Real time flows one way only, **Figure 5-2**. The future becomes the present and the present becomes the past, like a river that never passes the same spot twice. In the actual world, we can revisit the past only in memory.

Coherence

In our world, time affects everything identically, so everything moves together. Here, time is *not* like a river, in which some water drops flow along faster than others. In the world as we perceive it, everything is always at exactly the same "point in time."

At any one moment, people in different parts of the world may be at different hours of the day or even days of the week; but in the flow of time, all of us are always at exactly the same point.

Figure 5-2 Like a river, real time always flows in one direction.

Figure 5-3 Slow motion can make a hummingbird's wings visible.

(Credit: Dean E. Biggins/U.S. Fish and Wildlife Service)

Time in the video world, however, is the complete opposite of real-world time because nothing about it is constant. Its speed, flow, direction, and coherence are always controlled and continually changed by the director and editor.

Video Time Speed

You can control the speed at which time passes in your program, expanding an eye blink or compressing a hundred-yard stroll, so that each event fills exactly the same amount of screen time.

Screen time is the length of real-world time in which a video program (or a designated piece of it) is actually displayed on-screen.

There are two ways to change time speed:
- Alter the *actual* speed within individual shots.
- Alter the *apparent* speed of the action in editing different shots together.

The two methods of varying time speed have markedly different effects on the viewer.

Altering Actual Speed within a Shot

Altering speed within a shot is achieved by recording the shot at a slower or faster rate than normal. The result is an obvious change in the speed of time on the screen. For example,
- In slow motion, the lovers float lazily toward each other across a flowery field;

the whirring wings of a hummingbird seem to flap as deliberately as a buzzard's, **Figure 5-3**.
- In fast motion, the clowns skitter around the screen like doodlebugs; the flower's night-long opening unfolds in two brief seconds.

Fast, Slow, and Normal Speed

The rate at which video is recorded or displayed is called the ***frame rate***. In North America, video is displayed at 30 frames per second (fps).

For technical reasons, the actual frame rate is 29.97 fps.

So, material recorded at 15 fps and displayed at the standard 30 fps is seen in half the normal length of time, and the result is ***fast motion***. In the same way, material recorded at 60 fps and displayed at the normal 30 fps results in ***slow motion***. These are only examples. In fact, fast motion recordings can be made with frame rates as slow as, perhaps, one per hour, and slow motion shots can be recorded at up to several thousand fps.

Depending on the medium (video or film) and the location (such as the U.S. or the UK), the "normal" (standard) display rate may be 24, 25, or 30 fps. Fortunately, the shift to digital postproduction has simplified conversions from one standard to another.

Slow and fast motion can also be created in postproduction by changing the frame rate digitally.

As these examples suggest, different degrees of slow or fast motion serve different purposes.

Moderate slow motion imparts a dream-like effect. The feeling may be pleasant or romantic, as with the loping lovers; it may be nightmarish, as when a victim falls sedately toward certain death—and slowly falls and falls and falls. The mood may be nostalgic, as when an ex-football star recalls the 80-yard touchdown run that was the highlight of his career.

More pronounced slow motion has a different effect: it allows us to see otherwise invisible actions. It can divide a horseshoe pitch into many separate parts for analysis, or reveal the complex movement of the hummingbird's wings. Extreme slow motion lets us watch a droplet rebound from the water surface, **Figure 5-4**.

With fast motion, the differences are similar, though of course the effect is the opposite. Moderate fast motion alters the mood, usually making an ordinary action look funny. Silent comedy chases were often filmed in moderate fast motion, to heighten the comic effect.

More pronounced fast motion condenses action to make its overall shape or pattern easier to see. It can show us a sky full of small clouds boiling up into a full-fledged thunderhead right before our eyes. Extremely fast motion goes further by shortening actions that are so long that we cannot normally see them at all, like the passage of the moon across the sky, **Figure 5-5**, or the blooming of a flower. In its most extreme

Figure 5-5 Extreme slow motion can condense hours into seconds.

(Sue Stinson)

form, fast motion can compress whole months into a few seconds, as when the audience sees an office building go from vacant lot to finished structure in what seems like a single shot.

Popular editing software includes stop-motion commands that allow you to create time-lapse photography by automatically selecting, say, one frame for every minute of source footage—speeding up the motion 1800 times.

In short, slow motion expands the screen time during which an action happens, and fast motion contracts it. In both cases, the effect is obvious to the audience. When you watch a swinging baseball bat float majestically across the screen or the moon leap up in the sky, you realize that time has been stretched or condensed, and you know that the appearance of the action is not realistic.

Filmed model shots, such as miniature trains crashing, are generally recorded in slow motion to make the action look more realistic, while car chases are often shot in mild fast motion to increase the apparent speed.

Both slow and fast motion are used extensively in educational video programs, where their ability to present actions for analysis is valuable. They are also employed extensively in documentaries, to recreate past events or to shorten lengthy processes.

Figure 5-4 Specialized cameras can take thousands of frames per second.

(Michael Melgar)

In fiction programs, the movie makers usually want you to believe that the story is real. Since slow or fast motion shots remind you that you are watching a video or film, these effects tend to be used sparingly in fiction programs.

Fiction and nonfiction programs alike rely mainly on a second method of controlling video time. This method is editing.

Altering Apparent Speed in Editing

Slow and fast motion techniques are always visible because they vary time speed within each shot. In controlling time through editing, however, the action within every shot unrolls at a perfectly normal, realistic pace. It is *between* shots that editors expand or condense the passage of time. Because the manipulations occur between shots instead of within them, they are usually invisible to the viewer. Editors control video time speed in two ways: they slow time down by *overlapping* action, and speed it up by *omitting* action.

Overlapping and Omitting

To slow an action down, editors show parts of it more than once. For example, suppose you want to dramatize the feelings of a character who is wearily descending a flight of stairs. To do this, the director first covers the action from above, and then again from below.

Suppose that the staircase has 15 steps. To extend the time required to descend it, the editor shows the actor walking down steps 1 through 10 from the rear, and then cuts to the front angle. Here is the trick: even though the first shot ends on step 10, the second shot starts back up on step 5. Because the shots have such different points of view, the audience does not notice the overlap and the editor has extended the action by five steps (**Figure 5-6**).

To speed an action up, editors use the opposite technique: they cut out repetitive or otherwise uninteresting action. Imagine the same staircase scene, only this time the character is jogging briskly down the steps. To enhance the impetuous feeling of the scene, the director uses exactly the same camera placements as before, but this time the editor ends the first shot with the actor on step 5 and starts the second shot on step 10. This shortens the time required to descend the stairs and, as before, the audience does not notice that the

Cutting Out vs. Speeding Up

In the previous chapter, we showed how space can be shortened by editing (Figures A and B), and, of course, time can be condensed by the same process. In addition, it is now common to see shots that begin and end at normal speed, but are accelerated in the middle (Figure C). Here, our courtyard scene is shortened by each of these two methods.

Screen time can be shortened by editing… …or by speeding up part of the action.

Figure A

Figure B

Figure C

Which method is best? Removing the unwanted space and time by editing is more nearly invisible to the audience. However, where the space itself is important, speeding up the shot preserves the grand scale of the courtyard, while shortening the time required to cross it.

Figure 5-6 Extending action by overlapping two opposite angles.

Overlapping Action

Shot One

step 1

step 10

Shot Two

step 5

step 15

character has mysteriously skipped steps 5 through 9 (**Figure 5-7**).

In both speeding the action up and slowing it down, the key is invisibility. The editor is able to compress or expand video time by adjusting it *between shots*, where the audience cannot see what is happening.

Video Time Flow

Omitting or overlapping action changes not only the speed of video time, but its flow as well. Real-world time flows continuously, without so much as a moment repeated or cut. But, as we have seen, time flow in the very different world of video is constantly interrupted to control its speed. We adjust video time flow for other

purposes, too. To see how this works, we need to consider the three main forms of video time progression: *serial, parallel,* and *disjointed*.

Serial Time Flow

Up to this point, we have been discussing time flow in *serial* form: a series of events moving forward in a single stream. For example, a subject first showers, then gets dressed, then eats breakfast, then leaves for work. This serial flow of actions is the most common depiction of time in video, because that is how we perceive time in real life.

Single-stream time flow can be compressed to an amazing degree. A story can span three or four generations in two hours. A documentary video may cover hundreds of years of history.

Figure 5-7 Condensing action by shortening opposite angles.

Omitting Action

Shot One

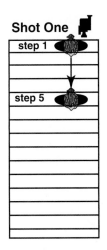

step 1	
step 5	

Shot Two

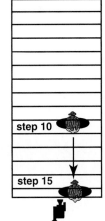

step 10	
step 15	

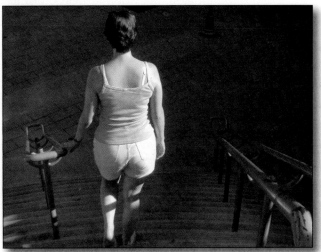

Despite its many advantages, however, a strictly serial time flow is limiting because you can show only one set of related actions. When you want to show two or more sequences of events at a time, you need to establish multiple streams of video time, flowing parallel to one another.

Parallel Time Flow

In the video world, we often see two or more sequences of action that are happening at the same time. This kind of time progression is called *parallel time flow*. To use a melodramatic cliché as an example, we might see the hero hanging from a railing around the roof of a building, while the heroine races up an endless stairwell to rescue him (**Figure 5-8**). The editor

switches back and forth among these parallel sequences in what is called *cross-cutting*. The audience never notices that in watching the movie, they are seeing the same moments of time twice-over (though in two different locations).

Cross-cutting is so common in movie storytelling that the silent movie title, "Meanwhile, back at the ranch..." has become a cliché.

In our example, video time is flowing in just two parallel streams. But, skillful directors and editors can keep three, four, or even more parallel time streams going at once.

The record for parallel cutting may be held by the film *The Longest Day*, which follows several people at once through the D-Day invasion of Normandy in World War II.

Altering Video Time to Edit Performance

Editors routinely adjust video time in order to fine-tune the performances of actors. To see how this works, imagine a romantic comedy in which a young woman is taking her boyfriend home to meet her parents, who believe their daughter is marrying a wealthy man. In two successive shots she says, "Don't forget: you're worth 50 million dollars," and he replies, "I'm not worth 50 dollars."

In the first version, his line follows hers immediately.

WOMAN: Don't forget: you're worth 50 million dollars.

MAN: I'm not worth 50 dollars.

In the second version, the editor has begun the boyfriend's shot about 1.5 seconds before he speaks.

WOMAN: Don't forget: you're worth 50 million dollars.

(No dialog)

MAN: I'm not worth 50 dollars.

Though the actor is simply looking at his partner, the added time makes him appear to be thinking about the hazards of this lie.

Disjointed Time Flow

If serial and parallel time flow are alternate methods of presenting sequences on the screen, *disjointed time flow* describes the way those sequences are usually recorded in production. Most programs are shot out of chronological order, and then reorganized into proper sequence during the editing process.

To save time and money, a production will typically shoot all the scenes that take place at one location (or on one set) at the same time, regardless of their positions in the edited program. For example, imagine that we see the parents of a naive young woman as they put her on a train for the big city. Four years later, in story time, they return to the same station to welcome back their now successful and sophisticated daughter. Typically, those scenes will be shot with different costumes and props, back-to-back in real-world time, often on the same day, to avoid the expense of a second trip back to the station location.

Figure 5-8 Two sequences take place simultaneously—the hero hangs from the railing, as the heroine races to save him.

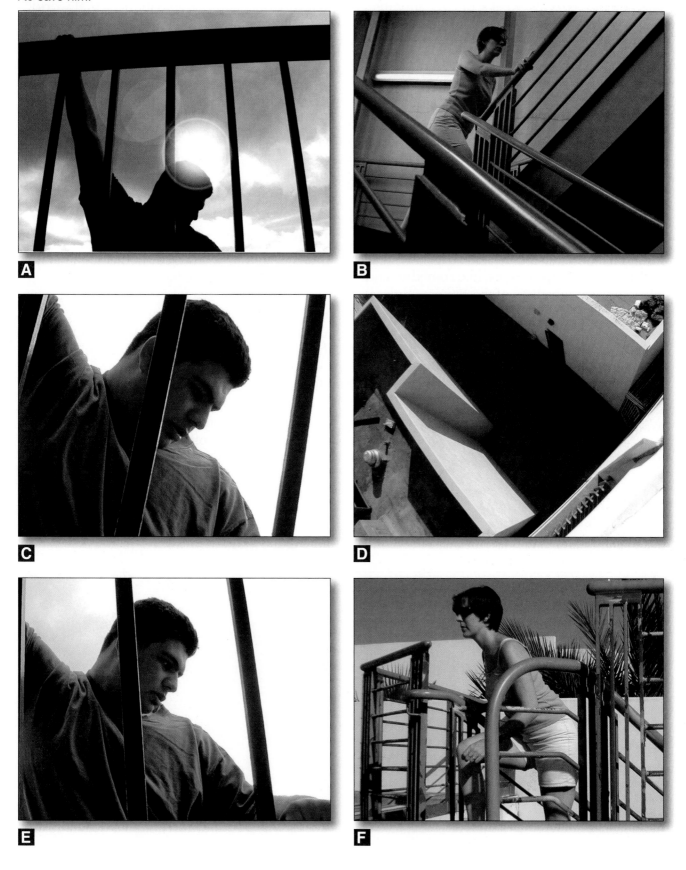

On the other hand, many sequences that seem continuous on the screen are shot in separate pieces, days, weeks, or even months apart. The most common example involves moving from exteriors to interiors and back. Often, the outside of the door is a real door on location, and the inside is part of a set. Though the actor's passage from outside to inside seems continuous in screen time, the two shots of that passage were made at very different points in real time (**Figure 5-9**).

In fact, *every* seemingly continuous video sequence is actually built from individual shots made at different points in real-world time. For example, a dialogue sequence that plays for two minutes on the screen may take a whole production day to shoot. (This is not true of multi-camera TV programs, like comedies and talk shows, in which all of the camera angles are recorded simultaneously by multiple cameras.) Different points in real-world time can even be combined in single video shots, as when actors play multiple roles via double exposure or are composited into backgrounds photographed at other times.

As long ago as 1921, Buster Keaton populated an entire theater with copies of himself in *The Playhouse*. At one point, he played nine different actors in the same shot!

Whether video time is manipulated to adjust pacing, to show parallel actions simultaneously, or to combine multiple real-world moments into single video time sequences, its flow is routinely stopped, started, and moved around in a way that is impossible in the actual world.

Video Time Direction

In the real world, we perceive time as moving always in one direction, as the future becomes the present and the present becomes the past. Though events in the video world usually move in the same way, the rules of this different universe also allow us to reverse time direction and to jump forward and backward in time.

Time Running Backward

The simplest way to reverse video time is by playing a recording backward, either for comedy or to achieve a special effect. We have all seen shots that show things like pie fragments leaving an actor's face and reassembling themselves into a complete pie, or milk rising like a small water spout to reenter a milk bottle. Like fast motion, reverse motion within a shot often has a comic effect.

Reverse motion can also be used to simulate actions that would otherwise be impossible. For example, recall the two men, introduced in Chapter 3, who make an impossible leap from the ground to the top of a wall six feet high, **Figure 5-10**. To accomplish this feat, they were recorded jumping *backward*, from the top of the

Figure 5-9 The two sides of a movie door are often not the same door.

On location outside

Studio set inside

Figure 5-10 Shots of two men leaping from the ground to the top of a six-foot wall.

The actors approach the wall and crouch to jump.

They leap upward.

They appear to jump to the top of the wall...

...and land there.

wall to the ground. The editor then reversed the shot in postproduction, **Figure 5-11**.

The same trick can be used to simulate a circus knife-thrower's blades striking dangerously close to an assistant's head. As displayed, the shot shows several knives flying into the frame and imbedding their points in the board behind the assistant. In reality, the shot is made with the knives already in place—their handles tied to invisible fishing lines. During the shot, off-camera crew members yank the knives out of the frame, one after another. By running the shot backward, the editor creates the thrilling effect of the knives flying into the shot.

Like slow and fast motion, reverse motion shots tend to be used sparingly because when they are noticeable, they look obviously unnatural and contrived.

Flashing Backward (and Forward)

Though audiences do not really believe that time itself can run backward, they readily accept the idea that a video program can jump backward in time, and then return to its place in the "present" and resume its forward direction. A time jump like this is called a *flashback*. The flashback is such an old and familiar technique that two examples of it will illustrate it sufficiently.

- At a 25th high school reunion, the now-sedate participants remember the crazy stunts they pulled as teenagers. As each participant's story unfolds, we jump back 25 years to watch it happen.
- As she takes the oath of office, the new president thinks of her long climb to get to this moment and we begin reliving that struggle with her.

Figure 5-11 The jumping shot as originally recorded.

**The actors jump *backward*
off the top of the wall...**

...and drop toward the ground.

**The ending shot of the jump is
omitted in the final sequence.**

In the first example, we jump back to the past in order to *contrast it* with the present. In the second, we explore the past in order to *explain* the present. In both cases, the movie makers flash us backward in time—a leap that is impossible in the actual world.

Video Time Coherence

Video makers manage time in one additional way—they pull moments in time apart by splitting their video and audio components, and mixing them with video and

audio of other moments. A common application of this technique is voice-over narration. When a boy on the screen delivers newspapers from his bicycle, as an adult voice on the sound track says, "Little did I know what would happen on that spring day so long ago...," the picture is unfolding at one point in time, while the sound is at a point years later.

Split Edits

An equally useful time-splitting technique is called, appropriately, a ***split edit***: a transition from one shot to another in which the video and audio portions do not change simultaneously. Split edits take two forms: video changes first or audio changes first.

To illustrate both, imagine a young woman planning a trip. In the first scene, she estimates her driving time and in the second, she tries to fix her broken-down car.

Split Edit: Video Leads

When the video changes first, the old audio continues off-screen over the new video (**Figure 5-12**).

Here, and throughout this book, you will encounter progressions of graphics that appear identical. Usually, the differences lie in the altered captions, as careful study will show you.

Split Edit: Audio Leads

When the audio changes first, the sound from the new scene begins over the preceding video, **Figure 5-13**.

Playing with Time

It is not a flashback when a character in a program actually travels through time, as people often do in science fiction stories. In a flashback, it is the *story*, rather than the character, whose position in time changes.

Of course, we can also flash *forward*, jumping into the future for a look, before returning to the main narrative in the present. In this case, time changes direction not when we make the leap, but when we jump backward again to return to the "present."

Today's audiences have become accustomed to unusual time schemes. In various movies, for example:

● The same day repeats endlessly.
● Time runs backward.
● Two characters who live in two different calendar years communicate through letters placed in a mailbox.
● A young woman's life splits into two different time streams.

Figure 5-12 Video leading split edit.

WOMAN: I can get there in an hour.

Whichever way the edit is split, the time of one element is overlapped with the time of the other.

About Time

It is evident that video time behaves very differently from actual time, because nothing about it is fixed—neither its speed, flow, direction, nor coherence. The skillful video maker uses the changeability of video time, manipulating it to condense the action, unfold the narrative, and create special effects.

Our discussion of time has, necessarily, been somewhat abstract and theoretical. In actual production, most video makers deal with space and time instinctively, and so can you. However, by taking the trouble to understand the very different laws that rule time in the video world you will be able to use them to greater advantage in your own programs.

Now that we have considered video time in this chapter and video space in Chapter 4, we are ready to look at other aspects of video communication. Chapters 6 and 7 concern communicating with video images—through composition in Chapter 6 and through video language in Chapter 7. Our survey concludes with communicating with audio in Chapter 8.

SOUND EFFECTS: Clanking of tools.
WOMAN (off-screen): I've got time to spare!

Figure 5-13 Audio leading split edit.

WOMAN: I can get there in an hour. I've got time to spare!
SOUND EFFECTS: R-R-R-R of starter motor.

SOUND EFFECTS: Clanking of tools.

Summary

- Real-world time always behaves in exactly the same way with respect to its speed, flow, direction, and coherence.
- The characteristics of time in the video world are always controlled and continually changed by the director and editor.
- Altering the recording rate of a shot results in an obvious change in the speed of time on the screen.
- Editors can slow time down by overlapping action, and speed it up by omitting action.
- Three main forms of video time progression are serial time flow, parallel time flow, and disjointed time flow.
- Rules of the video universe also allow us to reverse time direction and to jump forward and backward in time.

Technical Terms

Cross-cutting: An editing technique that allows two actions to be shown at once by alternating back and forth between them—presenting part of one, then part of the other, then back to the first, and so on. Can be used to show three or even more parallel actions. Also called *intercutting*.

Disjointed time flow: The way in which video sequences are usually recorded in production; most programs are shot out of chronological order and reorganized during the editing process to follow script order in the finished video.

Fast motion: A video effect that contracts the screen time during which an action happens; created by recording a shot at a slower rate or by changing the frame rate digitally in postproduction. A typical fast motion recording speed would be 15 fps. Played back at the normal 30 fps cuts the apparent time in half.

Flashback: A sequence that takes place earlier in time in the story than the sequence that precedes it.

Frame rate: The number of frames (images) per second at which video is recorded or displayed. Expressed as frames per second (fps).

Parallel time flow: Form of video time progression in which two or more sequences of action happen at the same time.

Screen time: The length of real-world time during which a sequence is displayed on the screen (in contrast to the length of video-world time that apparently passes during that sequence).

Serial time flow: Form of video time progression in which a single series of events moves forward in a single stream.

Slow motion: A video effect that expands the screen time during which an action happens; created by recording a shot at a faster rate or by changing the frame rate digitally in postproduction.

Split edit: An edit in which the audio and video of a new shot do not begin simultaneously. When "video leads," the sound from the preceding shot continues over the visual of new shot. When "audio leads" the sound from the new shot begins over the end of the preceding visual.

Review Questions

Answer the following questions on a separate piece of paper. Do not write in this book.

1. What is the difference between "screen time" and "real-world time?"
2. Moderate _____ motion in a shot imparts a dream-like or nostalgic effect.
3. *True or False?* Slow motion contracts the screen time during which an action happens.
4. How can editors control time in a way that is invisible to viewers?
5. Identify and explain the three main forms of video time progression.
6. Using _____ motion in a shot can simulate actions that would otherwise be impossible.
7. A backward jump in time is called a(n) _____.
8. What is a *split edit*?

STEM and Academic Activities

STEM

1. **Engineering.** Document the steps or actions in one of your daily routines. These steps illustrate serial time flow. Create a plan to develop the events in your routine into an interesting parallel time flow progression for a video program.

2. **Mathematics.** When you watch a video, 30 individual pictures are presented per second. This rate is expressed as 30 fps (frames per second). How many frames are presented in a 2-minute program? If the recording rate is changed to 20 fps, what is the percentage of change in the recording rate?

3. **Language Arts.** Watch several scenes from movies or television programs that make use of split edits (both video leading and audio leading). For each scene, explain how the split edit is used and describe the effect created by using this technique in the program.

(Sue Stinson)

(Michael Melgar)

CHAPTER 6 Video Composition

Objectives

After studying this chapter, you will be able to:

- Identify the key principles of video composition.
- Explain how contrast is used for emphasis.
- Illustrate techniques for creating depth in compositions.
- Summarize effective methods for directing the viewer's eye.
- Identify the principles involved in creating widescreen compositions.

About Video Composition

The word *composition* (literally, "placing together") has many different, though related, meanings. In video, *composition is the purposeful arrangement of the components of a visual image.* These components, often called pictorial elements, may be objects, such as people, buildings, or trees. They may also be abstract elements, such as lines, shapes, degrees of brightness, or colors.

What is the purpose of composition? One easy answer is that it somehow renders a picture more "pleasing" or more "artistic." But, that reply makes composition seem rather unimportant—nice to have, but not really necessary. In fact, composition is absolutely essential in video because it helps images communicate with viewers more quickly, efficiently, and powerfully. Composition does this by:

- Organizing pictorial elements so that viewers can quickly sort them out and identify them.
- Adding emphasis to direct attention to the most important elements on the screen.
- Creating an illusory third dimension in a two-dimensional image.

Overall, composition is such a powerful communication tool that every video maker needs to master its fundamentals. So, let us see how composition creates visual organization, guides viewer attention through emphasis, and enhances the illusion of depth.

The opening image on the previous page exploits all six perspective techniques for achieving apparent depth.

Organization in Composition

Composition organizes the elements within an image to help viewers decode it.

Decoding an image means assigning meaning to the elements in it, and identifying a relationship among them.

For example, the first image in **Figure 6-1** shows a number of small tools. Dropped in a disorganized heap, they are hard to identify. However, tools that are arranged in a composition are easy to spot. Viewers need composition to help them

Figure 6-1   Composition helps viewers "read" the image.

Disorganized presentation

Organized composition

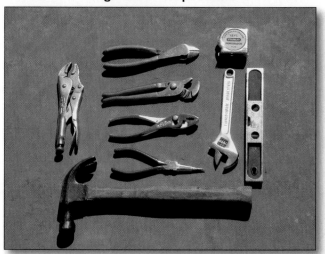

decode pictures, because they are unable to organize screen images as easily as they organize visual information from the real world.

Simplicity, Order, Balance

In organizing the pictorial elements in images, composition follows the key principles of simplicity, order, and balance (**Figure 6-2**).

Simplicity

Simplicity means eliminating unnecessary objects from the image, so that the viewer has fewer things to identify. Notice that the "Simplicity" picture in **Figure 6-2** is already easier to decode than the "Organized composition" image in **Figure 6-1**, merely because the number of tools in it has been reduced.

Principles of composition.

Simplicity   **Order**   **Balance**

   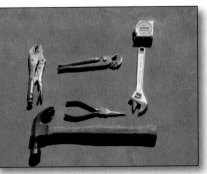   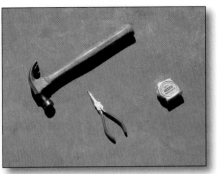

Order

Order means arranging objects in the image. In the "Order" picture in **Figure 6-2**, the tools are even easier to recognize because they have been lined up neatly.

Balance

Balance means distributing the objects in a way that gives about equal visual "weight" to each section of the image. Notice that arranging the large hammer in the "Order" picture in **Figure 6-2** so that it balances the group of smaller tools, makes it stand out in the composition.

Emphasis in Composition

Organization helps viewers make sense of the elements in an image, but by itself, it does not confer special importance on any of them. That is the function of emphasis. *Emphasis* is any technique that attracts the viewer's attention to one part of a composition. Techniques of emphasis include *position*, *relationship*, *significance*, and *contrast*.

Position

You can emphasize a compositional element by the way in which you position it within the frame. In any composition, the upper-left quadrant attracts the eye first, because that is where we automatically look to start reading a new page of text.

Obviously, this is not true of everyone. Readers of Chinese or Hebrew begin the page at other points.

Psychological experiments have confirmed that in a composition, the upper-left quadrant is most attractive, the upper-right one next, the lower-left quadrant third, and the lower-right one is the least attractive. See **Figure 6-3**.

Relationship

An element can be emphasized by placing it in relationship to other elements in the composition. These other components lead the eye to the center of interest. You can see this in **Figure 6-4**. Since the elements of video images are usually moving, a good director can exploit their ever-changing relationships.

Figure 6-3 In Western cultures, the upper-left quadrant is the most powerful part of the image.

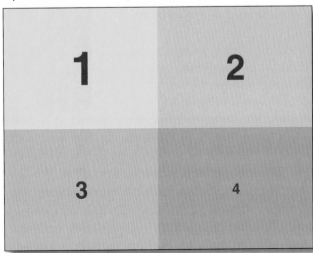

Composition and the Frame

We have noted that the frame around the image forms the border of the video world. At the same time, the frame acts as the most powerful component of visual composition. In fact, it is the frame that makes composition possible. A composition is the visual organization of a defined area and that definition is provided by the frame.

To see how this works, look at photo A. Though the edges of this picture do at least form a border, the image as a whole lacks an overall sense of composition. However, as you can see from photos B, C, and D, we can create several different compositions, according to how we place frames around different components of the original picture. (The composition in photo D will continue to improve as the animals on the left move across the frame.)

A The frame is too loosely composed.

B Composition 1

C Composition 2

D Composition 3

In addition to defining an area, the frame tells you what is horizontal (parallel to the horizon) and what is vertical. In these two portraits of former Vice President Al Gore, for example, the subject is positioned upright in both compositions, but seems to be tilted forward in the second one. In fact, it is the frame and not the person that has tilted in the second shot. But, because we derive our sense of vertical and horizontal from the frame, the picture continues to appear level, while the person within it seems tilted.

Straight frame

Tilted frame

Significance

Some compositional elements attract attention simply because they mean something to the audience. For instance, when you look at a photo of a celebrity, as in **Figure 6-5**, your eye goes immediately to the famous person because you recognize the face. It means something to you. Significance in a composition is widely used in Hollywood movies because the highly recognizable stars automatically attract viewers' attention.

Contrast

Contrast is the technique of emphasizing one element in a picture by making it look different from the other elements. An element can be

Figure 6-4   Diagonal lines point to the center of interest.

Figure 6-6   Contrast: size.

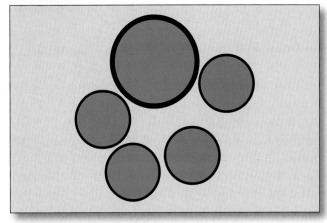

contrasted by adjusting its traits, including size, shape, brightness, color, and focus.

In cinematography, the word *contrast* is also used in a more technical sense, meaning difference in luminance (brightness) between the lightest and darkest picture elements.

Contrasting Size

The bigger something is, the more it calls attention to itself (**Figure 6-6**). It is simple to make one object appear bigger than other objects of a similar size: just place it closer to the camera. Not surprisingly, important action is typically staged in the foreground, where the actors are largest.

Figure 6-5   President Obama stands out even though he is off-center and surrounded by other people.

(Steve Adamson/Shutterstock)

Contrasting Shape

The rectangle in **Figure 6-7** attracts attention because its shape contrasts with the circles around it.

Contrasting Brightness

Light elements attract the eye more than dark ones (**Figure 6-8**). Much of the craft of video lighting involves emphasizing important elements through careful control of brightness.

Contrasting Color

Perhaps the simplest form of contrast is color. The viewer's eye is attracted to elements colored differently from the prevailing hues of the overall image. In **Figure 6-9**, the red circle stands out.

Contrasting Focus

Focus refers to the visual sharpness (clarity) of a pictorial element. Sharp, well-detailed

Figure 6-7   Contrast: shape.

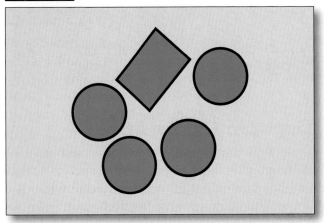

Figure 6-8 Contrast: brightness.

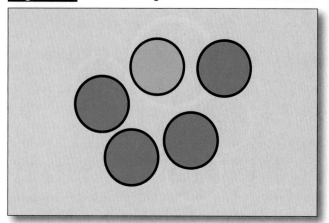

Figure 6-9 Contrast: color.

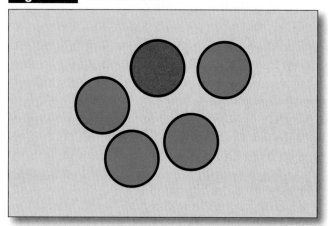

Figure 6-10 Contrast: focus.

elements attract viewers' attention away from soft, out-of-focus elements. Because video images are rarely equally crisp throughout their depth, cinematographers set the focus on the elements they want to emphasize. In **Figure 6-10**, the distracting background greenery is controlled by throwing it out of focus. Sometimes you will see the focus change in the middle of a shot to shift attention from one element to another.

Great painters, like Renoir, often achieved the effect of selective focus by painting important picture elements in fine detail and leaving less important ones sketchy and less distinct.

Depth in Composition

As we have noted, the video world has only two true dimensions: height and breadth. On the essentially flat monitor screen, what appears to be depth is only an illusion. Since most video programs are intended to mirror the actual world, depth-enhancing techniques play an important role in pictorial composition. Collectively, these techniques are called *perspective*.

Perspective

Perspective techniques are used to suggest the presence of depth on a two-dimensional surface, such as a painting, a photograph, or a video screen. Over many centuries, artists have evolved several ways to represent three dimensions on two-dimensional surfaces. These effects will appear in your compositions whether or not you employ them intentionally. But, if you purposely apply them, you can greatly enhance the sense of depth in your images. Perspective techniques include *size, overlap, convergence, vertical position, sharpness,* and *color intensity*.

Size

In the actual world, the closer an object is, the larger it appears (**Figure 6-11**). So, in a two-dimensional image, apparent size is a powerful clue to distance.

Overlap

In the actual world, one object will mask part of another if it is in front of it. So, in two-dimensional compositions, an object that overlaps another is perceived to be closer (**Figure 6-12**).

Figure 6-11 The young woman's much smaller image size tells viewers that she is farther away.

Convergence

In the actual world, parallel lines, such as railroad tracks or the edges of roads, seem to gradually come together as they recede into the distance. In two-dimensional compositions, converging lines convey a powerful sense of depth (**Figure 6-13**).

Vertical Position

In the actual world, the farther away objects are, the higher they usually appear in the field of view. In **Figure 6-14**, notice that the subjects in the background are higher on the screen than the foreground objects.

Figure 6-12 Partly masked by the older woman, the young woman appears to be farther away.

Figure 6-13 The converging lines add a strong feeling of depth.

Figure 6-14 The background subjects are higher in the frame.

Sharpness

The farther away objects are, the more indistinct they appear, for two reasons. First, their fine details are smaller and harder to distinguish. Second, distant objects are softened by the amount of air between them and the viewer. Because the effect is created by air, this is often called *atmospheric perspective*. This type of perspective is evident in a picture of foothills with mountains behind them. Each ridge appears softer and less distinct than the one in front of it (**Figure 6-15**).

Color Intensity

The same atmosphere that softens the appearance of objects also reduces the intensity of their colors. The farther away something is, the paler or more pastel its colors appear. The orange color of the runner's shirt is much more intense in **Figure 6-16B** than in **Figure 6-16A**.

As you can see, most images include several tools of perspective at once. For instance, **Figure 6-16A** is also a good example of distance indicated by size and vertical position.

Composing Video Images

Once you understand the ideas behind pictorial composition, you can make them work for you by creating compositions, directing the viewer's eye, and controlling apparent depth in your images. The resulting images will be both better-looking and more effective.

Figure 6-15 Each line of hills is less sharp than the one in front of it.

Figure 6-16 The color of the runner's shirt grows more intense as he comes closer.

Compositional Schemes

As you gain experience in framing video shots, you will instinctively use the principles of simplicity, order, balance, emphasis, and contrast to design your images. You do this to enhance their visual coherence by applying one of several compositional schemes.

The *American Heritage Dictionary* defines coherence as "...a logical, orderly, and aesthetically consistent relationship of parts."

Asymmetrical Balance

The most common of these schemes is called *asymmetrical balance*, in which visual elements are not equally opposed, like two children on a level seesaw, but are distributed less formally to give an overall impression of balance. Generally speaking, compositions are more pleasing when their elements appear to have the same visual "weight." But, when visual elements are balanced symmetrically, the result is often static and dull. It is usually better to organize them so that their characteristics are different, and their positions are not evenly opposed.

To see the difference between symmetrical and asymmetrical compositions, look at **Figure 6-17**. In the first image, the church is planted squarely in the middle of the frame, so that the resulting composition is not very interesting. In the second version, the church is smaller and placed far down in the lower-left corner, for a more striking composition. (Notice, too, how it is given emphasis by its contrasting brightness, color, and position.)

In **Figure 6-18**, the composition is strongly asymmetrical, with the archway and the bright sky leading the eye strongly to the right-half of the frame. Notice how the eye is then drawn to the large door on the left side, as if someone were about to come through it.

In **Figure 6-19**, the smaller, brighter subject on the right is balanced by the dark mass of the column on the left.

Figure 6-17 Balance in a composition.

Symmetrical balance is static.

Asymmetrical balance makes a more interesting composition.

(Sue Stinson)

Figure 6-18 The small figure balances the door.

(Sue Stinson)

Figure 6-19 A large dark mass can be balanced by a smaller, brighter object.

The Rule of Thirds

Perhaps the easiest way to design asymmetrical compositions is by using the "rule of thirds." To do this, mentally fill the camera viewfinder with a tic-tac-toe grid that divides the frame into three equal vertical and horizontal sections. Then, create compositions in which the important pictorial elements match the lines and intersections on the grid. It is not necessary to match *all* parts of the grid, (**Figure 6-20**).

The rule of thirds can be applied to most kinds of subjects, **Figure 6-21**. The rule of thirds is a useful guideline, especially when you are composing images quickly to record events as they happen.

The rule of thirds organization is now so common that many still and video cameras can superimpose a tic-tac-toe grid on their view screens.

Other Organizational Techniques

The rule of thirds is by no means a rigid rule. When you have time to arrange your subjects as well as your frame, it is often useful to ignore the rule of thirds, as you can see from **Figure 6-22**. Though each composition is balanced, it does not align on a tic-tac-toe grid.

Figure 6-20 Applying the rule of thirds. A—The scene to be composed. B—A grid imposed onto the image. C—The image reframed to align better with the grid.

A

B

C

(Corel)

Figure 6-21 Notice that most compositions do not align visual elements with every line or intersection on the grid.

(Sue Stinson)

Figure 6-22 None of these compositions follows the rule of thirds. The canoe builder is close, but note how the teapot and the large cylinder on the right side do not fit anywhere on a grid.

(Corel) (Sue Stinson)

Directing the Viewer's Eye

In addition to creating a coherent image, the second job of composition is placing the attention of viewers where you want it. To do this, you use emphasis and leading lines.

The most powerful way to attract the eye is through movement within the image, but that lies outside the subject of composition.

Emphasis

Emphasis, as noted earlier, can be added by adjusting subject position, relationship, significance, and above all, contrasting size, shape, brightness, color, and/or focus. **Figure 6-23** shows two compositions that lead the eye to distinctive shapes.

Figure 6-23 Emphasis. A—Though tiny in the frame, the human shapes are distinctive. B—The sand dollar draws the eye.

A **B** (Adobe)

Many compositions add emphasis using several techniques at once. In **Figure 6-24**, the first image directs the eye with a combination of size, position, and color. The second depends on size, shape, and color.

Leading Lines

The most straightforward way to emphasize a subject is simply to point to it using one or more lines in the composition, **Figure 6-25**. *Leading lines* can also be used more indirectly, **Figure 6-26**.

Controlling the Third Dimension

Composition plays a key role in creating the illusion of a third dimension on the two-dimensional video screen. In most cases, the goal is to enhance apparent depth using several different methods.

Staging Action in Depth

First, try to place subjects and camera setups to suggest depth in the image. In the second image in **Figure 6-27**, several tricks are used to enhance apparent depth. The revised camera placement makes the left subject larger than the

---

Figure 6-24 Adding emphasis using many techniques.

Size, position, and color.

Size, shape, and color.

(Corel)

(Adobe)

Figure 6-25 Leading lines. A—The subject is pointed to by no less than four leading lines. B—Here, the subject is so small that he would be unnoticed without the leading lines of the fence rails.

A

B

Figure 6-26 Indirect use of leading lines. A—The skywriter's smoke leads to the small biplane. B—The dark cow is emphasized by both horizontal and vertical lines, as well as contrasting color.

A

(Corel)

B

(Sue Stinson)

right one, the building windows diminish in size, and the clapboard plank lines converge. Notice also that the nearer subject "breaks the frame"—that is, parts of her head and left side extend beyond the image border.

Nineteenth century French artists learned this technique from Japanese prints and the then-new art of photography.

Figure 6-28 shows two other ways to stage action in depth. In the first image, the long street curves away out of sight in the background, suggesting that it continues for some distance.

The second image exploits apparent scale, placing the large boulder in the left foreground and the subjects so far away that they appear tiny.

Exploiting Perspective

Staging in depth uses tricks of perspective, each of which creates the illusion of a third dimension, **Figure 6-29**. And, wide angle lens settings and low camera angles enhance depth, **Figure 6-30**, especially with prominent foreground figures.

Figure 6-27 Staging for depth.

Flat staging.

Staging emphasizes depth.

Figure 6-28 Staging in depth. A—The street continues out of the frame. B—Using apparent scale.

A

B

(Sue Stinson)

Framing the Image

To add extra depth to an image, it often helps to frame it with foreground elements. The most common framing devices include tree branches and archways, **Figure 6-31**. Architectural elements, such as overpasses and roofs, also work well (**Figure 6-32**).

Windows make excellent framing elements. Try zooming in until the window is excluded and the subject fills the frame, **Figure 6-33**.

Also, by editing the shot together with an exterior shot of any building you select, you can establish that the view is from that building—even if the two real-world locations are entirely different.

Figure 6-29 Tricks of perspective.

Size

Overlap

Convergence

(Corel)

Vertical position

Sharpness

Color intensity

(Sue Stinson)

Figure 6-30 Using the camera to exploit perspective.

Wide angle lens

(Adobe)

Low camera angle

(Sue Stinson)

Figure 6-31 Framing with foreground elements.

Tree branches

Archway

(Sue Stinson)

Figure 6-32 Framing with architectural elements.

Overpass

Roof overhang

A

B

Working on the Picture Plane

Though you usually try to enhance the feeling of depth, there are times when you want the opposite effect: an image composed on the flat plane of the screen, called the *picture plane*. (Commercials and travel videos frequently use this technique.) Extreme telephoto lenses are often used to create this flat appearance, **Figure 6-34**. You can achieve the same effect without long lenses by carefully omitting all the perspective "clues" that you would normally include, if you were aiming to enhance the feeling of depth (**Figure 6-35**).

Managing Camera Movement

Professional-looking camera moves should be planned so that they both start and end with strong compositions.

Panning and Tilting

Without planning, moving shots often end with weak compositions. **Figure 6-36** shows the courtyard of a palace with groups of tourists on the right and left sides. If the shot is composed for the right-side group and then panned to frame the left group, the movement will end

Figure 6-34 Flattening the perspective. A—A telephoto lens compresses the "depth," placing all three racers on almost the same plane. B—The same technique turns this wall and building into a flat design.

A

(Corel)

B

Figure 6-35 Omitting clues to perspective. A—Without a horizon or information about relative ship sizes, the image lacks a sense of perspective. B—This semi-silhouette provides little help in determining perspective.

A

B

Figure 6-36 A palace courtyard.

(Sue Stinson)

unsatisfactorily because the size and shape of the second group are different, **Figure 6-37**.

If you have time to rehearse the shot, it is better to set the ending composition first, then set the starting composition, and make the pan;

zooming and reframing during the move to create the ending composition (**Figure 6-38**).

When *tilting* the camera up or down, the procedure is exactly the same (**Figure 6-39**).

Moving the Camera

The same concept applies when making a moving shot—whether hand-held or on a camera dolly. Rehearse both the ending and starting compositions, then make the move; adjusting camera angle and lens focal length setting as needed.

Of course, the camera adds to the feeling of depth when it moves forward or backward, into or out of a composition. But, it also enhances depth when moving *sideways*, parallel to the picture plane. That is because we notice even tiny changes in the relationships between objects, and use those changes to calculate depth.

Figure 6-37 Courtyard pan.

Start of the pan.

End of the pan.

Beginning and ending frames.

Figure 6-38 Set end composition first in a pan.

Start of the pan.	End of the pan.	Ending composition requires tilting up and zooming in.

Figure 6-39 Camera tilting up.

Start of the tilt up.	End of the tilt up.	Ending composition requires tilting up and zooming out.

(Corel)

In the first image in **Figure 6-40**, the two pilings overlap and only the tip of the distant island is visible at the right edge of the frame. In the second image, the camera has moved only about 10 feet to the left, while framing the same composition. Now the pilings have separated, and much more of the island has appeared. Our brains use these changes as clues to distance. That is why people who have lost the use of one eye can still perceive depth, even though they lack stereo vision.

In **Figure 6-41A**, the bridge structure flashing by adds a strong sense of movement. In addition, the brain can notice and interpret even the tiny apparent movement of the distant dock, enhancing the feeling of depth (**Figure 6-41B**).

Composition for Widescreen Video

Traditional video images are proportioned 1.33 to 1, meaning that the width of the frame is one and one-third the length of its height. All HDTV images, many DVD programs, and most TV shows and commercials are now recorded in a wider format, proportioned 1.78 to 1—commonly known as 16 to 9 or *widescreen video* (**Figure 6-42**).

Figure 6-40 Enhancing depth by moving parallel to the picture plane.

Starting from its first position...

...the camera dollies to the left.

Figure 6-41 Movement and depth. A—The bridge structure adds a feeling of movement. B—Note the changed position of the dock (marked by the red lines). The camera is dollying to the right.

A

B

(Sue Stinson)

Figure 6-42 Video frame proportions. The traditional video frame is proportioned 4 to 3. Widescreen video is 16 to 9.

Traditional (4 to 3)

Widescreen (16 to 9)

4

3

16

9

The Pros and Cons of Widescreen

There is no longer any competition between the two formats. With digital broadcast and the widespread replacement of older TV sets, the 16 to 9 screen ratio may be considered standard. That is not to say that it is necessarily an improvement over the old shape. Much of TV consists of closeups (also called "talking heads"), which are easier to frame pleasingly in the old format, as we will see.

Widescreen Advantages

For sports and other action, however, the widescreen format is obviously superior. In general, 16 to 9 frames work very well whenever the image is strongly horizontal (**Figure 6-43**).

Widescreen Drawbacks

16 to 9 frames are not without problems, however. As previously noted, they make closeups more difficult, as you can see from **Figure 6-44**.

Widescreen images can be less suitable for publishing on the Internet, where many sites continue to use the old 4 to 3 format. Widescreen images can also be difficult to display on traditional TV screens. If the program is to be shown on a 4 to 3 screen without some form of compensation, the edges of the composition will be clipped (**Figure 6-45**).

On older TVs, widescreen images are sometimes displayed *letterboxed*—with black bands on either side of the picture (**Figure 6-46A**). The problem is complicated by the need to transmit both standard and high-definition video images. Using one solution, the 16 to 9 high-definition image is displayed on standard sets with black on all four sides (**Figure 6-46B**). Another solution is to confine the essential information to the central 4 to 3 screen area, either by including bands of "filler" on the sides (**Figure 6-46C** and **Figure 6-46D**), or by composing the widescreen image so that it can be safely

---

Figure 6-43   Best uses of widescreen.

Widescreen lends itself to action...

...and to scenes with strong horizontal shapes.

Figure 6-44   Closeups in the widescreen format. A—Framing a closeup leaves a lot of empty space. B—Framing to utilize the screen area cuts off part of the subject's head.

A

B

Figure 6-45 A widescreen image displayed on a traditional TV screen.

A 16 to 9 composition...

...loses quality when clipped to 4 to 3 proportions.

(Goodluz/Shutterstock)

cropped (**Figure 6-46E** and **Figure 6-46F**). Some 4 to 3 images are adapted for widescreen display by both widening the image and cropping top and bottom (**Figure 6-46G**). Unfortunately, this both distorts the subject and degrades the composition.

In the case of widescreen feature films (which may be proportioned up to 21:9), "pan-and-scan" copying techniques are sometimes used to recompose the image for re-recording in 12 to 9 proportions.

Figure 6-46 A—Standard image letterboxed on a widescreen set. B—Widescreen image reduced to fit on a standard set. C—Standard screen image with added side filler visible on a widescreen set. D—Side filler is not visible on a standard set.

A

B

C

D

(Continued)

Figure 6-46 *(Continued)* E—Widescreen image composed for a widescreen set. F—Preserving a pleasing composition on a standard set. G—Filling the frame by horizontally stretching and vertically cropping the image causes distortion.

E

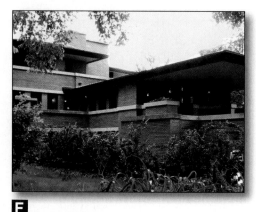

F

G

Composing Widescreen Images

16 to 9 frames will yield effective compositions, if you keep a few basic principles in mind.

Exploiting Horizontals

Since widescreen is a horizontal format, look for camera angles that can exploit a wide view. Obviously, horizontal subjects can be easily framed (**Figure 6-47**). Low angles and wide angle lenses can produce strong feelings of space and speed, **Figure 6-48**.

Using the Rule of Thirds

The rule of thirds works as well with widescreen compositions as it does with regular ones (**Figure 6-49**).

Composing at the Frame Borders

When the rule of thirds will not yield a good composition, try the opposite approach by placing important elements near the edges of the frame (**Figure 6-50**).

Trust Yourself

Although the craft of creating coherent, effective images appears complicated and technical, you can rely on your own taste and instincts to guide you to good compositions. Creating pictorial composition is like driving a car. When you are just learning, you have to remember every part of a complex process. But, before you know it, driving becomes so instinctive that you do it without conscious thought.

The same thing happens as you make video programs. You gradually internalize the principles of composition until you can use them naturally and instinctively.

Figure 6-47 Framing a horizontal subject.

(Corel)

Figure 6-48 Low angles can be effective in widescreen.

Figure 6-49 Applying the rule of thirds to widescreen images.

(Goodluz/Shutterstock)

Figure 6-50 Positioning important elements near the frame edges.

(Corel)

Framing Subjects

Generally speaking, widescreen images can be framed like traditional images, with a few notable exceptions. Obtaining large closeups requires cutting off part(s) of the subject's head. In these cases, framing off the chin produces an unattractive effect (**Figure 6-51A**). It is better to exclude more of the top of the head instead. Leave extra room below the chin so that it remains in the frame when the subject is speaking, **Figure 6-51B**.

Also, the horizontal format of widescreen allows more "lead room" in front of moving subjects. In **Figure 6-52**, notice that the plane formation has ample sky in front of it.

Figure 6-51 A widescreen large closeup. A—With her chin excluded, the subject looks cramped. B—Framing the chin and losing a bit more hair solves the problem.

A

B

(Corel)

Figure 6-52 Allow ample lead room for moving subjects.

(Corel)

From Image to Program

This chapter on composition has focused on the design of individual images. A video program consists of many thousands of these single frames, organized into shots, scenes, and sequences. Chapter 7, *Video Language* presents the principles behind this organization.

Summary

- Composition creates visual organization, guides viewer attention through emphasis, and enhances the illusion of depth.
- Organization in a composition helps viewers decode the image.
- An element in an image can be contrasted by adjusting its size, shape, brightness, color, and focus.
- Perspective techniques are used to suggest depth on a two-dimensional surface. Perspective techniques include size, overlap, convergence, vertical position, sharpness, and color intensity.
- Emphasis techniques and leading lines in a composition direct the viewers' eye where you want it.
- When recording in widescreen format, keep in mind the unique characteristics of the video format to create effective compositions.

Technical Terms

Asymmetrical balance: A composition in which dissimilar elements have equal "visual weight."

Brightness: The position of a pictorial element on a scale from black to white.

Contrast: (1) A technique of emphasizing one element in a picture by making it look different from the other elements. (2) The ratio of the brightest part of an image to the darkest.

Decoding: The process of identifying and understanding the elements in a composition.

Emphasis: Calling attention to a pictorial element in a composition.

Focus: (1) The part of an image (measured from near to far) that appears sharp and clear. (2) The object of a viewer's attention.

Leading lines: Lines on the picture plane that emphasize an element by pointing to it.

Letterboxed image: A widescreen image displayed in the center of a regular TV screen, with black bands above and below it filling the frame.

Picture plane: The actual two-dimensional image.

Staging in depth: Positioning subjects and camera to exploit perspective in the image.

Symmetrical balance: A composition in which visual elements are evenly placed and opposed.

Tilting: Vertical camera movement to point the camera lens up or down.

Widescreen video: A video image or video display screen proportioned 16 to 9, in contrast to the traditional TV screen's 4 to 3.

Review Questions

Answer the following questions on a separate piece of paper. Do not write in this book.

1. What is video *composition*?
2. In organizing the pictorial elements in an image, the composition follows the key principles of simplicity, order, and _____.
3. Any technique that attracts the viewer's attention to one part of a composition creates _____.
4. *True or False?* The rule of thirds is a technique used to create perspective.
5. *True or False?* The farther away an object is, the higher it usually appears in the field of view.
6. Describe the arrangement of items in an asymmetrically balanced composition.
7. What is the *rule of thirds*?
8. List ways to enhance the apparent depth in a video image.
9. Widescreen images have a _____ ratio.
10. List the principles involved in effectively framing 16 to 9 compositions.

STEM and Academic Activities

STEM

1. **Science.** Asymmetry is commonly used in video to create visual interest. However, asymmetry is also common in nature and biology. Investigate asymmetrical elements in nature and biology. Choose one example and explain the purpose or adaptation of the asymmetrical arrangement.
2. **Technology.** The DTV conversion occurred in the U.S. in June 2009. What are the benefits of digital television technology? What were the constraints of analog television technology?
3. **Language Arts.** Watch a program on DVD in both standard and widescreen formats. List the scenes that were different from one format to the other. Did the differences in scenes from one format to the other affect the message, plot, or action in the program?

7 Video Language

Objectives

After studying this chapter, you will be able to:

- Understand the application of video language.
- Identify the principal types of camera angles.
- Recognize the characteristics of camera angles.
- Understand how to create continuity of action among shots.
- Explain the use of transitions in a program and identify common transition effects.
- Recall how DVEs are used in a program.

About Video Language

As Chapter 3 explains, no matter how seamless videos may appear, they are really assemblies of individual pieces that are placed one after another in strings as long as one thousand units or more. These individual pieces are, of course, called *shots*.

A shot is a single, uninterrupted visual recording—a recording event during which the camera operates continuously. Of course, shots are not just strung together in random order. On the contrary, they are organized according to a set of rules. Collectively, these rules may be called the "language of video."

A classic book on film making titled *A Grammar of the Film* (Raymond Spottiswoode, University of California Press) was published as long ago as 1950.

Video Language Terms

Though the analogy between video and a language such as English is not perfect, it does provide a useful comparison. For example:

- A *frame* is like a unit of sound (a "phoneme" in verbal language); an essential building block that is usually too brief to deliver meaning by itself.
- An *image* is like a word. It is a, more or less, static segment of a shot that is perceived as a picture.
- A *shot* is like a sentence; the shortest assemblage that conveys a complete piece of information.
- A *scene* is like a paragraph. Composed of several shots (visual sentences), it conveys meaning about a single topic.
- A *sequence* is like a chapter. It assembles a number of scenes into a longer action that is also devoted to a single (though larger) part of the narrative.

The analogy between video and verbal language could be carried still further. In dramatic programs for TV, a major assembly of video sequences is called an *act* (just like an act in a stage play) and a complete video *program* could be compared to a book.

Why a Language of Video Is Needed

To continue the analogy with verbal language, imagine a series of words from which all the rules have been removed:

store back him waiting came he he she from would for the when be knew

Restore the grammatical rules, and the word series becomes:

He knew she would be waiting for him when he came back from the store.

Like verbal language, video language uses rules to combine individual pieces into larger units that communicate coherently.

Studying Video Language

Mastering the formal language of video can be quite intriguing. Individual shots are like the building blocks in a fascinating construction set that is very rewarding to experiment with. This chapter presents the basic rules for using that video construction set. Since we cannot discuss video components without identifying them, our guide begins with the naming of camera angles.

Camera Angle Names

A *camera angle* is a distinctive, identifiable way of framing subjects from a particular position at a particular image size. Most camera angles have specific names because they are relatively standardized setups that are used again and again, **Figure 7-1**. This is not because video

Figure 7-1 A "worm's eye" angle of a subject.

directors lack originality, but because a century of film and video production has developed a certain number of angles that convey information effectively and look attractive to viewers.

Technically, a *shot* is a single piece of video recording, a *take* is a single attempt to record a shot, an *angle* is a particular view of the subject, and a *setup* is a distinct camera position. In actual production however, these words are sometimes used interchangeably.

Usually, camera angles are named for one of several different sets of characteristics: *subject distance, horizontal camera position, vertical camera position, lens perspective, shot purpose,* and *shot population.*

Camera angle names are not standardized, so the labels used in the following discussion are only typical. For more on angle name variations, see the sidebar *What's in a Name?*

Subject Distance

The most common angle names describe the apparent distance between the camera and a standing adult human, see **Figures 7-2** through **7-11**.

In this discussion, all the figures are displayed in the classic 12 to 9 screen ratio. The shot names and definitions apply equally to widescreen images.

These camera angles are:
- *Extreme long shot*: The figure is tiny and indistinct in a very large area, **Figure 7-2**.

- *Long shot*: The figure is small in the frame (half the frame height or less) and slightly indistinct, **Figure 7-3**.
- *Medium long shot*: The standing subject is distinct and somewhat closer, but with considerable head and foot room, **Figure 7-4**.
- *Full shot*: The standing figure fills the screen from top to bottom, often with just a small amount of head and foot room, **Figure 7-5**.
- *Three-quarter shot*: Shows the subject from about the knees to the top of the head, **Figure 7-6**.
- *Medium shot*: Shows the subject from the belt line to the top of the head, **Figure 7-7**.
- *Medium closeup*: Shows the subject from about the solar plexus (abdomen) to the top of the head, **Figure 7-8**.

Figure 7-3 Long shot.

Figure 7-2 Extreme long shot.

Figure 7-4 Medium long shot.

Figure 7-5 Figure 7-5 Full shot.

Figure 7-6 Three-quarter shot.

Figure 7-7 Medium shot.

Figure 7-8 Medium closeup.

- *Closeup*: Shows the subject from the shoulders to the top of the head, **Figure 7-9**.
- *Big closeup*: Shows the subject from below the chin to the forehead or hairline, **Figure 7-10**.
- *Extreme closeup*: Shows the subject from the base of the nose to the eyebrows, **Figure 7-11**.

These names can be confusing. Although the names label the camera-to-subject distance, they *actually* refer to the subject size in relation to the frame.

Although these names all use a standing human for reference, they are also employed with both smaller and larger objects. For example, though a horse's head is obviously larger than a human's, a shot framing it is still called a closeup (**Figure 7-12**).

Figure 7-9 Closeup.

Figure 7-10 Big closeup.

Figure 7-12 A head shot is always called a closeup.

Figure 7-11 Extreme closeup.

Figure 7-13 Front angle.

Again, all the shot names discussed apply to both widescreen and traditional aspect ratio images.

Horizontal Position

After subject distance, the most frequently used angle names are based on horizontal camera positions. In our illustrations of angles named by subject distance (**Figures 7-2** through **7-11**), we always see the person from the front. But, the camera can view its subject from other directions, as well. The more common horizontal positions have specific names (**Figures 7-13** through **7-17**):

- *Front angle*: The camera faces the front of the subject, **Figure 7-13**.

- *Three-quarter angle*: The camera is placed between roughly 15° and 45° around to one side of the subject, **Figure 7-14**.
- *Profile angle*: The camera is positioned at right angles to the original front shot, **Figure 7-15**.
- *Three-quarter rear angle*: The camera is placed an additional 45° around the back side, so that the subject is now facing away, **Figure 7-16**.
- *Rear angle*: The camera is directly opposite its front position and fully behind the subject, **Figure 7-17**.

What's in a Name?

If you consult various books on cinematography, you will discover that there is no standard set of angle definitions—let alone the shot names that describe them. The reasons for this are simple: English language movie production developed independently in the United States, Canada, the United Kingdom, Australia, and New Zealand, and each country evolved its own terminology. Also, video, which began as broadcast television, developed its own vocabulary—borrowing some terms from the film medium and inventing others of its own. To illustrate this confusing situation, here are just some of the alternative names for common shot lengths. (The terms used in this book are in **bold** type.)

All these names apply to modern widescreen images, as well as to traditional ones.

Extreme long shot, extreme wide shot, very long shot, or extra long shot.

Long shot or wide shot.

Medium long shot, wide shot, medium wide shot, or full shot.

Full shot, long shot, full length shot, or medium long shot.

Three-quarter shot, loose single, knee shot, or mid shot.

Medium shot or waist shot.

Medium closeup, bust shot, chest shot, head and shoulders closeup, or closeup.

Closeup, close shot, or head shot.

Tight closeup, big closeup, choker, or extreme closeup.

Extreme closeup, detail shot, or eye shot.

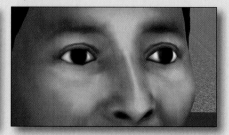

Of these alternative names, the most common variants are *extreme wide shot*, *wide shot*, and *medium wide shot*, instead of extreme long shot, long shot, and medium long shot.

Notice that the same terms are sometimes applied to different image sizes. For instance, some sources label a medium closeup as a closeup, and a tight closeup as an extreme closeup. One British source also includes an angle called a "mid shot," reaching from around hip level to the top of the head.

The cutoff levels of all these shots are based on two very different theories. One theory advises cinematographers to avoid cutting subjects at body joints because the results are said to look unattractive. The opposite theory, endorsed by this book, says that joints and other natural bending places, like the waist, are appropriate levels for framing the subject.

While reading this book, you need to understand the shot definitions and terminology. As you make videos, you will pick up the terms used in your particular production environment.

Angles cut the figure at natural bending points.

Figure 7-14 Three-quarter angle.

Figure 7-15 Profile angle.

Figure 7-16 Three-quarter rear angle.

Figure 7-17 Rear angle.

Taken together, horizontal camera angle and subject distance provide the most typical shot descriptions, for example, "three-quarter closeup." However, an equally important component of every angle is the camera's height.

Vertical Angle

Shots may also be labeled by the vertical angle from which the camera views the action. The easiest way to describe standard camera heights is by pretending that the camera is at the end of a clock hand and pointed at the center of the dial, **Figure 7-18**.

Using the clock for reference, here are some common camera height (vertical angle) names (**Figures 7-19** through **7-23**):

- *Bird's eye angle*: An extremely high camera position (around 1 o'clock) that simulates the view from a plane or high building, **Figure 7-19**.
- *High angle*: A shot in which the camera is evidently higher than the eye level of a human subject (between 1:30 and 2:30), **Figure 7-20**.
- *Neutral angle*: A shot in which the camera is more or less at the subject's eye level. (3 o'clock), **Figure 7-21**.
- *Low angle*: A shot in which the camera is evidently below eye level (between 3:30 and 4:30), **Figure 7-22**.

Figure 7-19 Bird's eye angle.

Figure 7-20 High angle.

Figure 7-21 Neutral angle.

Figure 7-18 Vertical angle shot names.

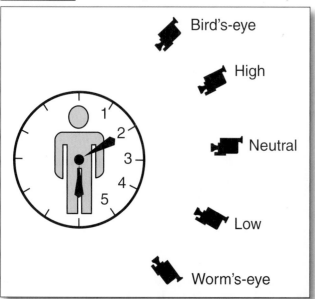

Figure 7-22 Low angle.

- *Worm's eye angle*: An extremely low camera position, looking dramatically upward (5 o'clock), **Figure 7-23**.

The height of an angle names the position of the camera, not the subject. For instance, the position of a dog may be very close to the ground, but in recording the pet from a standing position, you are shooting downward from a high angle.

Lens Perspective

Another type of shot name is based on the appearance created by a particular camera lens. Directors will often call for a shot by the type of lens to be used when they want a stylized rendering of perspective (**Figures 7-24, 7-25, and 7-26**).

Figure 7-23 Worm's eye angle.

- *Wide angle*: Wide angle lenses exaggerate apparent depth and dramatize movement toward and away from the camera, **Figure 7-24**. Directors often call for a wide angle shot when recording a chase, a fight, or some other sequence full of dynamic action.
- **Normal:** Normal angle lenses render perspective approximately the way human vision perceives it—neither increasing nor reducing the apparent depth, **Figure 7-25**.
- *Telephoto*: Telephoto lenses compress apparent depth and de-emphasize movement toward and away from the camera. Telephoto lenses are used to dramatize congestion and to intensify composition on the two-dimensional screen, **Figure 7-26**.

Figure 7-24 Wide angle lens shot.

Figure 7-25 Normal angle lens shot.

Figure 7-26 Telephoto lens shot.

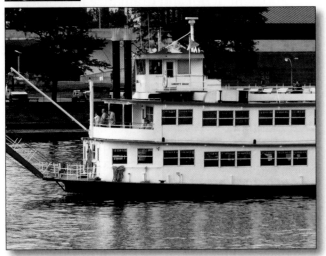

Figure 7-27 Master shot.

Strictly speaking, the opposite of a "wide" angle lens ought to be called a "narrow" angle lens. But in real-world production work, the term "telephoto" is used instead.

Shot Purpose

In addition to camera position, camera height, and lens type, shots are often named for the purpose they serve in the program.

Master Shot

The purpose of a master shot is to record all or most of an entire sequence in a full shot or even longer; capturing the action from beginning to end. After the master has been recorded, matching closer shots of the same action are made to replace parts of the master in the final edited version, **Figure 7-27**.

Today, shooting complete master shots of entire scenes is generally considered old-fashioned.

Establishing Shot

The purpose of an establishing shot is to orient viewers to the general scene and the performers in it. Though beginning a sequence with an establishing shot is also considered old-fashioned, it is often helpful to introduce one early in a sequence.

Since an establishing shot shows all the important parts and players in a scene, it is sometimes the beginning section of a master shot.

Reverse Angle

The purpose of a reverse angle is to show the action from a point-of-view nearly opposite that of the previous camera position. The position is *nearly*, but not fully, the opposite point-of-view because the camera usually stays on the same side of an imaginary action line to preserve screen direction (**Figure 7-28**).

Screen direction is discussed in Chapter 19, *Directing for Form*.

Over-the-Shoulder Shot

The purpose of an over-the-shoulder shot is to include part of one performer in the foreground while focusing on another performer. In addition to controlling emphasis, an over-the-shoulder shot enhances the feeling of depth, **Figure 7-29**.

These are often called "over-the-shoulder two-shots."

Cutaway Shot

The purpose of a cutaway shot is to show the audience something outside the principal action, or to reveal something from an on-screen person's *point-of-view*. For example, if we see several angles of a man mowing his lawn and then a shot of a dog lying in the uncut grass, the dog is a cutaway shot (**Figure 7-30**).

Insert

The purpose of an insert is to show a small detail of the action, often from the point-of-view of a person on the screen. The second image in

Figure 7-28 Reverse shot.

The original angle.   **The reverse angle.**

Figure 7-29 An over-the-shoulder shot.

Figure 7-31 is an insert because it shows the steering wheel from the driver's point-of-view. Because an insert often serves the same functions as a cutaway, it is sometimes called a "cut in."

POV Shot

The purpose of a POV (point-of-view) shot is to show the audience what someone on the screen is seeing. The insert shown in **Figure 7-31** is from the driver's POV

Glance-Object Pair

Cutaway, insert, and POV shots are often used with other angles to make combinations that are often called glance-object pairs. This term means just what its name implies. In the first (glance) shot, a performer looks at something off screen. In the second (object) shot, we see what the performer was looking at (**Figure 7-32**).

"Glance-object-pair" is a term used more by media writers and critics than by production people.

Shot Population

Finally, shots may be labeled according to how many performers are in them (**Figure 7-33**):
- **Single:** A shot showing one person.

Figure 7-30 Example of a cutaway shot in a scene.

Shot 1   Shot 2   Cutaway shot   Shot 3

Figure 7-31 The second image in this sequence is the insert. Because it is shown as the driver would see it, it is also a POV shot.

Shot 1   **Insert shot**   **Shot 2**

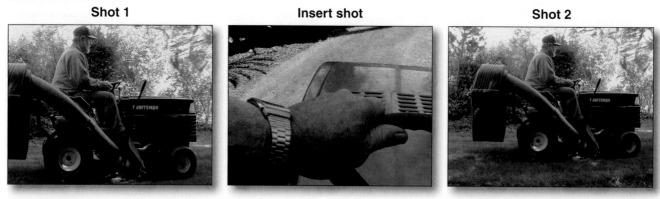

Figure 7-32 The subject glances offscreen (A) at the sundial (B) and then checks his watch against it (C). A and C are the same shot with B cut into it as an insert.

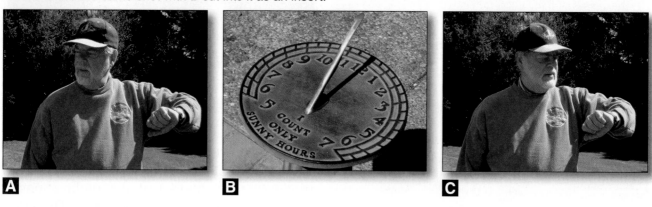

A   B   C

Figure 7-33 Shot population naming.

Single   **Two-shot**   **Three-shot**

- **Two-shot:** A shot showing two people.
- **Three-shot:** A shot showing three people.

By custom, a one-person angle is not called a "one-shot," but simply a "single."

Note that these terms number only the principal performers. For example, a shot of two people having dinner with other restaurant diners in the background could still be called a two-shot.

Shot Name Confusion

Shot names can be confusing at first, for several reasons. Many shot names are synonymous with the camera angles from which they are made ("low angle"), but others are not ("reverse"). Shots are often given multiple names. For instance, a director might call for a "neutral over-the-shoulder two-shot," or a "high angle POV closeup." Shots may be identified by different names. For example, if you have just made an "over-the-shoulder two-shot," asking for a "reverse" indicates that you want another over-the-shoulder two-shot, from the opposite point-of-view. As previously noted, shots often have different names in different countries (U.S. and U.K.), in different media (video, film), in different production centers (New York, Hollywood), or even in different studios in the same area. In practice, you will soon pick up the shot names preferred in the environment where you work.

To add to the confusion of names, some people use "scene" interchangeably with "shot." In this book, a *shot* is a single camera recording, from start to stop. A *scene* is a short, single piece of action, usually conveyed through a number of consecutive shots.

Creating Continuity

If the "words" in the language of video are shots, then video "sentences" are scenes made of several successive shots. Why break the action of a scene into separate shots? Why not simply record everything from start to finish in one continuous piece of video? The most important reason is to show each individual part of the action from the best possible position. This would be impossible to do if the camera were limited to one continuous take. Separate shots also enhance the interest of the material by providing visual variety (**Figure 7-34**).

In most programs, you wish to conceal the fact that you are constantly changing camera angles. Every time viewers notice a new angle, they are briefly distracted from the program content. So, the goal is to make the action appear to happen in a single, continuous flow. The process of doing this is called *continuity*. In creating visual continuity, you stage each shot so that it can be combined unobtrusively with the shots that precede and follow it. The key requirement for doing this is *matching action*. In matching action, you make the incoming shot appear to begin at precisely the point where the outgoing shot ends.

Of course, the action is never truly continuous (unless you are editing shots from two cameras recording simultaneously). In fact, the appearance of uninterrupted action is an illusion created by collaboration between the director and the editor.

The continuity described here is the classic Hollywood style that became the worldwide standard for commercial production. Today, however, these rules are treated flexibly.

Varying Shots

It might seem that the best way to conceal a change in shots would be to make the new angle as much as possible like the old one, as in **Figure 7-35**. But, as you can see, the opposite is true: instead of two shots that cut together invisibly, the result looks like a single shot with a piece chopped out of its middle. This is called a *jump cut* because the subject seems to jump abruptly on the screen.

Editors use the term **cut together** to describe how successfully one shot follows another.

To make a smooth edit, the technique is to match the action closely, while decisively changing

---

Figure 7-34 On screen, seemingly continuous action is constructed of multiple shots.

Shot 1   **Shot 2**   **Shot 3**

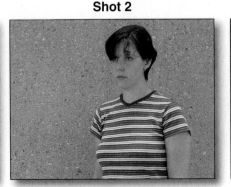
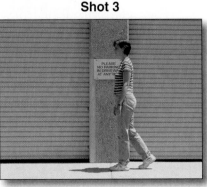

Figure 7-35 Using a camera angle that is too similar makes an obvious cut.

Shot 1

Shot 2

the camera angle, **Figure 7-36**. Generally, this means changing at least two of the angle's three major characteristics: camera position, camera height, and subject size. Unlike **Figure 7-35**, **Figure 7-37** and **Figure 7-38** do change one trait out of three. However, altering only a single trait can still result in an undesirable jump cut that calls attention to the edit. To achieve a smooth edit, a new camera setup should usually change two angle traits—most often the image size, plus either the camera position or height.

A **setup** is a completely new camera location, including the lighting and the microphone position.

Examples of effective angle combinations are shown in **Figure 7-39** and **Figure 7-40**.

Matching Action

While the angle should change decisively from one shot to the next, the action should be closely matched, to make the second shot seem to begin exactly where the first shot ends. There are two basic ways to match action: (1) have the actor repeat part of the action and (2) the editor can conceal the match point in postproduction.

Figure 7-36 A change in camera angle conceals the cut.

Shot 1

Shot 2

Figure 7-37 Changing only the vertical angle creates a jump cut.

Shot 1

Shot 2

Figure 7-38 Changing only the image size also creates a jump cut. However, a strong change in image size is sometimes enough change to conceal the edit.

Shot 1

Shot 2

Figure 7-39 Changing to a new image size and horizontal angle.

Shot 1

Shot 2

Figure 7-40 Changing to a new image size and vertical angle.

Shot 1

Shot 2

A performer can repeat part of the action from the first shot in the second shot. In **Figure 7-41**, for example, the subject pours milk into a mixing bowl and repeats this action in each of two shots. In the establishing shot (A), she raises the milk carton over the bowl. In the second shot (B), she pours the milk (the image shows the end of this shot). In the third shot (C), the subject tilts the carton upright and sets it down. If the edits are made correctly, the three different shots will appear to display a single continuous action.

A second (and easier) way to synchronize action is by concealing the *match point*.

A match point in two shots that record the same action is any point at which the editor can cut from one shot to the other, continuing the action without an apparent break.

In the first shot (A) of **Figure 7-42**, the performer exits the frame, leaving the shot empty. The next shot (B) starts with an empty shot, and then the performer enters the frame in the third shot (C). Though the shots may have been made in completely different locations, the action appears continuous. The performer does not have to both exit the first shot and enter the second one. Typically, an actor will leave the first shot and then be shown again as the second shot begins, or vice versa.

Making Transitions

Up to this point, we have discussed the language of video at the level of "sentences" (shots) and "paragraphs" (scenes). Now, we

Figure 7-41 The performer repeats part of the action in multiple shots. A—The subject begins pouring. B—The subject finishes pouring. C—The subject sets carton down.

A

B

C

Video Program Components

We have identified shots, scenes, sequences, and acts as components of video programs. And, we have established that a shot is a single, uninterrupted recording. But, what do the other terms mean? To obtain a feel for these different program components, imagine a dramatic hour-long TV show about a daring museum robbery.

Program

The program includes the entire story: planning the robbery, committing the deed, fleeing with the loot, and finally being captured and jailed.

Act

Typically around 10–12 minutes in length in a one hour TV program, an act is devoted to each of the four main parts of the story outlined above. In Act Two, for example, we see the entire museum robbery, from approach to escape. An act contains several (even many) separate sequences.

Sequence

A sequence is major component of an act, and has its own beginning, middle, and end. One sequence in the museum robbery act might be breaking in through the roof. First, the robbers cut a hole in the roof, then they anchor a climbing rope to a chimney, and drop the rope through the hole. Two men climb down the rope, while a third robber remains on the roof as lookout. Notice that this sequence is composed of several different actions, each of which is a scene.

Scene

In the sequence of breaking in through the roof, one scene shows the anchoring of the climbing rope. This scene is built up from several individual shots.

Shot

A shot is a single, uninterrupted recording. The rope anchoring scene might require three separate shots. In a medium shot, a robber loops the rope

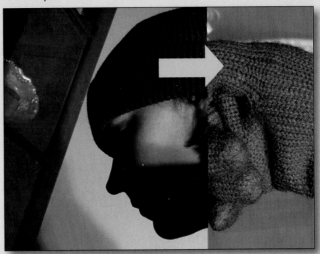

Sequences are sometimes linked by transitions, like this wipe.

(Sue Stinson)

around the chimney. In a closeup insert, he ties a knot in the rope. In a full shot, he tosses the other end of the rope out of frame. The camera pans to follow the rope, revealing a second robber dropping its end through the roof hole. (Remember: a "take" is not a program component, but only a single attempt to record a shot.)

Image

The last shot in the rope tying scene contains two distinct pictures: (1) the man throwing the rope, and (2) the second man lowering it down the hole. Since the shot includes two quite different pieces of visual information, we may say (informally) that it contains two images. A long moving shot may consist of a great many identifiably individual images.

Frame

For completeness, we should include the still picture called the frame. Each second of video consists of 30 frames (in the North American NTSC format). However, a *frame* is too brief to be used as a distinct video component, except in very special cases.

Figure 7-42 Synchronize action by concealing the match point. In the first shot (A), the performer exits the frame, leaving it empty (B) and then enters the empty frame in the third shot (C).

need to add a few rules for making transitions between "chapters." In video language, a chapter is a sequence—a program component assembled from several related scenes. The conventions for transitions from one sequence to another were established long ago and were strictly observed in classic Hollywood movies. Today, however, these rules are broken almost as often as they are observed.

Classic Hollywood Transitions

With few exceptions, traditional movies have signaled transitions with two effects: fades and dissolves.

Silent films also used an effect called an *iris out*, in which a black screen gradually revealed an image in an ever-expanding circle. An *iris in* reversed the direction—a circle gradually closed, leaving a black screen.

A *fade in* begins with a black screen, which lightens gradually until the image reaches full brightness (**Figure 7-43**, first sequence). A *fade out* (**Figure 7-43**, second sequence) is the exact opposite—the image on the screen darkens until the screen is completely black. Fades mark the beginning and ending of major program segments. For example, most commercial TV programs fade out before each commercial break, and then resume by fading in again.

Figure 7-43 Fade in and fade out transitions are exact opposites.

Fade in sequence.

(Sue Stinson)

Fade out sequence.

(Sue Stinson)

A fade out is sometimes called a "fade to black."

A *dissolve* (sometimes called a "mix" in video) is an effect in which one image is gradually replaced by another, **Figure 7-44**. A dissolve, in fact, is a fade in superimposed over a fade out. Traditionally, a dissolve is used instead of a fade to indicate a shorter break in the action, or a less decisive transition in the story.

A *wipe* is an effect in which a line moves steadily across the screen, progressively replacing the old image ahead of it with the new image behind it (**Figure 7-45**). Though a wipe may move along any axis, it is often a vertical line that progresses from left to right. After falling out of fashion for some years, wipes are once again quite popular.

Modern Transitions

Today's audiences are so well-trained to follow story continuity, that fades and dissolves are used less often—though fades almost always begin and end a program. Instead, editors use straight cuts or, sometimes, digital video effects.

Today, fades and dissolves, which were once created in printing the film, are also produced digitally.

Digital Video Effects (DVEs)

A *digital video effect (DVE)* is any computer-generated transition from one shot to another. Modern editing programs offer numberless different types, of which the most traditional are:

- *Flips.* The screen appears to revolve, as if to show its other side.
- *Pushes.* The incoming shot shoves the outgoing shot off the screen.
- *Fly-ins.* The incoming picture swoops to the center of the screen, growing from a tiny dot until it fills the frame completely.
- *Rotations.* Geometric forms with the outgoing shot on one face revolve to reveal the incoming shot on another face, **Figure 7-46**.

There are now so many hundreds of DVEs, many with proprietary names, that assigning them agreed-upon labels is impractical.

Figure 7-44 A dissolve gradually replaces one image with another.

Shot 1

Dissolve

Shot 2

Figure 7-45 A wipe is an old transition that is once again popular.

Shot 1

Wipe in progress.

Shot 2

(Sue Stinson)

Figure 7-46 A rotation effect.

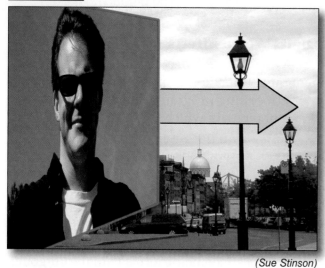

(Sue Stinson)

Digital video effects defy the main rule of classical editing—that transitions should be unobtrusive. On the contrary, DVEs purposely emphasize transitions by dramatizing them. Used judiciously, these arresting visuals add excitement and punch to a program. If they are overused, however, they can grow tiresome and annoying.

DVEs are lavishly employed in TV commercials and music videos. Commercials are very brief, sometimes as short as 5 seconds, and need to get viewers' attention quickly. DVEs are effective attention-grabbers. Because commercials are so short, their rapid-fire effects don't last long enough to wear viewers out.

As for music videos, these programs often use the music as an excuse to explore the possibilities of the video medium. This means that they don't always play by the rules of video grammar. At their best, music videos and other experimental programs may be compared to poems. Just as poets take "poetic license" with the rules of verbal language, video artists sometimes use camera angles, editing continuity, and transition effects without regard to the normal rules of the video medium.

The Frame in Motion

Today, transitions are only one of the tasks performed by DVEs. Now, you see them as continuously moving backgrounds behind news presenters and other "talking heads." The theory seems to be that even meaningless movement is more interesting than none at all. Video organizations use sophisticated software to create the moving backgrounds, and companies sell prefabricated patterns and effects that can be adapted to the users' needs.

Often, subjects on camera are positioned in front of blue or green screens, and the moving patterns are composited behind them (**Figure 7-47**). Compositing digitally replaces only the green or blue color with a different image.

Flashy transitions and other DVEs are often accompanied by "swooshes" and other dramatic sound effects.

Figure 7-47 Creating a DVE background.

A subject in front of a green screen.

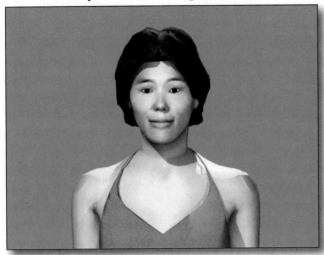

A moving background is composited behind the subject.

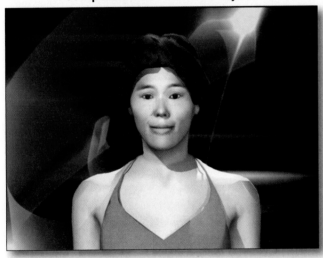

Playing by the Rules?

Nowadays, viewers have become so accustomed to digital effects, jump cuts, and breaks in continuity that they accept them readily, even in mainstream feature films. This means that the frequency with which you break the classic rules covered in this chapter—and the extent to which you break them—depends mainly on the style of your program and the effects you want to achieve. Only you (and your viewers) can determine when the rules have been ignored so thoroughly that your program begins to lose its effectiveness.

But no matter how much you decide to break the rules of video grammar, you need to learn and practice the language of video until you are thoroughly at home with it. Remember that even the most experimental art is grounded in a mastery of classical technique.

Though the artist Picasso usually created "unrealistic" images, he was a master draftsman who could execute realistic works when he chose to do so.

Video Language in Action

This simple sequence demonstrates the wide variety of shot names commonly used in making videos. To include every name would require three times as many images.

Establishing shot

Two shot

Over-the-shoulder shot

Reverse
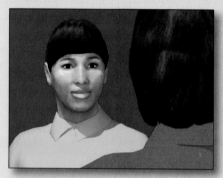

Insert
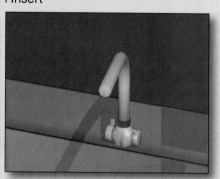

Profile shot
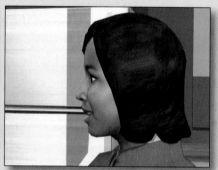

Medium closeup

Single

High angle

Glance

Object (cutaway)

Big closeup
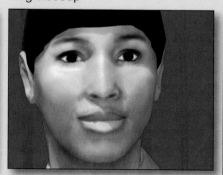

Summary

- Video language uses rules to combine individual pieces into larger units that communicate coherently.
- Camera angles are usually named for one of several different sets of characteristics: subject distance, horizontal camera position, vertical camera position, lens perspective, shot purpose, and shot population.
- In creating visual continuity, stage each shot so that it can be combined unobtrusively with the shots that precede and follow it.
- Video program components include frames, images, shots, scenes, sequences, and acts.
- Transitions are methods used to move from one sequence to another in a program.
- Fades and dissolves are effects that have traditionally been used to signal transitions in a program.
- DVE transition effects include flips, pushes, fly-ins, and rotations.

Technical Terms

Act: A major section (usually between 10–45 minutes) of a longer program.

Bird's eye angle: An extremely high, vertical camera angle that simulates the view from a plane or high building.

Camera angle: The position from which a shot is taken, described by horizontal angle, vertical angle, and subject distance.

Continuity: The organization of video material into a coherent presentation with continuous flow.

Cut together: To follow one shot with another. Also, two shots are said to "cut together" when the edit is not apparent to the viewer.

Digital Video Effect (DVE): Any digitally created transitional device, other than a fade or dissolve. However, fades and dissolves are usually digital, too.

Dissolve: A fade in that coincides with a fade out; the incoming shot gradually replaces the outgoing shot. Typically used as a transition between sequences that are fairly closely related.

Fade in: A transition in which the image begins as pure black and gradually lightens to full brightness. Used to signal the start of a major section, such as an act or an entire program.

Fade out: A transition in which the image begins at full brightness and gradually darkens to pure black. Used to signal the end of a major section, such as an act or an entire program.

Frame: A single still picture; 30 frames make a second of NTSC video.

High angle: Vertical camera angle in which the camera is positioned noticeably higher than human eye level.

Image: A single set of visual information. An image may last for many frames, until the subject, the camera, or both, create a new image by moving. Most shots contain several identifiable images.

Jump cut: An edit in which the incoming shot is visually too similar to the outgoing shot.

Low angle: A vertical camera angle in which the camera is noticeably positioned below human eye level.

Match point: The specific places in two shots with matching action, where the shots can be cut together to make the action appear continuous.

Neutral angle: Vertical camera angle in which the camera is positioned more or less at human eye level.

Point-of-view: A vantage point from which the camera records a shot. Usually abbreviated "POV."

Profile angle: Camera angle in which the camera is placed at a right angle to the front angle shot.

Program: Any complete video presentation, from a five-second commercial to a movie two or more hours long.

Rear angle: Camera angle in which the camera is placed directly opposite its front position and fully behind the subject.

Scene: A short segment of program content, usually made up of several related shots that occur in a single place at a single time.

Sequence: A longer segment of program content, usually consisting of several related scenes.

Telephoto lens: A lens (or a setting on a zoom lens) that magnifies subjects and minimizes apparent depth by filling the frame with a narrow angle of view.

Three-quarter angle: Camera angle in which the camera is placed roughly between 15°–45° around to one side in front of the subject.

Three-quarter rear angle: Camera angle in which the camera is placed about 45° around the back side of the subject, so the subject is facing away from the camera.

Wide angle lens: A lens (or a setting on a zoom lens) that minimizes subjects and magnifies apparent depth by filling the frame with a wide angle of view.

Wipe: A transition between sequences in which a line moves across the screen, progressively covering the outgoing shot before it with the incoming shot behind it.

Worm's eye angle: An extremely low camera angle in which the camera "looks" dramatically upward at the subject.

Review Questions

Answer the following questions on a separate piece of paper. Do not write in this book.

1. What is a *shot*?
2. *True or False?* An act is a major assembly of sequences in a program, usually 10–45 minutes in length.
3. A(n) _____ is a single attempt to record a shot.
4. Identify the six characteristics reflected in camera angle naming conventions.
5. In a(n) _____ shot, a standing figure fills the screen from top to bottom.
6. Explain the difference between a closeup shot and a big closeup shot.
7. What is a *worm's eye* camera angle?
8. How does a telephoto lens change the appearance of an image?

9. What is the purpose of an establishing shot?

10. The purpose of a(n) _____ shot is to include part of one performer in the foreground, while focusing on another performer.

11. "POV" is an acronym for _____.

12. *True or False?* Shot population naming conventions include both principal actors and background performs visible in the shot.

13. The process of organizing video material to create continuous flow is called _____.

14. Describe two ways to match action between shots.

15. Identify three traditional transitions and describe each.

16. DVE stands for _____.

STEM and Academic Activities   STEM

1. **Technology.** The technology behind blue or green screen effects is called *chroma key*. How do chroma key effects work? How are they applied? What are some common applications of chroma key effects?

2. **Social Science.** The video industry is not the only profession that has developed its own "language." What other professions use unique phrases or apply different meanings to common words when communicating? Provide some examples of the profession's "language" and the corresponding meanings.

Although cuts from extreme long shots to tight closeups are highly dramatic, their shock value wears off if used too often.

CHAPTER 8 Video Sound

Objectives

After studying this chapter, you will be able to:

- Identify the functions of audio (sound) tracks.
- Explain the relationship between sound and video in programs.
- Recognize the various roles of sound effects in a video sequence.
- Identify the layers of sound in a typical composite sound track.

About Video Sound

Sound is vitally important in the medium of video. It communicates so powerfully that you can create successful programs of audio without video (which, of course, is called radio), but not video without audio.

Even silent films were not truly *soundless*. Because stories unfolding in absolute silence conveyed an eerie feeling, even the smallest silent movie theaters showed their programs with musical accompaniment.

And yet, sound is often undervalued in video production. In amateur programs especially, the finished audio track may be nothing more than whatever the camera's microphone happened to pick up.

Strictly speaking, **sound** is what you record and hear, while **audio** is the audible half of a program (as opposed to "video"). In practice, the terms *sound* and *audio* are often used interchangeably.

Quality audio depends on both a mastery of the technical processes and an understanding of how sound contributes to the effect of a video program. The technical processes are covered in later chapters on sound recording and editing. In this chapter, we focus on how to use sound as a tool to communicate information and stir emotion. Along the way, we will explain the principal components of a typical sound track, consider the role of music in video, and look briefly at the history of audio recording. We will begin with sound's most important role: delivering information.

Delivering Information

The biggest task that sound performs is delivering information, most often through dialogue spoken by on-screen performers or the words of an unseen narrator. The audio track also provides information through sound effects, and, as we will see, these humble noises that are laid down by sound editors form a remarkably eloquent (and economical) method of communicating.

To "lay down a track" means to record one or more audio components. To "lay in" an individual piece of audio means to add it to an audio track in postproduction.

Dialogue

Video is, essentially, a visual medium, but part of good script writing lies in knowing when it is more efficient to tell something than to show it. Since reading lengthy on-screen titles is tiresome, "telling" typically involves some form of audible speech—usually *dialogue*, which is the on-screen conversations between characters (**Figure 8-1**).

Dialogue often delivers information explicitly. "I'm going to the zoo" tells us, in a straightforward way, where the speaker is headed. Dialogue may also contain implicit information. "I'm going back to the zoo," tells us that the speaker has come from the zoo, while "I'm going to the zoo again" indicates that the speaker has been there at some time in the past. If a second character says, "You're not going to the zoo again?" we learn three things: that the first character is going to the zoo, that he or she has been there often, and that the second character disapproves of these repeated visits. This dialogue has delivered three pieces of information in seven words and two seconds—working far more economically than visuals could in this case.

In a well-directed story program, dialogue also provides information about emotions and character. If "You're not going to the zoo again?" is said with fond amusement, it conveys certain things about the speaker's feelings. But, if the line is delivered with weary resignation instead, or perhaps in a resentful whine, the speaker's feelings appear quite different.

Figure 8-1 Dialogue is on-screen speech.

Dialogue is also important in many non-fiction videos, where it often takes the form of interviews. Some interviews are presented as conversations between two people on screen. (This approach is common in news and discussion programs.) In other cases, the interview is conducted and edited so that the interviewer is invisible and the questions are unheard. The interviewee seems to be speaking only to the viewer, **Figure 8-2**.

Narration

Often, a person being interviewed begins on screen. But, after a few moments the visual component of the interview is replaced by footage that illustrates what the person is talking about, while his or her voice continues on the sound track. This is called *voice-over*, (usually abbreviated "V.O." in scripts).

The terms "off screen," "off camera," and "V.O." are essentially interchangeable.

Voice-over interviews are widely used in documentary programs, where they may alternate with fully-scripted narration. Spoken by professional voice-over talent, narration is by far the most efficient way of delivering verbal information. A good script can pack a great deal into a short space, as you can see by the brief script excerpt shown in **Figure 8-3**. If you take a moment to study the narrator's

Figure 8-2 In the contemporary interview style, only the subject appears on screen.

(Adobe)

Figure 8-3 An excerpt from a script.

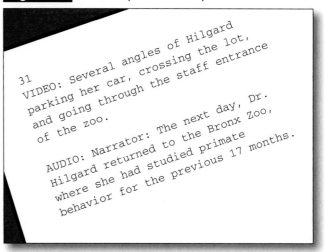

31
VIDEO: Several angles of Hilgard parking her car, crossing the lot, and going through the staff entrance of the zoo.

AUDIO: Narrator: The next day, Dr. Hilgard returned to the Bronx Zoo, where she had studied primate behavior for the previous 17 months.

sentence, you will discover that just 22 words deliver 10 pieces of information:
1. The subject is female.
2. Her name is Hilgard.
3. She has a Doctorate.
4. She went to the zoo.
5. It was the Bronx Zoo.
6. She had been there previously.
7. She had studied there.
8. Her study was in primate behavior.
9. Her study lasted 17 months.
10. The study period was quite recent.

The visuals, on the other hand, are essentially meaningless filler included to occupy the screen while the narration is spoken.

Filler visuals, often called "wallpaper," are covered more thoroughly in Chapter 10, *Program Creation*.

Another reason for narration's power lies in the skill of the professional narrator. Using vocal pace, inflection, and emphasis, a narrator organizes and highlights text in the process of speaking it. This makes the narration easier to understand and more interesting to listen to. If you have ever winced at the sound of an amateur reading a text in a droning monotone, you will appreciate the complex techniques involved in speaking narration with professional skill.

Sound Effects

Sound effects are the noises that come from the world on the screen. Some are recorded with the video and consist of the actual sounds produced by the on-screen action. Many others are "laid in" by the sound editor in postproduction, **Figure 8-4**.

Typically, the sound effects synchronized with on-screen action are designed to heighten the sense of reality. To see how this works, try muting your TV sound during a sequence of vigorous action, such as a car chase or a fight. Notice how the on-screen activities suddenly grow remote and uninvolving.

Although sound effects are typically used to heighten the realism of visual action, they can also deliver information independently of the video. For example, imagine a car chase during which our hero's brakes suddenly fail. The director communicates this information with an insert—a close shot of a foot repeatedly punching a brake pedal. The problem with this is that the foot movement looks no different from that of heavy braking, so the shot does not convey the essential information that the car's brakes have failed.

To supply this information, the sound editor lays in repeated sound effects of the pedal clanking on the floorboard of the car. It is that sound, not the picture, that tells the audience that the brakes are useless (**Figure 8-5**).

In pre-digital film editing, each sound track component—such as dialogue, effects, or music—was on a separate reel of magnetic film. These physical components were labeled "sound track *elements*."

Figure 8-4 A digital sound editing program.

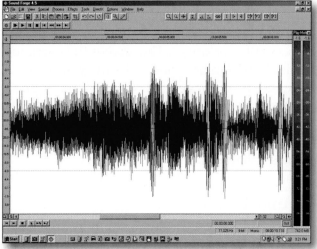

(Sound Forge)

Figure 8-5 By itself, the video communicates only "braking." With an added sound effect, it communicates "brakes failed!"

No sound effects.

Sound effect added.

CLANK!

Conveying Implications

Although sound is often used to supplement the video, its most important informational role is to tell you about things that are not on screen—things that exist only by implication. Remember the first law of video space: what is outside the frame does not exist, *unless that existence is implied*. If you provide the audience with audio clues that something is there, even though they cannot see it, they will believe you.

Sound Implies Existence

By far, the easiest way to imply the presence of things off screen is through sound. In **Figure 8-6**, for example, a mother says goodbye to her

The Perception of Sound

Sound does not really exist in nature. What we think of as "a sound" is actually a disturbance of the air, moving in a pattern of waves. These waves apply rapidly shifting air pressures to our ears, which create nerve impulses, which our brains organize and interpret as "sound."

Sound in the Actual World

Human hearing is marvelously acute and subtle, because our brains are skilled at organizing, processing, and interpreting the audio babble that continuously assaults our ears. We can regulate overall volume, adjusting our tolerance for loudness to suit the environment. For instance, imagine driving at 65 miles per hour with the car radio playing. When you slow to 30 mph, the radio suddenly sounds uncomfortably loud, even though the volume has not increased. The noisy radio did not seem too loud at 65 mph, because your brain can adjust for the higher overall sound levels of high-speed driving.

We can also listen selectively; processing some parts of the overall sound, but ignoring others. For example, most people quickly filter out the drone of an air conditioner until they don't notice it at all. On the other hand, parents can often hear the faint cries of a baby two rooms away, even through the sound of a TV program. With the air conditioning, they suppress and ignore the sound; but with the baby, they isolate and recognize the sound, despite competition from other noise. Even though people *hear* everything, they *listen* selectively.

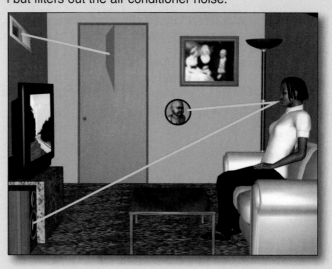

The listener hears the TV sound and the baby's cry, but filters out the air conditioner noise.

Sound from Video Programs

Unfortunately, we do not seem able to listen to an audio track as selectively as we listen to real-world sound. This creates both a problem and an opportunity for video makers. The problem is that the sound editor must do the job that our brains do with real-world sound: regulate volume and filter out unwanted sounds, while isolating and emphasizing wanted ones. The opportunity lies in the brain's ability to recognize and process many sounds at once. This allows editors to build audio programs out of many layers of audio. In complex, big-budget productions, individual sound track elements can number in the dozens.

Historically, complex audio was uncommon in video programs because the hardware required to play back, synchronize, and mix audio tracks on videotape was very expensive. The alternative of *mixing* a few tracks at a time and then remixing the subassemblies, resulted in loss of audio quality. (Larger studios could build up multiple tracks on one- or two-inch audiotape.) With the establishment of digital recording, however, these difficulties have disappeared, and audio accompanying video can now match the sophistication of audio for film.

daughter and watches her ride away. But what, exactly, is the daughter riding in or on? We can't tell from the visuals because her transportation remains outside the frame.

Now suppose we add sound effects to the sequence, as in **Figure 8-7**.

In video scripts, "SFX" is the standard abbreviation for "sound effect."

With just a honk, a hiss, and an engine roar, we have constructed a complete bus (**Figure 8-8**) and added it to the scene—a bus whose existence is only implied, and implied exclusively by sounds.

Figure 8-6 A sequence without sound effects.

Mother and daughter embrace...

...the daughter leaves the frame...

...and her mother watches her ride away.

Figure 8-7 Sound effects added to the mother and daughter sequence.

SFX: A bus horn honks impatiently.
DAUGHTER: G'bye, Momma.
MOTHER: Honey, you take care, y'hear?

SFX: The hiss of bus doors opening...

SFX: ...and then closing; then the diesel rumble of a bus, fading in the distance.

Figure 8-8 A phantom bus is created entirely by sound effects.

(Sue Stinson)

We could change the young woman's transportation just by substituting different effects:

- A brass bell chimes over an idling speedboat motor. The motor revs up and then gradually fades, leaving the sound of gulls and water lapping around a pier.
- A conductor calls, "Booooard!" and we hear the first slow chuffs of a big steam locomotive. The puffing grows louder and faster, then diminishes in the distance, with the wail of a whistle.
- The motor of a single-engine plane idles smoothly. After a moment, the door chunks shut, the engine revs, and the sound moves farther away as the plane starts to taxi down the runway.

By using sound effects alone, we could supply the daughter with four completely different modes of transportation.

Sound Implies Locale

With the bus and the airplane, we can only assume that the scene is set on a road or an airstrip. In the case of the train, however, the conductor's call supplies the train station for us. In the same way, the gulls and lapping water inform us that we are on a boat dock.

You can easily think of the locations that are established by sounds like the crackle and roar of a forest fire, the hiss of rain and rattle of thunder, or the deafening roar of a steel mill (**Figure 8-9**). We could multiply examples, but the point is clear: sound is psychologically powerful in evoking a sense of place.

Sound Creates Relationships

Sound is very useful in bringing together and unifying program components that were originally shot separately. For example, look carefully at the sound elements supporting the visuals in **Figure 8-10**. The exterior shots of the

Figure 8-9 The sounds of a steel mill are as overwhelming as its images.

(Corel)

truck were probably made on location with stunt doubles doing the driving. The interiors with Bob and Bill were shot separately, with their pickup towed safely behind a camera truck.

Figure 8-10

BOB: Yer goin' too fast, Bill!
SFX: *(muted)* **Engine roar.**

BILL: *(O.S., faint)* **Not fast enough!**
SFX: *(loud)* **Engine roar, gravel spraying.**

BILL: *(muffled by window)* **Hey, there's a shortcut.**
SFX: *(loud)* **Engine roar continues.**

BOB: *(O.S., faint)* **That's a field!**
SFX: *(loud)* **Engine roar and gravel spraying continue.**

BOB: *(muffled by windshield)* **Now ya done it!**
SFX: *(loud)* **Engine, springs creaking, rocks hitting metal.**

BOB: *(O.S., faint)* **We're stuck.**
SFX: *(loud, close)* **Engine, hiss of broken radiator.**

Layers of Sound

The audio part of a video program is usually built up from several different sound tracks. These individual components can vary considerably in both their level of reality and their function in the overall sound track.

Levels of Reality

Some sounds are absolutely authentic, some are completely artificial, and some fall in between these two extremes. Listed in order of reality, we find:

- **Live sound.** This audio is recorded simultaneously with the video and is often called *production sound*. Live sound includes spoken dialogue, the noises created by the action, and the ambient sounds of the shooting location.
- **Sweetened sound.** This is live audio that has been processed in postproduction to improve its quality. In some cases, a process called *equalization* strengthens certain sound frequencies (pitches), while relatively weakening others. Often, extra live sound is added to conceal the audio differences between different shots.
- **Foley sound.** This is a track of sound effects recorded live in a special studio while the video is displayed, so that Foley technicians can synchronize the sounds they are creating with the program in real time. For example, a Foley studio may have several sections of floor covered variously with gravel, asphalt, concrete, or tile, for use in recording footstep sound effects synchronized with the footfalls of on-screen actors walking on similar material.

- **Library sound.** This is audio that is not created specifically for a particular program, but is stored as digital files for use where needed. Some sound libraries contain thousands of different effects. General background sounds, such as waves, forest noises, or city traffic, are often designed as digital "loops" that will play continuously.
- **Synthesized sound.** This is digital audio that is created on a computer, instead of being recorded acoustically.

Audio Functions

Audio tracks can also be classified by their functions. Here are some of the most common uses for sound:

- **Production sound.** As noted previously, this is sound that is recorded with the video.
- **Presence.** Often called "room tone," presence is recorded at a shooting location while the cast and crew remain perfectly still. The result is a recording of the location's ongoing background sounds, whether birds, distant traffic, air conditioners, or anything else. Presence is used to fill gaps left in the production sound track when unwanted sounds are cut out.
- **Background sound.** This is an ongoing sound effect, like the chatter of diners and the tinkle of china and silver in a restaurant. When the background track is mixed with the live production tracks, shot-to-shot changes in sound quality are smoothed over. Like

Part of a Foley setup for recording footsteps on different surfaces.

Thousands of sound effects are available in digital libraries.

presence, background sound smoothes out differences in the production sound track. But, it is also intended to supply sound that was not present during production recording.

Background sound may be recorded during production or selected later from a library.

- **Sound effects.** Sound effects (often abbreviated "SFX") are specific sounds, such as walking, gunfire, or car engines that are added in postproduction, whether from Foley recording or from a sound effects library.
- **Dialogue replacement.** This is a process by which actors re-record dialogue in a studio, synchronizing it to the lip movements of performers as played back on a monitor or screen. The procedure is still often called "looping" because in pre-digital days it was accomplished by splicing the film containing a few lines of dialogue into an endless, head-to-tail loop, which could be projected continuously while actors practiced synchronizing replacement dialogue with lip movements on the screen.
- **Music.** Music is added to the sound track to enhance continuity and communicate feeling. Music composed for a specific program may be timed right down to the frame. Library music (pre-composed and stored selections) may be selected by length, or else faded in, faded out, or both, in order to time it to the picture. Digital music loops contain embedded information that allows them to be adjusted precisely to fit any length required.

The following are some typical uses for Foley sound effects:

Microphone too far away to record actual sound.

Actual sound does not sound the way it should.

Exaggerated sound needed for emotional effect.

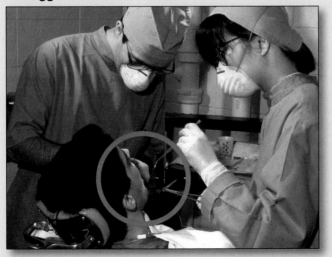

Actual sound obscured by background noise.

Incidentally, notice in the last frame that the "wreck" has been created by tilting the camera off-level.

To meld these different times and places into a single sequence, the sound editors used three common techniques:

- They laid in some of the actors' dialogue as off-screen voice-overs accompanying the exteriors.
- They continued the car sound effects over shots of the truck's interior.
- They varied the volume, perspective, and quality of the dialogue and sound effects from shot to shot to match the different viewpoints of the audience, inside or outside the truck.

The combination of audio styling and matched-action cutting reinforces the illusion of a single, continuous sequence.

Strengthening Continuity

Sound is a great smoother of rough connections. In the example of Bill and Bob (**Figure 8-10**), those connections are between shots in a sequence. Sound is also used to tie separate sequences together through a technique called a *split edit*.

The Split Edit

In Chapter 5, *Video Time*, we discussed the split edit. We return to it here because a split edit is impossible without audio. It is a transition from one shot to another in which video and audio do not change simultaneously. Either the incoming video is accompanied by a few extra moments of the outgoing audio or, conversely, the incoming audio is laid over the last part of the outgoing video. Split edits are usually, though not always, used during transitions between two sequences. To see how split edits work, we return to Bob and Bill, still stuck in their disabled car.

Since the next four trios of pictures (**Figure 8-11** through **Figure 8-14**) are identical, pay close attention to their captions, which demonstrate the different options for split edits.

Straight Cut Version

For reference, **Figure 8-11** shows a *straight cut* (without a split edit). Notice that Bill's picture and sound conclude together. In this straight cut version, the joke is plain but heavy-handed.

Split Edit, Video Leading

In **Figure 8-12** the video changes to the new sequence ahead of the audio. We cut to Bob sitting in the darkness (third sequence in **Figure 8-12**) as Bill's line from the preceding scene is completed in voice-over. Placing the old line over the new shot adds an ironic comment.

Split Edit, Audio Leading

Figure 8-13 shows the opposite form of split edit. In this version, we hear the obvious sounds of night creatures as Bill completes his line in the first sequence. Instead of conveying a sense of irony, like **Figure 8-12**, this opposite approach allows the audience to anticipate the joke.

Figure 8-11 Straight cut version.

BOB: *(O.S., faint)* **We're stuck!**
SFX: *(loud, close)* **Engine, hiss of broken radiator.**

BILL: **Aw, simmer down. Someone'll come along any minute.**

SFX: **Night frog/cricket sounds: ribbit, ribbit, ribbit…**

Figure 8-12 Split edit, video leading.

BOB: *(O.S., faint)* **We're stuck!**
SFX: *(loud, close)* **Engine, hiss of
broken radiator.**

BILL: **Aw, simmer down.**

BILL: *(O.S.)* **Someone'll come
along any minute.**
SFX: **Night frog/cricket sounds:
ribbit, ribbit, ribbit…**

Figure 8-13 Split edit, audio leading.

BOB: *(O.S., faint)* **We're stuck!**
SFX: *(loud, close)* **Engine, hiss of
broken radiator.**

BILL: **Aw, simmer down.**
SFX: **Night frog/cricket sounds:
ribbit, ribbit, ribbit…**
BILL: **Someone'll come along
any minute.**

SFX: **Night sounds continue.**

Two-Way Split Edit

Figure 8-14 shows the audio split both forward and backward. The sounds from the second sequence begin over the last shot of the first one, and the dialogue from the first sequence continues (voice-over) in the second.

Which version is best? It depends on the dramatic needs of the program.

Evoking Feelings

In addition to performing all its other tasks, sound is very useful for conveying and enhancing feelings, both momentary responses and overall moods. Music is the most powerful mood setter, as we will see, but sound effects and background noises can also stimulate viewers' emotions.

Sound and Mood

Certain background sounds are useful for setting the overall mood or tone of a video sequence. For example:

- Gentle surf and steady rain sound peaceful and soothing.
- Howling wind sends a message of desolate emptiness.
- Steady mechanical vibration imparts a mood of tenseness. (That is why the Death Star megaship in the original *Star Wars* film is always accompanied by an ominous, almost subsonic, rumble.)
- The sounds of busy city streets convey feelings of energy.

Figure 8-14   Split edit, two-way.

BOB: *(O.S., faint)* **We're stuck!**
SFX: *(loud, close)* **Engine, hiss of
    broken radiator.**

BILL: **Aw, simmer down.**
SFX: **Night frog/cricket sounds:
    ribbit, ribbit, ribbit…**

BILL: *(O.S.)* **Someone'll come
    along any minute.**
SFX: **Night sounds continue.**

The History of Video Sound

The recording of sound has a longer history than the recording of visuals. In fact, in the late 19th century, Thomas A. Edison worked to develop moving pictures; not as a new form of entertainment in its own right, but as an accompaniment for the phonograph that he had recently invented.

Radio

The first broadcast medium to use audio was radio. And radio, of course, was sound only. It was here, in this "theater of the mind," that sound technicians invented ways to suggest places and events through sound effects. Background noises, such as theater audiences or rushing rivers, were recorded on phonograph records.

Sound effects teams developed "live" sounds—operating miniature doors, "walking" shoes across different surfaces, and firing blank shots in real time—while the story unfolded. Once sound movies arrived, these live sound effects operations were adopted by the Foley recording studios of movies. "Foley recording" is named for Jack Foley, the Hollywood technician credited with inventing this process for use in sound films.

Sound effects technician Tom Keith rehearses for the *Prairie Home Companion* radio program.

(Minnesota Public Radio photo by Frederic Petters)

Early Sound Films

At first, movie audio was necessarily simple: a live production sound track "sweetened" with some effects and underscored with music. Because the emotional power of music was well understood, early sound movies used music lavishly—in some cases, almost continuously.

Modern Films

After World War II, movie sound advanced again with the introduction of magnetic recording on both tape and film. (Movie sound had previously been recorded optically by photographing the changing light outputs of photoelectric cells.) Magnetic recording offered improved quality, flexibility, and economy, so that the sound track elements used to make a composite audio track could multiply from fewer than a dozen, to as many as fifty or more.

Originally, film audio was an actual "track" running beside the images.

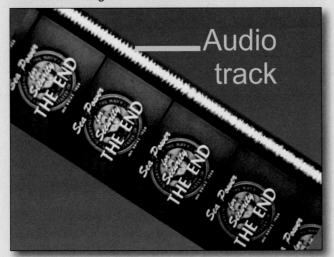

Optical sound recording gave way to magnetic recording.

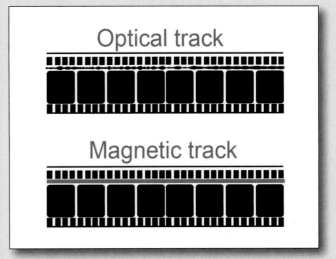

Magnetic recording brought another benefit too: it allowed sound technicians to lay down *control tracks*—electrical signals that gave each "frame" of sound (1/24 second) a unique identification, like a street address. This made it easier to synchronize sound tracks with one another (and with the picture) without mechanical connection. As the techniques of film and video gradually merged, film sound engineers began using the time code method, in which the length of a recording is measured in hours, minutes, seconds, and frames, and the time "address" of every single frame is recorded with it.

Television

In the early days of TV, all sound was "live," as it had typically been in TV's parent, radio. Music and background effects were usually prerecorded on audiotape, but individual sound effects were often produced in real time. The arrival of videotape permitted more sophisticated sound editing and the development of time code made it practical to lay in individual synchronized sound effects. Nonetheless, the process was time-consuming and editors tended to avoid elaborately layered tracks.

Video Sound Today

Video sound came of age with the development of computer-based editing, in which audio components (like video components) are digitized and all editing is performed on the computer before final archiving (storage) digitally or on film. Nonlinear sound editing allows almost instant access to all sound components, and that speeds up editing substantially. Also, the computer can handle dozens of tracks at once. As a result, the possibilities for creating video sound tracks are almost limitless.

- Jet engine background sounds deliver a feeling of power, **Figure 8-15**.
- The crackling of logs in an open fire suggests cozy safety.

These sound effects are mood setting, partly because they are more or less continuous, running in the background of the audio track.

Sound and Emotional Response

Other sounds can evoke more specific emotional responses. This is often because they are associated with particular actions or objects. For instance:

- Echoing footsteps on concrete create suspense.
- The crackle of crunching bones and the pulpy thud of smashed flesh can arouse disgust, and even pain.
- The metallic click when a rifle hammer is cocked can provoke a sense of dread.
- The chuckling sound of pouring champagne says warmth and romance.
- Fireworks explosions can signal triumph, **Figure 8-16**.

Figure 8-15 Jet engine sound effects convey a sense of power.

(Corel)

Figure 8-16 Triumphant fireworks are enhanced by sound explosions.

When any sound is weirdly distorted, it can impart a feeling of disorientation and even insanity. Listed on the printed page, these evocative sounds may seem like clichés; but when editors lay them skillfully to appropriate visuals, they can be remarkably effective.

Unrealistic Sounds

Some sound effects are frankly unrealistic, especially in comic programs. For example, as a running character skids around a corner, we hear the sound effect of squealing tires. And when he rushes past the camera, he is accompanied by the sound of an engine roar.

Unrealistic effects can also be used in serious programs. For instance, in a fine nature documentary titled *The Year of the Jaguar*, we hear the ominous rattle of a shell or bamboo wind chime whenever a jaguar spots some off-screen prey. Note the care with which this effect was selected. The sharp tinkling of a glass wind chime would sound out of place in this Central American rain forest. But, the more organic sound of shells or bamboo is acceptable, even though it is equally unrealistic.

Perhaps the most unrealistic effects of all are the sounds made by the spaceships in many science fiction programs. We all know that sound does not exist in the vacuum of space, but we accept it anyway, as an evocation of the power of these mighty vessels.

The Role of Music

The most powerful sound tool for evoking emotion is music, and music has been an integral part of movies since their very beginnings. Joy, sorrow, excitement, dread, and almost every other human feeling have been evoked by music. Building upon the innovations of opera composers, the creators of film and video music have evolved a wide variety of techniques for supporting and enhancing the visual track, **Figure 8-17**. Though music's emotional effect is evident in fiction programs, it is equally important in other types of videos.

Commercials and promotional videos rely heavily on emotions to sell their ideas or products. Even documentary and training programs utilize music to set an overall tone, create transitions between program sections, and supply a sense of closure at the end. Like production sound, sound effects, and background sound, music is treated by sound editors as a separate audio component to be blended with the other sound elements in creating a final composite sound track.

Figure 8-17 Library music compositions are offered in several versions.

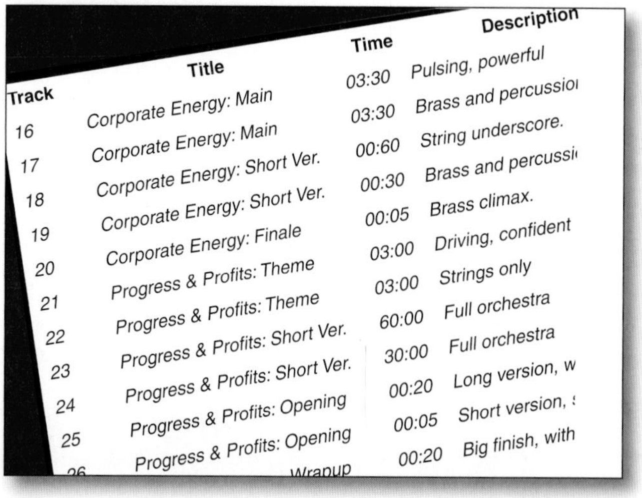

Track	Title	Time	Description
16	Corporate Energy: Main	03:30	Pulsing, powerful
17	Corporate Energy: Main	03:30	Brass and percussion
18	Corporate Energy: Short Ver.	00:60	String underscore.
19	Corporate Energy: Short Ver.	00:30	Brass and percussi
20	Corporate Energy: Finale	00:05	Brass climax.
21	Progress & Profits: Theme	03:00	Driving, confident
22	Progress & Profits: Theme	03:00	Strings only
23	Progress & Profits: Short Ver.	60:00	Full orchestra
24	Progress & Profits: Short Ver.	30:00	Full orchestra
25	Progress & Profits: Opening	00:20	Long version, w
	Progress & Profits: Opening	00:05	Short version,
26	Wrapup	00:20	Big finish, with

Communicating with Sound

As you can see, sound is a powerful resource in creating effective video programs. It can deliver abstract information far more efficiently than visuals can. It can suggest places, things, and even people that don't appear on screen. Sound helps bind the many separate shots that make up the visuals into a single, flowing whole. Sound has the power to arouse all kinds of emotions in viewers. And now, with digital postproduction, the full resources of sound are available to video makers. Digital editing offers the ability to weave subtle audio textures, creating composite tracks that are rich and complex.

Summary

- Sound is what you record and hear, while audio is the audible portion of a program.
- The functions of sound in video include delivering information, conveying implications, strengthening continuity, and evoking feelings.
- The audio of a video program is built from several different sound tracks, each track serving a specific purpose.
- Sound effects support the video portion of a program, but they also provide information that is not presented on the screen.
- The layers of sound in a typical composite sound track include production sound, presence, background sound, sound effects, dialogue replacement, and music.

Technical Terms

Audio: Collectively, the sound components of an audio-visual program.

Dialogue: Speech by performers on-screen; often as conversation between two or more people.

Equalization: The adjustment of the volume levels of various sound frequencies to balance the overall mixture of sounds.

Mixing: The blending of separate audio tracks together, either in a computer or through a sound mixing board.

Production sound: The "live" sound recorded with the video.

Sound: The noises recorded as audio.

Sound effects: Specific noises added to a sound track.

Straight cut: (1) An edit in which audio and video change simultaneously. (2) An edit that does not include an effect, such as a fade or dissolve.

Voice-over: Narration or dialogue that is recorded independently and then paired with related video.

Review Questions

Answer the following questions on a separate piece of paper. Do not write in this book.

1. What is the difference between *sound* and *audio*?
2. The on-screen speech of program characters is called _____.
3. *True or False?* Narration is the most efficient way of delivering verbal information.
4. When synchronized with on-screen action, sound effects heighten a program's sense of _____.
5. How does sound convey implications?
6. What is the *production sound track*?
7. Explain the difference between *presence* and *background sound*.
8. What is the typical script abbreviation for "sound effects?"
9. *True or False?* In a split edit with audio leading, the sound changes before the picture.
10. Give some examples of how sound can set the mood or evoke emotion in a sequence.

STEM and Academic Activities

STEM

1. **Science.** Investigate how sound frequency is measured. What is the audible range of sound frequencies?
2. **Engineering.** Develop an illustration that graphically illustrates the concept of a split edit. Choose one type of split edit to illustrate.
3. **Mathematics.** Research three different sound effects library products/companies. Chart the subscription plan details for each sound effects library. What is the cost per download for each sound effects library? Which product/company offers the best value?

The sound effect "whooosh!" of a vehicle passing at high speed would not be realistic for a foot race, but it could be effective—especially in comedy.

(Rafal Olkis / Shutterstock.com)

CHAPTER 9 Project Development

Objectives

After studying this chapter, you will be able to:

- Identify the steps involved in video project development and the tasks included in each step.
- Summarize a video project in terms of its subject, objectives, audience, delivery system, length, concept, and genre.
- Compare different types of delivery systems.
- Recognize the various levels of program treatments.
- Explain the use of a storyboard in project development.
- Recall the appropriate applications of common script formats.

159

About Project Development

Project Development is the first of three chapters on the preproduction phase of video. The second chapter, Chapter 10, *Program Creation*, addresses the creative process of writing scripts (or storyboards) for popular video genres, such as fiction, instructional, and documentary. The third chapter, Chapter 11, *Production Planning*, covers the nuts and bolts of preparing for actual shooting.

In this chapter, Chapter 9, we are concerned with the first steps in the preproduction phase: the process of developing a video project—of any type—for scripting and eventual production.

In Hollywood, a movie being prepared for production is said to be "in development." Why not just say that it is being "scripted" or simply "written"? The reason is that no matter how important a script may be, it is only an end-product in a much larger creative process—the process of project development.

For simplicity, "project," as used here covers the development process of all types of videos. "Production," the subject of the following chapter, is used in creating specific types of videos.

This chapter covers the fundamentals of this vital process. We will start with a common sense approach to defining the video project you wish to produce. Then, we will look at various ways to create a design for your project—a design intended to guide you and your colleagues through production and postproduction.

If you propose to shoot a short simple video, you may be tempted to skip the development process. If, for example, you are making a music video about a garage band, you may feel that all you have to do is record the group playing a song and then edit the results. But, this approach is almost guaranteed to produce a less than satisfactory program. Even a simple project like that garage band video can benefit from the process of development. You begin this development process by defining the project.

Even spontaneous videos shot with mobile phones or other portable devices will benefit from some planning, as explained in Chapter 19, *Directing for Form*.

Defining the Project

The first step in the development process is to describe exactly the video that you want to make. This may sound obvious, but far too many programs suffer because they were poorly defined to begin with. By developing a detailed blueprint of your proposed program, you provide the information you need to plan, shoot, and edit your video.

To illustrate the steps in defining a project we will use a promotional video for a fictional product. Imagine that you have been hired by Acme Power Tools, Inc. to produce a short program that features their new cordless electric drill, the Sidewinder. Acme Power Tools, Inc. will **publish** your video on its corporate website and on general video websites, as well.

At first, the approach appears obvious: simply show what the Acme Sidewinder drill can do. But, there is much more than that to defining this program. To draw a detailed blueprint, we need to specify the program's *subject*, *objectives*, *audience*, *delivery system*, and *length*.

Defining a program also involves selecting a **genre** (such as story, training, documentary, etc.). Program genres are covered in Chapter 10, *Program Creation*.

Subject

The first step is to identify the subject matter, and the easiest way to do this is by assigning your project a temporary working title that announces its topic, such as *Our Camping Trip*, *Ed and Darlene Get Married*, *Warehouse Operations*, or *The Lions Club in Your Community*. Each of these titles summarizes the content of the proposed video.

For your finished program, you may want to replace the literal working title with a more imaginative alternative.

Later, you will further limit and refine your subject; but even at this first step, your title may suggest that your topic is too unfocused. For instance, the title *Warehouse Operations* may tip you off that your topic is too broad for a single training program. To reduce it to a manageable

size, you might select one important part of warehouse operations and plan a program titled *How to Drive a Fork Lift*. In the case of Acme Power Tools, your imaginary client, the subject/title is obvious: *The Sidewinder Cordless Drill* (**Figure 9-1**).

Objectives

The next step is to identify the client's objectives: what does Acme want this program to achieve? The obvious answer is *convince people to buy Sidewinder drills*. But, that is only a start. Some potential buyers may not even think that cordless drills are very useful. Others may understand that such drills are worth considering, but they are leaning toward drills made by Acme's competitors. Still others do not think that they want *any* type of drill. These facts suggest two obvious objectives for your program—persuade customers that:

- A cordless drill is a tool worth buying.
- The Acme Sidewinder is the best cordless drill to buy.

Notice that these objectives are clear, specific, and simple; and they are phrased in terms of their effect on viewers. An objective like, "to demonstrate cordless drills" merely seeks to present information. By contrast, "to persuade customers that a cordless drill is worth buying," is an improvement because the objective in this form intends to actively move the viewer.

Figure 9-1 The Sidewinder cordless drill.

Even a personal video benefits from at least one clear objective. For instance, the objective for a child's birthday party video might be to communicate the love and care that went into creating the event.

These program objectives are few in number and they fit together logically. To see why limiting objectives is important, imagine that Acme Power Tools initially wants two additional objectives:

- To increase customer interest in do-it-yourself projects in general.
- To develop positive feelings toward the Acme Power Tools, Inc.

But, you realize that general "do-it-yourselfing" would make the subject too broad to cover properly, and that the Acme corporate public image is not directly relevant to product usefulness. So, you convince your client to limit the program's objectives specifically to those that will sell the Sidewinder drill.

Audience

Now: to whom do you want to sell the drill? Different audiences will watch your video with different interests, different prejudices, different amounts and types of knowledge. To communicate effectively, you must identify your **target audience** and address your video directly to them. To see how this works, imagine that you are preparing two programs: *How to Choose a Retirement Community* (an informational video for older people) and *Power on the Water* (a commercial aimed at speedboat buyers), **Figure 9-2**. The senior audience will probably appreciate a straightforward, low-pressure presentation of their retirement options. The potential boat racers will respond to dynamic, exciting shots of action on the water, backed by pulsing music and the howl of big engines.

To see how this works, imagine what would happen if you delivered a loud and frantic video to the mature viewers, and a steady, systematic presentation to the boat racers. Both groups would reject your message. To succeed, your video must be carefully tailored for its target audience.

Figure 9-2   Retirement community vs. power boating.

(Corel)

To return to the Sidewinder cordless drill, pretend that in discussing the project with the Acme marketing department, you discover at least four different groups of potential viewers:

- *Buyers for hardware and building supply chains.* These people will decide whether to offer the Sidewinder for sale, and if so, how to promote it against drills made by Acme's competitors.
- *Potential customers.* These people will decide whether they want to buy a cordless drill, and if so, whether the Acme Sidewinder is their best choice.
- *Female customers.* Unlike some males, these potential buyers are not dependably attracted to appliances merely because they are intriguing adult toys. Instead, they want to know what the Sidewinder drill will do, how easily it will do it, and why doing this will be useful to them (**Figure 9-3**).
- *Canadian customers.* Some of these viewers prefer French to English.

Once you have identified your potential audience, you can make some informed decisions. First, you know that Acme wants to sell its drills in Canada, so you will need to plan for a second sound track in French. Secondly, store buyers have interests that are quite different from those of customers, so you eliminate the buyers as primary targets. Next, you realize that it would be impractical to provide a separate program for female viewers, but you

decide to address their interests and concerns as you aim your program at potential customers of both genders.

So, by comparing Acme's objectives to groups of possible viewers, you have decided that your primary audience consists of potential drill buyers of both genders in all of North America. As you continue developing your program, you will try to speak directly to this target audience.

Delivery System

The next question is, how and where will this audience see your video? The answer will determine many things about your program. Acme has asked for a video that it will publish

Figure 9-3   Female customers want to know how well the product works.

on its own and other websites. So, in this case, the delivery system is already selected.

But, suppose Acme had wanted what is called a "point-of-sale" video instead—a program to be shown in stores to customers who are ready to decide right then and there whether to buy a Sidewinder drill. That situation would require another method of displaying your program—a different *delivery system* (**Figure 9-4**). It would probably involve a flat screen video monitor placed at a prominent spot in a retail store, with customers strolling past it as they shop.

The website delivery system imposes certain requirements on your program:

- It must be vivid enough to keep viewers from clicking away from it.
- It must be simple, visually, so that people can see it clearly on the typically small website screen size.
- It must not depend too heavily on its sound track, because some people have only laptop speakers and others disable the audio playback.
- Above all, it must be short—brief enough so that viewers will not use their player controls to skip through it.

In contrast to a website program, a training video designed to teach people how to use the Sidewinder might be watched by an employee in a quiet room or by a do-it-yourself home owner in the comfort of a family room or den.

With those very different delivery systems, these restrictions would not apply. Whatever form it may take, your program's delivery system (the situation in which it will be shown) should have a strong influence on its approach and style.

Program Length

The next step is to determine your video's length. Program running time may be governed by one or more of several factors:

- *Standard time units.* In broadcast and cable TV, for example, programs and commercials alike are always produced in standardized lengths, **Figure 9-5**. Video websites often impose maximum running times on programs they accept.
- *Resources.* As a rule, the longer a video runs, the more it costs to produce. So, your program's length may be affected by the size of your budget.
- *Audience tolerance.* The length of the audience's attention span depends on the type of program you are making. Viewers might enjoy a movie for two hours or more. A TV infomercial about a line of products might hold them for 30 minutes at most. (Infomercials are covered in the following chapter.) Training videos are hard to sustain for longer than 10 or 15 minutes, and commercials are too intense to hold up much longer than 60 seconds.

Figure 9-4 Sidewinder video delivery systems.

Streaming video on the web.

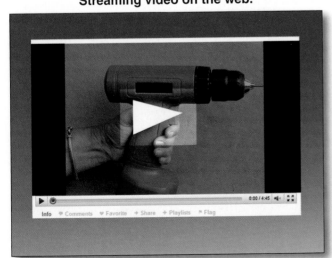

Point-of-sale display.

Figure 9-5 TV program segments are standard lengths to accommodate commercials.

Sale ends Monday!

(Adobe)

- *Subject matter.* Within the program types that are not standardized in length, the actual duration of a video may depend on the extensiveness of its content.

Superficially, subject matter might seem to be the first determiner of length, rather than the last. It would appear logical to allow a video to run as long as it takes to cover the chosen content with appropriate thoroughness. This is often how publishers set the lengths of nonfiction books, such as this one. A book may have 300 pages, or 600, or 1,200 depending on the extent of its subject. But, a book is a random access document. You can dip into it anywhere you please, read it at your own pace, pick it up and put it down when you feel like it. The length of a book is not critically important, because the reader controls how much of it to absorb at any one time.

The video viewer, by contrast, cannot control the speed at which information is delivered, and most programs are designed to be watched continuously, at a single sitting. For these reasons, the lengths of video programs are influenced less by their subject matter than by the other factors discussed above.

Considering how these factors influence your Acme Sidewinder video, you realize that impatient net surfers are unlikely to watch a promotional video for even five minutes. On the other hand, you cannot demonstrate the

major benefits of the new drill in a 30-second commercial. With this in mind, you set a program length of about three minutes.

Budget

Finally, you need to consider budget: the amount of money available to produce your program. Budget constraints will determine both *what* you can do and *how* you can do it.

What You Can Do

For the Sidewinder drill program, the original concept was to use fast motion techniques to show several women building an office building using Sidewinder drills. However, you soon realize that a lengthy location shoot in a real construction site would be far too expensive, **Figure 9-6**. Instead, you develop a less expensive concept involving three performers in a kitchen.

How You Can Do It

With any concept, there are often alternative ways to produce the program. For example, your first version of the kitchen concept might be to shoot the Sidewinder program in the elaborate kitchen of an expensive home. But, when you add up the costs of renting the location, lighting the kitchen, and transporting cast, crew, and equipment, you realize that it will probably cost less to design a kitchen set and build it on a sound stage instead. However, when you cost

Figure 9-6 Women constructing an office building would be too expensive to produce.

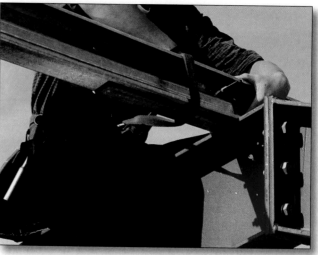

(Corel)

out this plan, you discover that a newly designed set would still be too expensive. Instead, you can rent a stock kitchen set inexpensively from the studio where you will shoot. And so, by repeatedly revising your concept and estimating costs, you arrive at a plan that fits your budget.

Even the most expensive Hollywood productions must usually adjust the production plans to fit the budget.

Selecting a Concept

When you have specified the objectives, the audience, the delivery system, the length, and the budget of your program, you have developed a nearly complete profile of your intended video. With these determining factors firmly in mind, you are *almost* ready to start writing your program. But, you still need to come up with a program **concept**. A concept is an organizing principle; an idea that gives shape and meaning to your video, **Figure 9-7**. It determines what you include and how you treat it.

You could say that the concept guides your approach to your subject, or your perspective on the subject, or simply your "angle." However you describe it, the concept of your video program enables you to make a coherent statement that your audience can understand and respond to. These program organizers are simple to demonstrate by examples. The sidebar *Sample*

Program Concepts suggests plausible concepts for various types of personal and professional videos.

In the case of the Sidewinder cordless drill, a possible concept is hidden in the already-selected objectives and audience. Acme wants to communicate the message that their drill is the best choice, and they want to appeal to female buyers, as well as male. Considering these two desires together, you develop a concept: *Get your own Sidewinder so your husband can have his back.*

With the concept in place, the whole program suggests itself: a group of brief scenes in which a husband keeps asking for his Sidewinder and his wife keeps promising to give it back as soon as she's finished with it. In each scene, she uses the drill in a different type of project, demonstrating its simplicity and versatility as she does so.

Preparing a Treatment

If you are planning a short, personal video, you can probably start production as soon as you have settled on the content, length, and (above all) concept. Most professional videos, by contrast, require that your informal design be transcribed onto paper as a *script*, a *storyboard*, or a *treatment*.

Uses for Program Treatments

The simplest transcription is called a **treatment** (**Figure 9-8**). A treatment is a few paragraphs that explain the program's concept,

---

Figure 9-7 The subhead reveals the underlying concept: "David slays Goliath."

McKinley Beats Fillmore High
Underdogs Rally to Defeat League Champs

Sparked by quarterback Charley Folsome, the McKinley Presidents rallied in the third quarter to power home two touchdowns that opened a lead Fillmore was never able to close.

When the ground attack pursued in the first half failed against the bigger Fillmore line, Folsome took to the air, completing three passes to teammates Bruce Petit, Mickey Fleschner, and Norm Simas.

Boosters Club Donates Scoreboard

At a ceremony during the pep rally before Saturday's game, McKinley Boosters Club President Amanda Wang turned on the all digital

---

Figure 9-8 A treatment is an outline in narrative form.

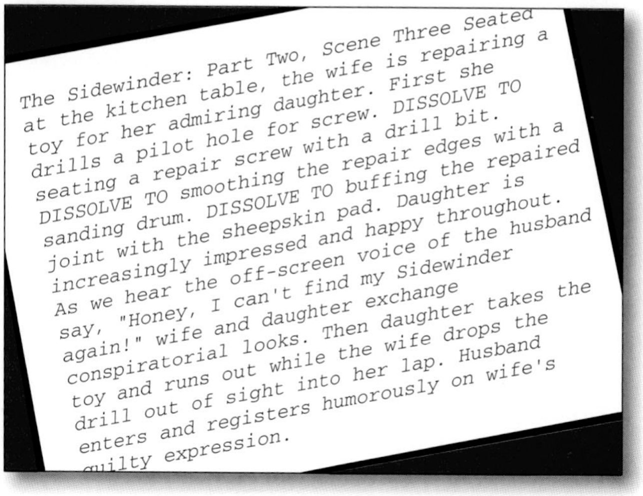

The Sidewinder: Part Two, Scene Three Seated at the kitchen table, the wife is repairing a toy for her admiring daughter. First she drills a pilot hole for screw. DISSOLVE TO seating a repair screw with a drill bit. DISSOLVE TO smoothing the repair edges with a sanding drum. DISSOLVE TO buffing the repaired joint with the sheepskin pad. Daughter is increasingly impressed and happy throughout. As we hear the off-screen voice of the husband say, "Honey, I can't find my Sidewinder again!" wife and daughter exchange conspiratorial looks. Then daughter takes the toy and runs out while the wife drops the drill out of sight into her lap. Husband enters and registers humorously on wife's guilty expression.

Sample Program Concepts

To provide a clear idea of what program concepts are and how they can guide the video maker's approach to developing programs, here are six examples: three personal programs and three professional ones.

Personal Programs

Program type: Vacation video
Subject/working title: *Lake Omigosh Vacation*
Concept: Triumphing over rain.
Summary: Video focuses on comic results of trying to camp out during two solid weeks of bad weather.

Program type: Holiday video
Subject/working title: *Easter Egg Hunt*
Concept: The great Easter egg deception.
Summary: Before letting the kids find the eggs, the parents repeatedly hide them in spots where the children have already looked.

Program type: Family oral history video
Subject/working title: *Grandfather Remembers*
Concept: Family connections survive despite long times and great distances.
Summary: Questions and Grandfather's answers focus on family continuity across three continents and two hundred years.

Professional Programs

Program type: Training video
Subject/working title: *Using Your Multi-line Business Phone*
Concept: Conquer your fear of buttons.
Summary: Humorous acknowledgment that business phones can be complex and frustrating, evolves into the idea that a little study clears up the confusion.

Grandfather Remembers

Program type: Community service promotional video
Subject/working title: *The McKinley Boosters*
Concept: Communities depend on volunteers.
Summary: The Boosters Club preserves and beautifies the town of McKinley.

Program type: Wedding video
Subject/working title: *The Anders/Goldstone Wedding*
Concept: As the glass flies apart, two lives come together.
Summary: Starting with a slow-motion shot of a goblet shattering under a cloth, the video moves from this symbol to documenting a traditional Jewish wedding.

its subject, its order of content presentation, and its style. A treatment serves two purposes: it allows you to see and evaluate the organization of your video, and it communicates your plan to others. By committing your ideas to paper, you can see whether they are appropriate to your program and whether they flow smoothly and logically. By examining your treatment, you can spot problems and opportunities that might otherwise be overlooked.

A treatment is also useful in communicating your vision to other people, especially the colleagues who will help you produce your program and the clients who will pay for it. Without a treatment's overview of the program, your crew can only make one blind shot after another, without knowing how they should fit together and what they should achieve. As for the clients, few if any will underwrite your production without a clear idea of the program you propose to deliver.

Most clients demand a full script, rather than just a summary treatment.

Levels of Treatment

Video program treatments have no fixed style or length. They can be a one-sentence statement of concept and content, or a multi-paragraph synopsis. They can also be an outline so detailed that it identifies every separate content component. Whatever the level of detail, program treatments attempt to convey the effect of finished videos. Here are samples of program treatments developed to three different levels of detail. Each is for the Sidewinder drill program.

Skeletal Treatment

A skeletal treatment covers all three parts of the Sidewinder video, in the briefest possible form:

The Sidewinder Drill
Part One: A succession of quick scenes shows the many jobs performed by the drill.

Part Two: Several vignettes in which a husband is frustrated because his wife is always using his Sidewinder drill.

Part Three: After he presents her with her own Sidewinder, the two of them collaborate happily on a construction project.

Summary Treatment

This excerpt from a summary-level treatment covers just one third (Part Two) of the skeletal treatment in greater detail:

The Sidewinder Drill, Part Two
Scene One: Husband asks where his Sidewinder is, as we see wife using it to repair kitchen cabinet hinge.

Scene Two: As wife assembles picnic table bench in backyard, husband appears and again asks where his Sidewinder is.

Scene Three: Wife is repairing child's toy at kitchen table when she hears husband asking where drill is. As he appears in kitchen doorway, she hides drill in her lap.

Detailed Treatment

This excerpt from a detailed treatment, covers only Part Two, Scene Three, as summarized in the previous version.

The Sidewinder Drill: Part Two, Scene Three
The daughter brings her broken sailboat to the mother, at the kitchen table. The mother says it will be simple to fix. She uses the sidewinder to drill a pilot hole for screw. The daughter watches, impressed. The mother hands the boat back, and the daughter says, "Great! Thanks, Mom." The husband starts into the kitchen saying, "Honey, I can't find my Sidewinder again!" Seeing that her husband is coming, she hides the drill under the table as the daughter takes the toy and runs out. Husband enters and registers humorously on wife's guilty expression.

The amount of detail in your own treatments will depend on how minutely you need to pre-visualize your program, and how completely you want to communicate it to clients and colleagues.

Creating a Storyboard

The old saying claims that one picture is worth a thousand words, and this is often true in developing video programs. In graphic-based project design, a succession of pictures resembling a graphic novel, sketches all the important moments in the program. This script in picture form is called a *storyboard*.

Storyboards got their name from the bulletin boards on which the drawings of scenes for animated cartoons are pinned for inspection and editing.

Storyboard Uses

Storyboards have two main uses: to help others visualize the look of the eventual program and to pre-plan complex sequences shot-by-shot.

Visualization

Storyboards are particularly valuable for communicating content to clients and crew,

A Storyboard Sequence

Here is part of a storyboard sequence as an advertising agency might create it for the Sidewinder Drill program.

A storyboard sheet.

GIRL: Mom, the rudder broke!

GIRL: Can you fix it?

MOM: Sure! I'll get the drill...

SFX: Drill whirring.

MOM: I put in a bigger screw.

GIRL: Great! Thanks, Mom.

DAD: Honey, I can't find my sidewinder again.

DAD (O.S.): Every time I need that drill...

DAD: ...it disappears!

DAD (O.S.): You don't know...

DAD: ...where it went.

DAD (O.S.): Do you?

(FrameForge)

because they present concrete images instead of the abstract words that describe them. Some people have less talent than others for thinking graphically. The problem is that they are often unaware of their inability to visualize, and so they indicate understanding of written descriptions when they really can't imagine them. The result can be serious miscommunication. A storyboard presents the images in *pre-visualized* form, along with captions containing dialogue, sound effects, and descriptions of the action.

Shot Planning

Where complex visual sequences are involved, storyboarding can help you as well as others. By planning all camera shots in advance, you can see how well they will edit together and how

clearly they will communicate the content. Many directors make extensive use of storyboards in their productions, especially in laying out highly complex action sequences. Storyboarding is also useful for working out consistent screen direction in complex physical sequences, like fights and chases.

Managing screen direction is covered in Chapter 19, *Directing for Form.*

In today's productions, compositing and other techniques are often used to combine quite different visual elements. Storyboarding is essential so that the creators of live action, computer graphics, and other elements can design components that match and fit together in the final images.

Writing a Script

Another way to lay out a detailed production design is by writing a full *script*. A script describes every sequence in your program, including both video and audio components. A script is especially useful if the program contains dialogue to be memorized and spoken by actors and/or voice-over text to be read by an off-screen narrator. Scripts are also a common alternative to storyboards for presenting programs to clients. They are especially valuable for planning and budgeting a production, since they include every element of the program in a compact narrative form.

The Scripting Process

Writing a script can follow any system—or no system at all. In commercial and industrial production, however, the process of creating a script often breaks down into five stages:
1. Producer and client agree upon the program's content and concept.
2. A detailed content outline is written. This is often critiqued by the client and then revised by the writer.
3. A first draft script is written. This is the initial attempt to lay out a complete production script. Usually the client reviews this draft and orders revisions.

4. A revised draft of the script incorporates the client's changes.
5. The revised script draft is reviewed by the client and further changes are ordered. If all is going well, these changes do not require a complete third draft, but only a refinement of the second one.

Common Script Formats

In professional productions, video scripts have standardized formats that are adhered to rigidly—so rigidly, in fact, that failure to use the right format is considered the mark of an amateur.

Final Draft® software, discussed below, even has a template for the special format preferred by Warner Brothers Studios.

Fiction Script Format

The formatting of movie and TV scripts includes so many rules that at least one entire textbook has been written on this subject alone. However, an Internet search will offer several detailed guides on script formatting. Once you have learned the rules, you can then set up word processor paragraph styles to automate the process (**Figure 9-9**). The main styles include:

- *Action:* Margin-to-margin descriptions of people, places, and actions.
- *Character:* Names of characters speaking.

Figure 9-9 Word processor styles customized for screenplay writing.

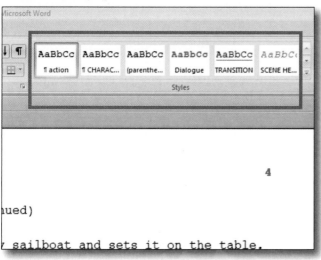

- *Parentheses:* Very brief descriptions of the dialogue that follows, such as *(loudly)*.
- *Dialogue:* Lines spoken by the characters.
- *Transitions:* Type of transitions between scenes, such as CUT TO or DISSOLVE TO.
- *Headers:* Identifications for each new scene, such as INT–THE KITCHEN–DAY.

Nonfiction Script Format

The traditional nonfiction script layout consists of two vertical columns with the visuals on one side of the page and the dialogue, narration, and other audio on the other side. The problem is that this side-by-side layout can be inconvenient to use in a word processing program.

Word processors with a "parallel protect" option can work in two column format, because they ensure that each paragraph on the audio side remains directly opposite its related shot on the video side.

For convenience, nonfiction scripts (like the following excerpt) are sometimes formatted over-and-under, instead of side-by-side (**Figure 9-10**).

Figure 9-10 Over-and-under nonfiction script style.

```
25.
VIDEO   TWO SHOT: The GIRL brings in a toy
        sailboat and sets it on the table.
AUDIO   GIRL: Mom, the rudder broke!
26.
VIDEO   OTS on the GIRL.
AUDIO   GIRL: Can you fix it?
27.
VIDEO   OTS on MOM
AUDIO   MOM: Sure! I'll get the drill.
28.
VIDEO   CU drilling a hole in the sailboat
        rudder.
AUDIO   SFX: Drill whirring.
```

Storyboard Software

If you search for "storyboard software" on the Internet you will find several examples—some of them freeware. Some programs are more like production planning boards; others let you import photo images or help organize your own informal sketches. With the most advanced types, you create images by loading standard components from the program's library—backgrounds, actors, and props—and then customizing them to fit your story. One of these programs is a program called FrameForge.

Like software for architects or CG animators, FrameForge creates environments in three dimensions so that the camera can move around in space.

The FrameForge work screen.

Views from different camera setups

Set ground plan

(FrameForge)

In this series, the camera dollies into the scene, reframing the subjects as it moves.

Using standard components, you can build a variety of environments—all of them fully three dimensional.

A restaurant.

A living room.

A city street.

An office.

And, you can populate these environments with characters customized to fit your story. Using "texture mapping," you can actually place the faces of actual actors on the characters.

You can select several ethnic types and dress characters in different costumes.

You can adjust characters' ages and change the colors of clothing items.

You can pose characters in different attitudes…

…because they can be adjusted at natural human bending points.

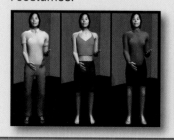 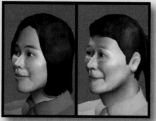  

By trying different camera setups, you can show different angles and see how they will cut together in editing.

Two shot.

Her medium closeup.

His closeup.

Her over-the-shoulder shot.

You can approximate the actual lighting on the set or at a location.

Character fully lit.

Character with lights "turned off."

The program can adjust depth of field to match different lens focal lengths, apertures, and distances from subjects.

After re-arranging shot orders, sometimes cutting back and forth between shots, you can produce storyboards complete with shot labels and descriptions. In addition to fully "three dimensional" views, frames can be rendered in a variety of more traditional 2-d formats.

Original three-dimensional version.

"Cartoon" rendering.

Pencil sketch.

Colored ink drawing.

Like all complex software, pre-visualizing storyboard programs take considerable time and effort to master. Even for a skilled user, assembling environments and props, selecting, modifying, dressing, and posing actors is time consuming. If a fiction program has, say, 8 major characters and takes place in 6 locations, you can build and store all the components beforehand and then quickly load them, as needed. But, if a commercial runs through 20 characters in 12 locations in 30 seconds, the work involved in using this program may not be cost-effective.

Nonetheless, this pre-visualizing (or "previs," as these programs are called) is so powerful that many directors and cinematographers learn and use it just to experiment.

Storyboards or Scripts: Which Method Is Best?

Treatments, storyboards, scripts—professional video makers employ all these forms of program development, sometimes mixing and matching them as needed. For example, a program may be documented completely in script form, which is supplemented by storyboards of action sequences or other activities that demand precise visual pre-planning.

When you make fairly short, simple programs, it is usually enough to write down your concept, develop a narrative treatment that covers the major components of your video, and perhaps storyboard critical sequences for camera angles and continuity. As your productions grow in scope and complexity, you will probably move up to fully scripted programs, with or without more extensive storyboarding. Whatever form or forms you choose, the result is the blueprint from which you create your video program.

Professional Scripting Software

Several word processing programs are designed exclusively to simplify script writing. Though there is no production industry standard, the most widely used software is *Final Draft®*, which is available in two different versions: a screenplay version for fiction scripts and an AV version for scripts demanding two-column format. *Final Draft* has two big advantages: it has pre-formatted styles for each type of paragraph, and its smart typing function remembers character names and standard commands. For example, if you have a character named Alexander, the second time you start to enter the name in a "character" paragraph, as soon as you type *Al* the program offers to enter *exander* automatically. If you also have a character named *Albert*, the program first offers you both names to choose from by clicking one. If you then type "a b," it offers just *Albert*. Since script writing involves so much repetitive typing, this feature saves a great amount of time. The same is true of scene locations and standard transitions.

Final Draft Screenplay

The figure below calls out the standard formatted paragraphs used by *Final Draft*. These include:
- A—Character name (always capitalized automatically)
- B—Parentheses for short descriptions of the speech below.
- C—Dialogue
- D—Action
- E—Transition
- F—Script header
- G—Scene description (also all-caps)

The black line indicates a page break.

Parts of two typical *Final Draft* screenplay pages.

(Final Draft)

Final Draft also includes features that allow writing partners to collaborate—even on different screens.

The next figure shows a few of the program's other features. The Navigator Window shows every scene in the script. Scenes can be sorted by number, location, or page color.

The Navigator Window and Scene Properties box.

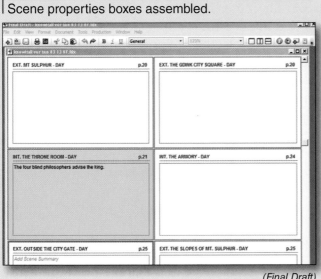

(Final Draft)

The Scene Properties box (on top of the Navigator Window) includes essential information and allows you to summarize the content of the screen. These boxes can be viewed collectively, either as an outline or as a group of index cards.

Scene properties boxes assembled.

(Final Draft)

This is especially useful because the index card view presents an outline of the entire story. In professional scripts each revision is assigned a different page color. Pages 21–23, for example, might go from white to blue to green colored paper as they are repeatedly revised. Notice the green tint on page 21, to match the script revision code.

On this view, you can drag scene cards to different locations and the script will automatically adjust scene order to match. Since fiction scripts depend so heavily on their dramatic structures, the ability to try different scene orders is amazingly valuable.

Final Draft AV

The audiovisual (**AV**) version of *Final Draft* is intended for scripts of commercials, documentaries, training programs, and other videos requiring a two-column layout. It is especially convenient because nonfiction videos use so many different formats and conventions, and because they are typically difficult to lay out.

The first page of an AV script.

Agency		Writer	
Client		Producer	
Project		Director	
Title		Art Director	
Subject		Medium	
Job #		Contact	
Code #		Draft	

VIDEO	AUDIO
TWO SHOT: the GIRL brings in a toy sailboat and sets it on the table.	GIRL: Mom, the rudder broke!
OTS on the GIRL.	GIRL: Can you fix it?
OTS on MOM.	MOM: Sure! I'll get the drill.
CU: drilling a hole in the sailboat rudder.	SFX: Drill whirring.
TWO SHOT favoring GIRL	MOM:

(Final Draft)

Summary

- The major phases of developing a video project include: defining the project, preparing a treatment, creating a storyboard, and writing a script.
- During project development, a program's subject, objectives, and audience are refined and are often more limited than in the beginning of the process.
- How and where the audience will see the program determines the delivery system.
- Budget constraints will determine both *what* you can do with a program and *how* you can do it.
- The goal of every program treatment level (skeletal treatment, summary treatment, and detailed treatment) is to convey the effect of finished video.
- Storyboards help others visualize the look of the eventual program, and assist in pre-planning complex sequences shot-by-shot.
- Video scripts have standardized formats that are adhered to rigidly. The script formats used for fiction and nonfiction programs have unique characteristics specific to the genre.

Technical Terms

AV: Abbreviation for "audiovisual," a catch-all term for all nonfiction video genres. Pronounced "a-vee."

Concept: The organizing principle behind an effective program. Often called an angle, perspective, or slant.

Delivery system: The method by which a program will be presented (such as website, TV monitor, or kiosk), as well as the situation in which it will be watched (alone at a desk, in a training room, in a crowded store, etc.).

Genre: A specific type of program, such as story, documentary, or training.

Pre-visualizing: The process of creating manual or computer images to plan shots and shot sequences prior to actually recording them. Often abbreviated as "previs."

Publish: To distribute a video program publicly by uploading it to a website.

Script: Full-written documentation of a program, including scenes, dialogue, narration, stage directions, and effects, that is formatted like a play.

Storyboard: Program documentation in graphic panels, like a comic book, with or without dialogue, narration, stage directions, and effects.

Target audience: A specific group of viewers for whom a program is designed.

Treatment: A written summary of a program that is formatted as narrative prose; may be as short as one paragraph or as long as a scene-by-scene description.

Review Questions

Answer the following questions on a separate sheet of paper. Do not write in this book.

1. *True or False?* Program development is needed only for larger video projects.
2. List the steps involved in project development.
3. The group of potential viewers is called the _____.
4. Identify possible delivery systems for video programs.
5. What is a video *concept*?
6. A(n) _____ is a few paragraphs that explain the program's concept, subject, order of content presentation, and style.
7. Describe the uses of storyboards.
8. Identify the five common stages of script creation.
9. A two-column script format is often used for _____ programs.

STEM and Academic Activities

STEM

1. Technology. Explain the impact of word processing programs on the process of writing a script. Include a comparison to the script writing process before the availability of word processing programs.
2. **Engineering.** Relate the steps in video project development to the development process for another type of project:
   - Define the project.
   - Prepare a treatment.
   - Create a storyboard.
   - Write a script.
3. Language Arts. Imagine that you are developing a training video for a company with international locations. List the items you must consider when making a video for an international market and briefly summarize how you would handle each item listed.

In producing corporate videos, meetings with clients are inescapable.

After studying this chapter, you will be able to:

- Summarize the basic components of story construction.
- Identify the principal nonfiction program genres.
- Understand how the three *T*s are used in training program scripts.
- Recognize appropriate video and audio resources for a documentary production.

About Program Creation

The previous chapter addressed project development in general because the basic process is much the same for many different types of videos. But, once you begin writing a particular script, the creative process will depend on the type of program you are creating. That is because each video *genre* is different.

The *American Heritage Dictionary* defines genre as "A category of artistic composition...marked by a distinctive style, form, or content."

How many video genres are there? As you can see from any week's TV listings, video programs come in many different varieties, including fictional series and movies, news and sports features, documentaries, instructional programs, and commercials, to name just some of the more common types. When you turn to popular Internet video sites, **Figure 10-1**, you can find every conceivable program genre and some videos that are quite unclassifiable.

Many videos published on websites seem so unstructured, so spontaneous, that they may not have been written at all. (This does not mean that all of them are inferior to more systematic efforts.) But, all professional programs are intended for specific purposes and carefully scripted to achieve them.

A program may be documented solely by a storyboard or treatment. For convenience, however, we will refer to all forms of program creation as "writing," and the end products as "scripts."

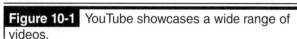

Figure 10-1 YouTube showcases a wide range of videos.

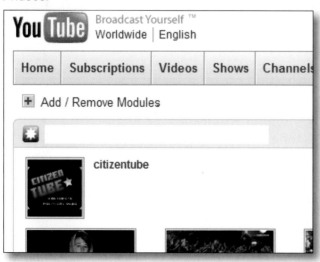

The scripts for these program genres have characteristic structures, each evolved through many years of trial and error. To thoroughly discuss each script genre would require a separate book for each one. We can, however, offer writing hints for some of the more common types.

Programs fall into two overall classes: fiction and nonfiction. Fiction genres include feature movies, TV miniseries and program episodes, and short fiction often intended for internet publishing (or simply for personal satisfaction). But, whatever their length and style (drama, comedy, science fiction, etc.), we can discuss them as a group because most story videos are constructed in much the same way.

Story Videos

Story videos are dramas (or comic dramas), and they typically employ *dramatic structures* that have been used for many centuries.

The Athenian Greeks had developed the art of dramatic structure by around 450 BC.

In general, dramatic story structure calls for three acts:
- **Opening act:** The story is introduced and launched.
- **Middle act:** The story unfolds and conflict intensifies.
- **Concluding act:** The conflict reaches a critical point and is finally resolved.

Though this three-act organization is not the only dramatic structure, Hollywood has found that it works equally well for two-hour movies and 20-minute TV comedies. So, when you write a story script, you might try using this structure.

Story Construction

Here is a breakdown of a typical three-act story structure, illustrated by our acquaintances, Bob and Bill.

Act One

Taking about one-fourth of the total story time, act one introduces the characters, the setting, and the situation or event that starts the *dramatic action*.

Bob and Bill are good friends, except that Bob is involved with Nancy and Bill would like to be. A big dance is announced. Bill decides to win Nancy away from Bob and take her to the dance.

The dramatic action begins when Bill makes that decision (**Figure 10-2**).

Act Two

Consuming about half the program time, act two shows the escalating *conflict* between opposing forces in the story.

Bill works to get Nancy to notice him. Bob discovers he has a rival and tries to get a commitment from Nancy. Taking advantage of the situation, Nancy starts playing Bob against Bill. Each one offers competing inducements, like a dinner before the dance, a stretch limo, etc. Step by step, the conflict grows more intense, all the way up to the day of the dance. Not wanting to hurt either Bob or Bill, Nancy says "yes" to both of them—but how can she attend the dance with both escorts?

Dramatic stories are almost always about *conflict*. Without a clash between opposite sides, the drama is missing from the story.

Act Three

The remaining one-fourth of the story leads to the climax of the conflict, and then shows how it is resolved.

Making an excuse to meet Bob at the dance, Nancy arrives with Bill, then uses one improvisation after another to shuttle back and forth between Bill and Bob. The crisis occurs when they discover the trick and confront each other. Going outside, Bob and Bill wrestle each other to a draw, ruining their prom clothes and inflicting considerable damage. Finally, with the honor of each one upheld, Bob and Bill go off as friends again, leaving Nancy alone at the dance.

In summary (**Figure 10-3**), then:
- Act one (¼) introduces the characters and the event that starts the conflict.
- Act two (½) intensifies the conflict with every event and builds up tension.
- Act three (¼) brings the conflict to a climax and then resolves the action into a satisfying conclusion.

If the Bob and Bill video were to run 20 minutes, then act one would take about 5 minutes, act two would need around 10 minutes, and act three would unfold in the final 5 minutes.

Though this popular three-act structure may seem to reduce stories to formulas, it is a perfectly sound way to organize a fiction video script. And, since it has worked well for 2,500 years, it has certainly proven its effectiveness.

Even so, the three-act structure is not the *only* way to organize a story, so do not worry if your script does not follow it perfectly.

Figure 10-2 Bob and Bill are called the protagonist and antagonist in the conflict.

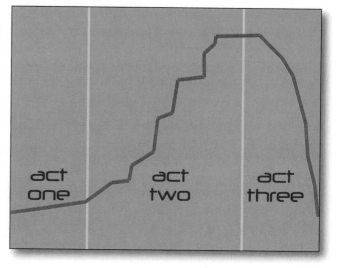

Figure 10-3 In graphic form, dramatic action might look something like this.

Other Factors

Structure is, arguably, the most important aspect of a fiction script, but other factors are also worth careful planning.

Stories for TV

Though TV stories follow the same basic beginning-middle-end format, they are more difficult to structure because of the constraints of television—especially interruptions for commercials.

Some programs precede the first section with a very short "teaser"—a brief, intriguing glimpse of what is to come in the story.

- TV part one is usually longer than average, to "hook" viewers by getting them involved in the story. In effect, all of the first act is delivered in this opening part. Typically, the initial event of act two ends the section in order to keep viewers during the commercials.
- TV parts two and three generally cover the remainder of the second act.
- TV part four corresponds to the third and final act.

Some programs end with a very brief epilogue ("tag"), but by this time the story is essentially over.

Character Arcs

Dramas, especially longer ones, are generally more interesting if at least some characters are changed by events in the story—if they grow in some way or gain new insights or simply change. These progressive changes are called *character arcs*, **Figure 10-4**. Like many devices for structuring stories, these character developments can be tired clichés when not handled well. But, like so many other ancient guidelines for writing drama, character arcs remain effective because they work as well as ever.

Short Stories

Until you are an established professional, your fiction videos will probably be short—perhaps 5–20 minutes long.

Some video websites will accept only short submissions.

Figure 10-4 By the time of these Roman theatrical masks, the concept of character arcs was well known.

Within that time frame, you can still follow a three-act structure, and even provide one or more character arcs. For example, suppose Nancy realizes that in trying not to hurt either Bob or Bill, she has hurt both of them, despite her good intentions. For a short story like this one, Nancy's insight completes a satisfactory character arc (**Figure 10-5**).

Getting Started

These tips for writing story videos are necessarily brief and basic. As noted previously, whole books and year-long writing courses are devoted to explaining the details of this demanding

Figure 10-5 Bob and Bill leave Nancy behind to think about what she has done.

(Corel)

craft. For short, simple projects, however, these hints can help you construct a stronger story.

Nonfiction Programs

By far the greatest percentage of videos are not stories, but programs that deal (in one way or another) with real people, places, and events. To verify this, look at the number of cable channels devoted exclusively to history, nature, science, medicine, cooking, shopping, antiques, golf, and just about any other interest you can think of. Also, corporations, educational institutions, and not-for-profit organizations use video programs constantly. Companies devoted exclusively to producing instructional programs advertise on the Internet; and both organizations and individuals freely contribute nonfiction videos to public video websites.

For example, a Google video search of the phrase "Lighting for Video Production" found well over 2,700 instructional programs. Eliminating titles that were not truly relevant still left over 100 entries—some of them 20 minutes or longer.

Despite their enormous variety in subject matter and treatment, most nonfiction programs belong to one of three broad types: instructional, promotional, and documentary.

Instructional Videos

Instructional videos are intended to teach certain sets of knowledge and skills. Typically, they are carefully targeted—that is, they are limited in scope and organized to cover specific topics. In this respect, they differ from documentaries on, say, science or history, which generally deliver a broader range of information. To put it simply, documentaries on academic subjects are typically for *education*; instructional videos are for *training*.

Promotional Videos

Promotional videos are designed to convince viewers of something—to persuade them to buy a product, vote for a candidate, or support an idea. Some have broader ambitions: to promote a cause, like conservation, or deliver a positive impression of a corporation.

Companies that help support non-commercial American public television stations are highly skilled at presenting "corporate identifications" that look remarkably like commercials (**Figure 10-6**).

Historically, many promotional programs were intended for classroom use. For example, *Rhapsody of Steel* (1959) was an animated program about steel production and uses sponsored by the United States Steel Company.

Of course, the most common promotional videos are commercials—short, vivid pieces that range in length from five seconds to a minute, or longer. In addition to selling products, commercials offer two incidental, but important benefits: they stimulate the invention of innovative video (and audio) techniques, and they serve as proving grounds for writers, designers, and directors, who often move on to TV programs and feature movies.

Documentary Videos

Documentary videos are programs intended to capture and present (to *document*) the actual world. They differ from instructional programs because they do not expect viewers to master specifically targeted information, but only to learn about it. Documentary forms include current events, "reality" programs, infomericals, and infotainment.

Figure 10-6 Sponsor identifications are intended as corporate promotions.

Current Events

Current events include news, news features, investigative reports, and sports. Most of these are segments of longer programs, either daily newscasts or weekly shows divided into self-contained stories (**Figure 10-7**).

"Reality" Programs

In "reality" programs, cameras follow groups, like families, police officers, or celebrities, documenting their lives as they unfold. However, many reality situations are set up specifically to be documented. For example, teams compete to complete tasks or survive in hostile environments; unlikely combinations of people are required to live together while the resulting difficulties are recorded.

Some purely fictional shows are designed to look like reality programs, with improvised dialogue, haphazard lighting and sound, and shaky camera work. They are sometimes called "mockumentaries." Despite their haphazard appearance, these programs are actually structured and rehearsed by the professional actors who perform in them. The mockumentary form is more often seen in motion pictures, such as *This Is Spinal Tap* and *Best in Show*, but they also appear on cable channels, especially as comedies.

In their original form, reality documentaries were often called "cinema verité," because they did not use music or narration, and relied solely on the footage they were able to capture.

Infomercials

Infomercials are TV programs of up to 30 minutes (or even more) that are designed to give the appearance of regular programs, but are actually intended to sell products. For example, an infomercial on cooking might spend most of its time selling the sponsor's brand of pots and pans, **Figure 10-8**.

Infomercials are easy to recognize on TV schedules, which generally list all of them as "paid programming."

Infotainment

Infotainment videos do indeed document reality, but they are pre-scripted, recorded, and edited to entertain viewers while informing them. Most of the programs on the many history, science, and nature channels are infotainments.

The Basic Types

The rest of this chapter focuses more closely on three types of nonfiction video: *instructional programs*, *news reports*, and *documentaries*. First, we will look at how to construct each type. Then, we will survey the resources you can use to build them.

As for promotional programs, video professionals are often called upon to make them, but we will not explore them further for two reasons: first, promotional programs seeking an overall favorable impression for their sponsors (like *Rhapsody of Steel*) are a form of documentary

Figure 10-7 A news feature about a street fair.

Figure 10-8 A "cooking show" infomercial may actually be selling sets of cookware.

(Adobe)

and are constructed the same way. Second, commercials and infomercials are just too varied in style and method to be covered in the space available here.

Instructional Programs

Instructional programs fall into two types: *documentary* style and *training* style. Essentially, documentary style instructional programs capture and present material covered in real time. Training programs are pre-scripted in great detail and depend heavily on the techniques of instructional design, **Figure 10-9**.

Many programs combine the two methods, positioning and reinforcing content with instructional techniques that punctuate major segments of documentary-style presentation. For simplicity, however, documentary-style and training-style programs are treated separately here.

Documentary-Style Instruction

Documentary-style instructional programs tend to be demonstrations—recordings of processes, with explanatory comments. For example, using two or three cameras, you might record a ceramics teacher in real time as she turns a lump of clay into a basic pot, explaining how she does it as she proceeds. Or, you might document a physics or chemistry instructor demonstrating a procedure on a lab bench.

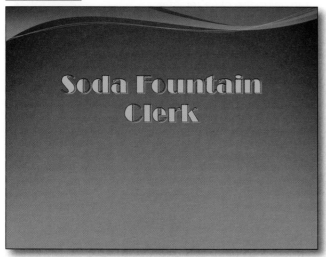

Figure 10-9   A typical training video.

Classroom demonstration tables are often supplied with overhead mirrors, angled so that students can look down on the table surface. One trick is to record the action as seen in the mirror, and then reverse it digitally ("flip it") in postproduction, so that viewers see it oriented correctly, as the instructor does.

Documentary-style instructional programs are relatively quick and simple to produce. (Many short programs published on the Internet belong to this type.) They also give the impression that viewers are watching actual events unfolding in real time. But, documentary-style instructional programs are typically not designed to supplement their basic information, reinforce learning, or review material covered. Those techniques require training videos.

Also, demonstration videos should not be confused with tutorials—interactive programs in which viewers actively participate. Though demonstrations may ask viewers to complete each step by themselves as it is presented in the video, the programs cannot adjust when the viewers' materials or situations are different. Sooner or later, viewers may grow frustrated as their own results diverge from the program.

Training Programs

Training programs are not tutorials either, but they are the products of what is called *instructional design*. That means that training programs are carefully structured to organize information logically, present it clearly, and reinforce and review it for better viewer understanding and retention. These programs are often simple to organize because the subject matter tends to do the job for you, **Figure 10-10**.

Some programs use chronological order for organization. If your topic is, say, how to assemble a model airplane, then the program will cover each construction step in the correct sequence. Other programs divide up the content by type. In a program on how to clerk at a soda fountain, topic types might include making drinks (like sodas and milkshakes), assembling sandwiches, and ringing up sales. Each topic would have its own section of the program.

Figure 10-10 Training videos are used to teach job skills.

The problem with training videos is that they cover the subject at the speed set by the director—a speed that may be too fast for some learners and two slow for others. With printed materials, learners can move at their own, most comfortable pace, slowing down for difficult passages, flipping back to review materials already covered, and skimming things they already know.

Videos, however, move at a certain fixed pace, and (despite DVD chapters and fast-forward and reverse controls) learners are generally forced to move at that pace. Also, it is difficult to keep the content of a training video organized in the learner's mind. At various moments, viewers worry: *What did we already cover? Where does this part fit in the whole? What is left to learn?*

Instructional Design

To address these difficulties, producers have developed three techniques so effective that they can be found in almost all training videos:

- *Introduce* the program by briefly listing the material to be covered.
- *Cover* each topic, constantly relating it to the topics before it and the topics to come, so that viewers understand where they are in the flow of information.
- *Summarize* the materials covered, to reinforce what viewers have learned.

These techniques are widely known as *the three Ts*. The three *T*s refer to the fundamental structure of the training program:

- **T**ell them what you are going to tell them.

- **T**ell them.
- **T**ell them what you have just told them.

Perhaps these would be better labeled the three *S*s, because whenever practical, **s**howing something in a video is more effective than just telling it.

Tell them what you are going to tell them. Begin the instructional body of the program by listing the topics and/or having them read aloud by the narrator (**Figure 10-11**). Because some viewers learn better aurally and others visually, it is generally best to use both titles and narration.

In **Figure 10-11**, notice that the narrator's language closely follows the words on the title. If the narration used different language, such as, "creating drinks and sandwiches and processing customer payments," viewers would have to reconcile the differing words on the screen and on the track—distracting their attention from the program content. Even when you read them on this page, below, it takes a moment to realize that the two different versions say the same thing:

NARRATOR: Making fountain drinks, making sandwiches, and using the cash register.

TITLE: Creating drinks and sandwiches and processing customer payments.

Tell them. When you have listed the topics of the program, start presenting the live-action coverage of the first topic and continue through the training content (**Figure 10-12**).

Figure 10-11 NARRATOR: In this program, you'll learn about making fountain drinks, making sandwiches, and using the cash register.

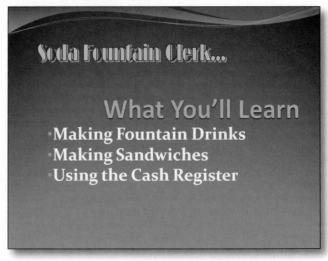

Figure 10-12 Note that the headlines of the topic titles match the language of the "What You'll Learn" organizer title.

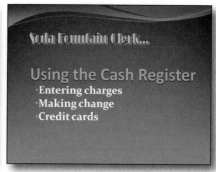

Tell them what you have told them. When you have finished presenting the training material, summarize what viewers have learned—often by repeating the opening organizer slide (**Figure 10-13**).

The three *T* system organizes the material for viewers and continually reminds them of where they are in the presentation. This organization works best when supported by two other techniques: *buildups* and *signposts*.

The title card graphics in this section use these techniques, and, as you can see, they exploit the use of on-screen lists. Training video lists are simply content outlines broken down and presented in separate parts to help viewers organize program information (**Figure 10-14**). Three kinds of lists are common:

- *Introductory lists:* Present the major topics.

- *Topic lists:* Break out the key points within each topic.
- *Summary lists:* Remind viewers of the topics just covered.

Figure 10-14 Introductory lists set out the major sections of the program content. Topic lists include the key points in single topics.

Introductory list

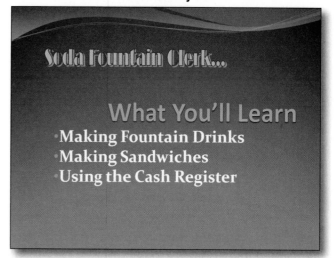

Topic list

Figure 10-13 NARRATOR: And so, we have covered making fountain drinks, making sandwiches, and using the cash register.

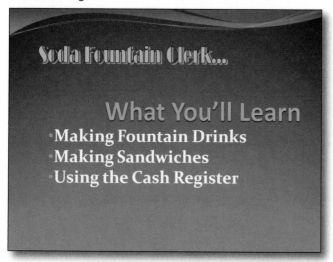

As always, the narrator also delivers the key information using the same language as the title on-screen at the time.

Buildups

All three types of lists employ *buildups*. These are lists that grow as a topic unfolds. This is the basic buildup technique:

- List the topics to be covered, in order, and in the same, relatively neutral color.
- List the first topic alone in a different, more vivid color.
- List the next topic in the same vivid color, while the previous topic remains above it, but reverts to its original, neutral color.

That way, each new subtopic (highlighted by a contrasting color) is seen in the context of the subtopics that have preceded it. See **Figure 10-15**.

These titles are easily created in presentation software, such as PowerPoint®, and then imported into the video during postproduction.

The advantage of buildup lists is that they continually remind viewers of where they are in the program as a whole.

Signposts

By retaining previous topics on the screen, buildups can remind viewers of what they have covered, but they cannot point to where viewers

Figure 10-15 Buildup technique; this sequence is repeated throughout the program. A—The video starts with the program organizer list. B—The program organizer list is repeated with just the first topic, which is highlighted in red. C—The first topic organizer list appears. D—Subtopics on sodas and malteds disappear, and the subtopic on milkshakes is highlighted.

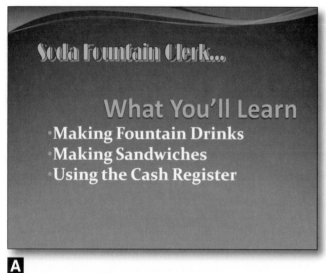

A

B

C

D

(Continued)

Figure 10-15 *(Continued)* E—At the transition to the next subtopic, the second subtopic title is added and highlighted, while the first subtopic reverts to white. F—The final subtopic appears, highlighted, and the previous ones are white. G—The program organizer list reappears with the second major topic added and highlighted. H—The second topic organizer appears.

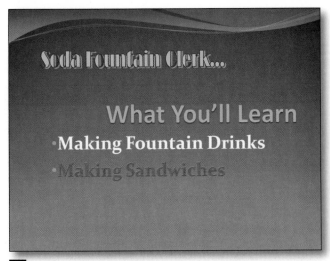

are going. *Signposts* can do both. Signposts are additional organizers that reorient viewers by reminding them of where they are in the large subject of the program (**Figure 10-16**). Typically, they are created by repeating the introductory title with the new topic highlighted (**Figure 10-17**). As always, the narrator reinforces the text on the screen.

Seen in isolation here on the printed page, lists with their buildups and signposts do not seem very exciting. In actual training videos though, they appear only at intervals, separated by more interesting live-action visuals. And, while it is true that some training videos do

Figure 10-16 Signposts within the program point in both directions.

Figure 10-17 NARRATOR: Now that we've covered fountain drinks, let's look at making sandwiches.

not lend themselves to this approach, it works well for many programs. For other subjects, the documentary style covered earlier may be more interesting and effective, or perhaps a well-planned mixture of both.

News Reports

News reports are mini-documentaries intended, mainly, for daily newscasts. Like some documentaries, news and sports features are often scripted before, during, and after shooting. This is especially true of the longer features made for "news magazines," like *60 Minutes*. To see how this works, we will follow the progress of a news feature about a county fair.

Before Shooting

Prior to sending out a news crew, the producer develops a concept for the feature. (For more on concepts, see Chapter 9, *Project Development*). The concept: *Ranching is the heart of our county economy*. The concept tells the reporter and crew to focus their shooting on the livestock events at the fair, **Figure 10-18**.

During Shooting

At the fair, the news crew discovers that the dairy cow competition is both photogenic and easy to shoot. So, the reporter writes (or at

Figure 10-18 The livestock barns at the county fair.

least outlines) and records standup narration to explain that dairy farming is a good example of local ranching (**Figure 10-19**).

After Shooting

Back in the editing suite, the producer determines that the best footage was shot during the judging of Jersey heifers. Furthermore, the competition generated suspense and human interest. So voice-over narration is written to focus attention on this (**Figure 10-20**).

If time and circumstances permit, the reporter is sent back to the fair to record on-camera narration and, if possible, interview the mother/daughter team that won the competition. A quicker alternative is to record studio narration for use as voice-over.

Figure 10-19 The reporter records scripted narration.

Figure 10-20 The recording focuses on the jersey heifer contest.

(Sue Stinson)

Though not uncommon for longer investigative reports, this **pickup** (supplementary) shooting is rarely possible for topical news features.

In this typical example then, a news feature script is roughed in before shooting, focused more narrowly during production according to the shooting opportunities offered, and finalized during postproduction to fit the concept as refined during editing. The result *looks* as if the production team originally set out to follow one competitor in the heifer judging, when, in fact, they did not know precisely what their feature was about until the postproduction phase (**Figure 10-21**).

Figure 10-21 The edited feature is now entirely about the jersey heifer contest.

up next...
lovely young ladies!

(Sue Stinson)

In *breaking news*, such as crimes, accidents, and weather disasters, the scripting process is even more informal. In these cases, the production team gathers whatever footage it can, a reporter delivers an on-the-spot standup narration to explain it, and the producer writes introductory and concluding copy to be read by newscasters on the air (**Figure 10-22**).

Documentaries

Documentaries are programs that select, organize, and present factual material on particular subjects. The makers of *verité* documentaries claim to show events exactly as they unfold before the camera, without any script at all.

As explained in Chapter 3, even the most nonjudgmental verité documentaries actually "script" their material informally by choosing what and what not to shoot. Then by selecting, trimming, and sequencing the footage that appears in the finished program.

Infotainment Documentaries

On the other hand, typical documentaries of the kind seen on cable channels and public television are usually fully written before shooting begins. These infotainment documentaries on history, biography, science and medicine, wildlife, travel, public affairs, and many other subjects are as tightly scripted as any story video

Figure 10-22 A professional reporter can improvise from notes made at the scene of the breaking news.

or training program, **Figure 10-23**. Since verité documentaries are not formally scripted, we will focus on informational documentaries—almost all of which are infotainments.

As you prepare to write a pre-scripted documentary, remember that most such programs are designed to be entertaining. In fact, except for videos designed specifically for instructional use, the entertainment component may be more important than the information.

Legal Cheating

A scripted documentary can employ many of the techniques developed for story programs, including the three-act structure discussed earlier. Some subjects lend themselves to conflict and suspense (*will the expedition make it across the hostile desert?*). Many subjects unfold over time, so you can organize them as if they were stories (*how a frog egg becomes a tadpole and then a frog*).

Be aware, however, that fiction techniques are dishonest when they falsify the facts of documentary subjects. For example, if that supposedly hostile desert is really only a two-hour journey across and the season is cool, it is misleading to present the desert as a dangerous antagonist.

Also, it is easy to falsify behavior. Suppose, for example, you show a seal suddenly raising her head and calling (**Figure 10-24**). If you follow

Figure 10-23 American history is a popular documentary subject.

(Corel)

Figure 10-24 The seal raises her head and calls.

(U.S. Fish and Wildlife Service)

that shot with a clip of a baby seal turning its head as if listening, you are probably honest. The two shots were not recorded in sequence, the pup may not be listening to anything important, and the baby may not even belong to this mother. Nonetheless, baby seals do respond to their mothers' calls (**Figure 10-25**).

If the mother and baby seal were photographed facing in the same direction, the editor might flip one shot to indicate visually that one is "talking to" the other.

But suppose instead, that you do not have a shot of the seal pup, so you write voice-over narration: *Sensing that she has been away too long, the mother seal suddenly calls out to her pup.* That is less than honest, because you do not really know what the mother was thinking, or even if she is capable of such thoughts.

The Three *T*s, Again

In organizing a documentary subject, it helps to orient viewers at the beginning, so that they understand what the program is about and why they are looking at it. The three-*T* approach often works well, if it is handled informally. Instead of displaying title cards with topic buildups, use voice-over narration for introductions and transitions from one topic to the next.

Because lengthy commercials can make details hard to recall, infotainment programs often begin each new segment by summarizing the previous one.

Figure 10-25 Falsifying behavior.

The mother's call...

...appears to be heard by the baby.

(U.S. Fish and Wildlife Service)

Cross-cutting can be particularly effective, especially when you are following two or more sets of subjects. Suppose, for example, your topic is a year with the wildlife on a certain stretch of seashore. The main script organizer is the seasons, but within each one, you could switch back and forth among the seals, birds, otters, and even the dolphins offshore (**Figure 10-26**).

Documentary Program Elements

Scripted documentary programs (and short news features, too) are constructed from several elements that have been developed through decades of production. These video, audio, and graphics resources have become relatively standardized.

Video Resources

Despite the vital roles of interviews and narration, video is essentially a visual medium. Wherever practical, the most important component should be what appears on the screen. Preparing a documentary script, you have six basic types of visuals at your disposal: production footage, standup commentary, on-camera presenter, library footage, reconstructions, and interviews.

Digital effects and rostrum camera footage are covered separately in the Graphic Resources section later in this chapter.

Production Footage

Live-action footage typically provides the most compelling visuals. Today's lightweight, high-quality, high-definition cameras encourage the use of production footage—not only because less equipment allows smaller crews, but also because newer storage systems are so much cheaper.

Nature documentaries recorded on film are expensive because they require so many hours of costly film and processing in order to capture outstanding sequences. Most nature programs are now video recordings.

Standup Commentary

As we have seen, the *standup* report delivered on-camera is one of the most common components used in both news features and documentaries, **Figure 10-27**. Like interviews, the subject often delivers the entire text on camera. In postproduction, most of his or her audio is used independently, as narration over other visuals.

Creating "Wallpaper"

Documentaries often need to convey abstract information—usually by voice-over narration. Since abstract concepts have no physical existence, they are impossible to record visually. But because the screen cannot be blank, the script writer must invent visuals that appear plausibly connected to the narration, even if they do not show what is actually being talked about on the sound track. This type of seemingly (but not actually) related visual is often called *wallpaper*.

To demonstrate wallpaper technique, here is part of a documentary on the Pacific island nation of Vanuatu.

NARRATOR: Lacking both natural resources and heavy manufacturing facilities…

…the people of Vanuatu have become experts at repairing and conserving…

…the infrastructure of their transportation system…

…which, in a nation distributed among dozens of islands…

…means ships and boats.

Here, workers restore a small inter-island freighter.

Obviously, the boat-painting sequence receives more attention than it deserves. But, its job is to use up screen time while the narration discusses the abstract concept of transportation infrastructure.

If you analyze today's infotainment documentaries, especially science programs, you will see that half (or even more) of their footage is wallpaper.

On-Camera Presenter

An on-camera *presenter* is very similar to a standup reporter, and the footage is used in the same ways. However, a presenter is often recorded in a fully-lit studio or location, and often reads narration from a teleprompter. Due to improvements in miking and recording technology, it is now common to see the presenter in the field and participating in the video—trekking through the rain forest or shooting the rapids.

Library Footage

Where live-action material is unobtainable (as in history documentaries), archival footage can often be found in stock film rental libraries.

Figure 10-26 The program switches back and forth among four groups of animals.

(Corel)

Documentaries about 20th century wars are typically assembled from *library footage* (**Figure 10-28**). Entering "stock footage libraries" on an Internet search engine will find a large number of companies who license film and video footage on every conceivable subject.

Reconstructions

Often it is possible to restage and record actions that happened in the past, called *reconstruction*. In some cases, participants in the original events repeat their original activities. In other cases, professional actors impersonate historical figures. When using this technique, the ethical producer may need to use one of several methods to stylize the footage to indicate that it is not real. (For more shooting on re-enacted events, see the sidebar *Shooting Re-Enacted Events*.)

Figure 10-27 A standup commentator reading narration from a teleprompter. An appropriate background will be composited onto the green screen.

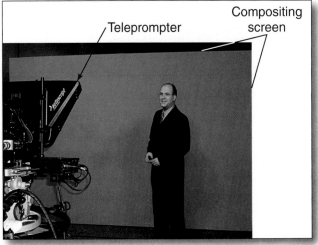

Teleprompter

Compositing screen

(JVC)

Shooting Re-Enacted Events

Documentaries often dramatize past events by restaging them. To make it clear that the results are not actual footage, editors typically distinguish them by altering them in postproduction. Here are just a few of the more popular methods of processing staged footage.

Sepia toned monochrome images evoke the past.

Setting the camera shutter speed slower than normal produces dreamlike images.

Purposeful over-exposure can stylize the footage.

(Sue Stinson)   *(Sue Stinson)*

The infotainment production industry has grown so large and experienced, that numerous additional techniques have been developed. The easiest way to discover them is by watching these programs.

In many historical documentaries, unmodified footage can be used because the participants are obviously impersonators. Viewers are unlikely to believe that images of Benjamin Franklin or Julius Caesar show those actual people.

Interviews

Interviews are the backbones of many documentaries, partly because they contain both video and audio components. Today, most interviews use an "invisible" interviewer—the interviewer is never shown and the questions are edited out (**Figure 10-29**). Interviewees are coached to give answers that do not sound like responses to questions. For example, the interviewer would not ask, "When did you start working with lions?" That question might

Figure 10-28 Library footage from World War II.

(Corel)

Figure 10-29 Most interviews now show only the subject.

(Corel)

elicit an answer like, "Ten years ago," which would make no sense when the question was removed in editing. Instead, the question is phrased, "Tell us how and when you started working with lions (and be sure to include *working with lions* in your answer)." Now the interviewee might respond, "I started working with lions ten years ago and I've studied them ever since." That answer will be clear, even when the question is removed.

As with standup reports, the interview often starts with the subject on camera, then switches to video footage of the topic while the interview responses continue on the sound track as voice-over narration.

In many cases, interviews are pre-scripted to obtain specific information for use at designated points in a program. To do this, the script includes sets of questions designed to elicit answers that fit the topic. The interviewer may have to phrase and rephrase requests for information until the interview subject produces answers that fit the needs of the script at that point.

Audio Resources

Audio is especially important in documentary and news programs. As we have repeatedly noted, the audio track is often more effective than the video—at least where delivering information is concerned.

Production Sound

Nothing adds more realism to visuals than live, synchronized sound. The problem is that documentary shooting does not always allow the time or the opportunity for obtaining optimal sound quality, though improvements in battery-powered wireless microphones now enable better audio in the field. Because of the unpredictability of documentary shooting and the wide differences in sound environments, it is especially important to record a track of general background sounds at each location. This track is used in smoothing out shot-to-shot differences in the production sound (**Figure 10-30**).

Narration

Whether the source is an interview subject, a reporter, a studio presenter, or a professional

Sit-Down or On the Fly?

In both straight documentaries and docudramas, conventional interviews are often called "sit-downs," for obvious reasons.

To impart a more dynamic feeling to the essentially static interview format, directors often shoot them as "on-the-fly" interviews, in which the subject talks while doing something else. In one common type of on-the-fly interview, the subject is driving while talking. This setting is preferred for two reasons:

- Outside shots of the moving vehicle provide quick and easy cutaways.
- Though the vehicle is moving, the subject and cinematographer (often including lighting and sound equipment) are conveniently locked together because both are sitting in the front seats.

If the subject were moving around a location instead, lighting, miking, and camera operations would be more difficult.

voice-over performer, narration is the most important audio component, often delivering the majority of information. (See an example of this under the "Narration" section in Chapter 8, *Video Sound*.)

In some cases, especially with historical subjects, the script is written as narration only. When it is complete, video and graphics resources are recorded or obtained to illustrate it. As

Figure 10-30 A background track of wind and wave sounds can tie together a variety of shots.

(Corel)

visuals are added, the narration is then adjusted to match the particular materials obtained for the program. In the sidebar *Storytelling with a Digital Rostrum Camera*, for example, the painting of the sea battle was studied to discover details that could tell the story. Then the basic narrative account of the battle was adjusted to match the frames selected.

Audio Reconstructions

Historical documentaries rely heavily on original sources, like letters, diaries, speeches, and public writings. Appropriate excerpts from these materials are recorded by professionals. Since we do not know how people sounded before the invention of voice recording, voice characterizations are generally developed that reinforce the public image of these characters. If you need to quote George Washington, for example, you develop an overall impression of the man and then use a voice that reinforces that impression.

Some poetic license is allowed. Actual recordings of Theodore Roosevelt reveal that this super-masculine western adventurer and leader in the Spanish-American War spoke in high, squeaky tones, which were quite at odds with his forceful image, **Figure 10-31**. Audio impersonators reading his words generally give him a more "appropriate" speaking voice.

Figure 10-31 Theodore Roosevelt's high voice did not match his forceful appearance.

Music

The key role of music is discussed in Chapter 8, *Video Sound*. Original music is very rarely composed and recorded for documentaries because of prohibitive costs. Instead, music libraries offer thousands of themes, and music arrangement software permits a remarkable range of customization. It is routine to select a musical theme from the recorded candidates, and then set its precise length, musical style (rock, country, classical, etc.), instrumentation, tempo, and other characteristics.

Music that sounds repetitious is a symptom of unskilled or hasty postproduction. Music that runs under most, if not all, of the video suggests a weakly structured program being "tied together" by its music track.

Graphics Resources

If you watch infotainment documentaries on cable channels, you know how much they depend on graphics: photographs, computer simulations, historical art, and other media. This is partly because past events are not available for recording and partly because certain types of information are best conveyed by graphics.

Rostrum Camera Footage

A *rostrum camera* is a rig that allows you to record photographs and other two-dimensional objects, imparting an appearance of movement by zooming in and out and panning from one picture detail to another. Specialized and expensive, rostrum setups are being replaced by software that enables you to digitize images, pan and zoom around them with your computer, and then import the finished sequences into your program (**Figure 10-32**). Today, many video editing programs include this "pan and scan" ability.

Maps and Diagrams

These common graphics are useful for orienting viewers and visualizing abstract ideas. As with photographs and artwork, these are now being digitized (or created in graphics software applications) and then set in motion by "virtual rostrum" software (**Figure 10-33**).

Figure 10-32 Panning and zooming on a still image.

From a wide shot of the painting…

…the camera pans and zooms…

…to frame the locomotive.

(Corel)

Computer Graphics

Because documentary programs are often necessarily slower-moving and short on dramatic live-action footage, lively computer graphics, titles, and transitions are often added to enhance the visual energy. In general, nonfiction programs of all types allow more varied and obvious effects than fiction programs, where excessive computer graphics distract viewer attention from the story (**Figure 10-34**).

Combining Program Elements

Mixing and matching all these video, audio, and graphic resources requires skill and practice. However, you can learn a great deal about the

Figure 10-33 A sense of motion created by "virtual rostrum" software.

A broad area of the Pacific shown on a map.

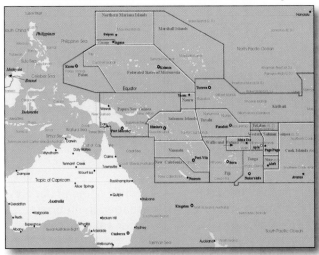

Zoom in to locate the nation of Vanuatu.

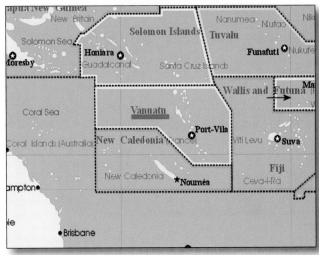

(Corel)

Storytelling with a Digital Rostrum Camera

Ingenious rostrum camera techniques can illustrate complex stories with a minimum of graphic resources. Here, an engraving of an early 19th century sea battle...

...is used with voice-over narration and supporting sound effects to create an entire sequence.

Imagine that in some cases, the camera pans and zooms from one image to the next, while in others, the change is by direct cut.

A 19th century engraving scanned into a computer.

(Corel)

Each red frame surrounds one of the images in the sequence.

NARRATOR: The sea battle lasted 7 hours and involved over 20 ships. SFX: Cannon fire continues throughout sequence.

The cannon fire... SFX: Cannon shot.

...from opposing frigates... SFX: Cannon shot.

...was deafening. SFX: Cannon shot.

Dismasted early in the melee, the U.S.S. Indomitable wallowed helplessly...

...while some of her crew escaped in lifeboats. Other sailors, however...

...were not so lucky.

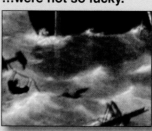

By dusk, the fleet had taken a terrible pounding... SFX: Cannon fire continues.

...but the Americans refused to strike their colors.

Notice that the last two images have been tinted blue in postproduction to suggest night. A sequence like this is completely scripted in advance, including narration, images, and digital "camera" movement.

Figure 10-34 Libraries of constantly moving patterns are commercially available; large production companies also create their own.

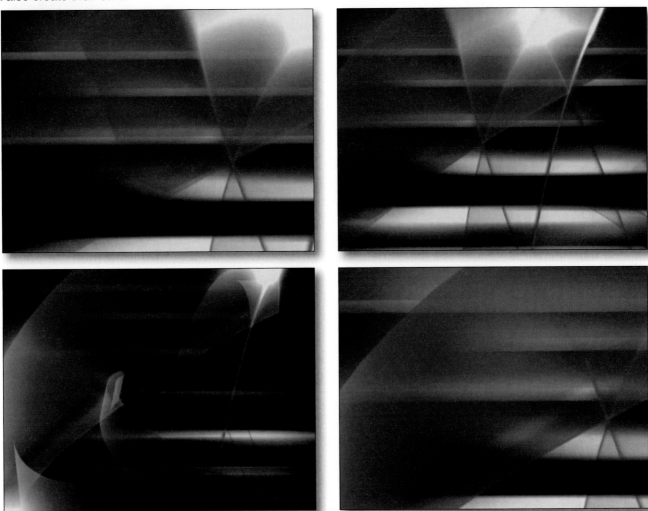

(Footage Firm)

process by watching documentaries on TV. But, you must watch them critically. No matter how interesting the subject may be, focus on the techniques used to present it. Notice how often live-action footage is recycled in different parts of the program. Whenever you see a lengthy shot of the landscape from a low-flying helicopter, ask whether the visual delivers information, or is it just wallpaper? Decide whether the content fills its allotted time, or whether it has been stretched and padded to achieve the right program length. Above all, decide whether the topic has been sensationalized for the sake of drama, as in "Our planet is doomed—*doomed* to destruction in only a brief *four billion years*!!"

From Development to Planning

With any type of video program, developing the script is only the first half of preproduction. Once the program has been written, the next phase is the thorough planning needed to ensure a successful shoot.

Summary

- In general, dramatic story structure calls for three acts: opening act, middle act, and concluding act.
- Most nonfiction programs belong to one of three broad types: instructional, promotional, and documentary.
- The three *T*s refer to the fundamental structure of a training program—**T**ell them what you are going to tell them; **T**ell them; **T**ell them what you have just told them.
- There are six basic types of visuals available when preparing a documentary script: production footage, standup commentary, on-camera presenter, library footage, reconstructions, and interviews.
- When delivering information, the audio track is often more effective than the video.

Technical Terms

Breaking news: News that is covered as it is happening, or soon after.

Buildup: A sequence of title cards, each one adding a new line of information.

Character arc: The growth, development, or simply change in a character during the course of a script.

Conflict: The struggle between opposing sides that creates dramatic action.

Dramatic action: The essential story line of a script; the plot.

Dramatic structure: The organization of a story to build interest and excitement.

Genre: A type of video, such as fiction, documentary, or training program.

Infomercial: A program-length commercial designed to resemble a regular TV program.

Instructional design: The craft of organizing an effective education or training script.

Library footage: Film or video collected, organized, and maintained to be rented for use in documentary programs. Also called *stock footage*.

Pickup: Video and/or audio material recorded later than the principal production, to add to or replace parts of material already recorded.

Presenter: An on-camera narrator who speaks directly to the viewer.

Reconstruction: A re-enactment of past events that is recorded and used in a program.

Rostrum camera: A camera rigged for moving around still images to record different details of them. (Now largely replaced by manipulating images scanned into computers.)

Signpost: In a training or documentary program, a reminder to viewers of what has been covered and what will come next.

Standup: A report presented on camera, usually by a reporter.

The Three Ts: Informal name for the basic organization of training programs—**T**ell them what you will tell them; **T**ell them; **T**ell them what you have told them.

Training program: An informational program designed to teach specific subjects.

Wallpaper: Footage intended to take up screen time while the narration presents material that cannot be shown.

Review Questions

Answer the following questions on a separate sheet of paper. Do not write in this book.

1. *True or False?* Programs fall into two overall classes: drama and comedy.

2. List the acts typically included in a dramatic story structure and identify the characteristics of each.

3. What is a *character arc*?

4. _____ videos are intended to teach certain sets of knowledge and skills.

5. What is the difference between *infomercial* programs and *infotainment* programs?

6. The two general types of instructional programs are _____ style and _____ style.

7. What are the three *T*s?

8. *True or False?* Buildups are lists that grow as a topic unfolds.

9. Explain how wallpaper is used in video programs.

10. How is a track of general background sounds used in a video program?

STEM and Academic Activities

STEM

1. **Technology.** Investigate the technology behind image manipulation software. How does image manipulation software edit or alter still photos?

2. **Language Arts.** Record an episode of a dramatic television program that you regularly watch. Analyze the episode and identify the opening act, the middle act, and the concluding act. Summarize the plot and action in each act of the program.

Underwater shooting is common in nature and science "docutaiments".

CHAPTER **11** Production Planning

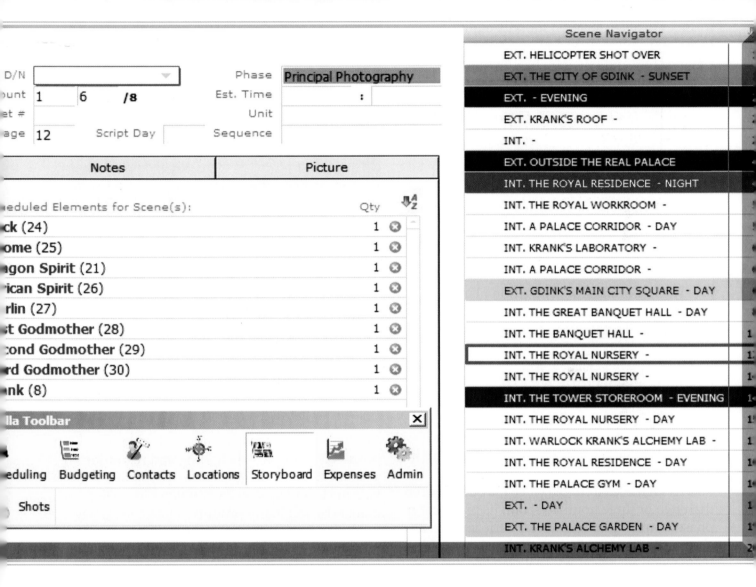

Objectives

After studying this chapter, you will be able to:

- Summarize the production requirements for a small-scale video production.
- Identify the people, places, and things required for a production.
- Create a production schedule.
- Create a production budget.

About Production Planning

When the content of your proposed video is well developed, you are ready for *production planning*—the process that will organize and manage the complexities of shooting. Planning is critically important for two reasons: time and money. For example, you may have permission to shoot in a key location for only eight hours (time). Or, you may be renting equipment or paying crew and talent by the day (money). In professional video production, the phrase "time is money" is literally true because nearly everything (and everyone) is paid for by the week, day, or hour.

Thorough production planning can make the difference between a smooth, successful shoot, and a disaster. To show you how to plan effectively, this chapter will take you through the process of preparing to produce a simple video.

The Big Moment

The importance of production planning becomes obvious if you think of an actual video shoot as similar to a theatrical performance. With a stage play, many people spend weeks conducting rehearsals, building sets, sewing costumes, and publicizing the show. All this lengthy preparation is for the two or three hours of the play's actual performance. During that critical period there should be no major mistakes, because all that preproduction effort pays off in a show that works.

Though even a simple video may take longer than two or three hours to shoot, its production phase is much like a theatrical performance: a short, intense period when all the planning and preparation are rewarded. That planning involves:
- *People:* The cast and crew of the production.
- *Places:* The locations and/or sets where shooting will take place.
- *Things:* The equipment used to shoot the program, the props and costumes in it, and miscellaneous items, like talent and location releases, **Figure 11-1**.
- *Plans:* The scheduling and other organizational processes that ensure a smooth shoot.
- *Budgets:* The predicted production costs.

The rest of this chapter examines each of these planning topics, in turn.

Figure 11-1 Finding, renting, and scheduling this vintage car for a sunset shoot in the desert required extensive planning.

(Corel)

People: Crew and Cast

In most professional video productions, the personnel involved demand the largest organizational effort and consume the majority of the budget. Some of these people appear in the production. Other equally important personnel work behind the camera.

The people who appear in video programs, whether actors, narrators, interviewees, or others, are often collectively called the **talent**.

If you make personal or small-scale professional programs, finding and recruiting people can be difficult because most amateurs lack the technical skills that video production demands, and many producers lack the money to pay professional-level salaries. The sidebar *Small Video Crews* presents some ideas for solving these problems when gathering people for low-budget videos.

Finding Production People

Even the simplest video project may be too much work to handle alone, so the first step is to find some assistants. The sidebar *Small Video Crews* describes three crews. The crew duties listed should give you an idea of how to assign responsibilities to your crew members.

Where do crew members come from? The first resource is always friends and family,

Small Video Crews

A video production *crew* can be as large as 20 members or more, each performing specialized tasks. Most small-scale projects, however, are crewed by just a few people—or even produced by a single person. For obvious reasons, members of small crews often do several jobs at once. The job divisions suggested are only typical.

2-Person Crew	
Title	**Responsibilities**
Producer/Director/ Videographer	Producing: Organizing and scheduling production, casting. Directing: Determining camera setups and shaping the performances of cast. Videography: Operating camera. Lighting: Designing and executing lighting with reflectors and/or lighting units.
Assistant	Preproduction management: Tracking cast members. Sound recording: Miking talent. Wielding the microphone boom. Controlling sound levels with production audio mixer. Camera/lighting assistance: Helping transport and set up video equipment. Setting and adjusting reflectors and lights as directed by producer. Continuity: Keeping track of shots and setups. Ensuring consistency of shot contents and action from shot to shot. Recording shot information for editor. Transportation: Taking people and equipment to and from locations, as necessary.

4-Person Crew	
Title	**Responsibilities**
Producer/Director	Producing: Organizing and scheduling production. Directing: Determining camera setups and shaping the performances of cast.
Videographer	Videography: Operating camera. Lighting: Designing and executing lighting with reflectors and/or lighting units.
Video Crew	Sound recording: Miking talent. Wielding the microphone boom. Controlling sound levels with production audio mixer. Camera/lighting assistance: Setting and adjusting reflectors and lights as directed by producer.
Production Manager	Continuity: Keeping track of shots and setups. Ensuring consistency of shot contents and action from shot to shot. Recording shot information for editor. Management: Organization and tracking of shooting schedule and all contributing components. Makeup/costumes/production design: Designing and applying cast makeup. Selecting, fitting, and managing cast costumes. Determining overall visual effect of production. Transportation: Taking people and equipment to and from locations, as necessary.

10-Person Crew	
Title	**Responsibilities**
Producer	Producing: Organizing and scheduling production.
Director	Directing: Determining camera setups and shaping the performances of cast.
Videographer	Videography: Operating camera. Lighting: Designing and executing lighting with reflectors and/or lighting units.
Gaffer	Lighting assistance: Setting and adjusting reflectors and lights as directed by producer.
Asst. Camera/Grip	Camera assistance: Help with camera setups, pull focus, perform scene carpentry.
Sound Recordist	Sound recording: Miking talent. Wielding the microphone boom. Controlling sound levels with production audio mixer.
Script Supervisor	Continuity: Keeping track of shots and setups. Ensuring consistency of shot contents and action from shot to shot. Recording shot information for editor.
Production Manager	Management: Organization and tracking of shooting schedule and all contributing components. Transportation: Taking people and equipment to and from locations, as necessary.
Makeup/Costume	Makeup: Designing and applying cast makeup. Costume: Selecting, fitting, and managing cast costumes.
Production Designer	Design: Determining overall visual effect of production. Props: Dressing location and/or set shooting environments with properties.

especially for jobs that do not require too much technical expertise. For example, a camera operator needs to know some techniques for obtaining good quality images (**Figure 11-2**), but a production manager can draw upon skills learned in almost any managerial position.

When you do need technical expertise, the first place to look is in community theater or local clubs that feature music groups. Both activities attract talented amateurs to design and operate lighting schemes and mix multi-track audio. If your project is for an organization (whether a business, a church, a school, or a community group), look for assistants within that organization. After all, group members are likely to be interested in the success of your project because their organization stands to benefit from it.

The Internet is another good source for crews:

- Post crew calls (and casting calls, too) on online classified ad services.
- Umbrella organizations, such as Meetup, may list video clubs in your area.
- Use creative search terms in Internet search engines—try: video + crew + volunteer, and possibly add your city.
- Online magazines, such as *Videomaker*, often list local sources.

In the case of the McKinley Boosters (introduced in the sidebar, *Case Study: The McKinley Booster Club*), you might find a member who is interested in audio recording to become your sound person. A local amateur theater

technician might want to learn video lighting. The business and professional members should offer any number of executives willing to manage the production (particularly if given a title, like "Producer"). Other members might be called on to help provide transportation, hold reflectors during outdoor shooting, and perform other nontechnical duties. Still other club members, of course, will provide most of your on-camera talent: the *cast*, **Figure 11-3**.

Casting the Production

In personal and entry-level professional productions, you may not have much choice in casting the parts in your production. Your strategy is to make the most of the choices that you do have. To cast your show successfully:

- Do not always cast the obvious person. For roles that are not gender-specific (unlike "Father" or "Mother"), cast males and females even-handedly.
- In casting a real-world character, do not limit your choice to the actual person. For example, the current president of the McKinley Booster Club is a nice man, but not a very dynamic speaker. The former president, however, is a woman with a fine platform presence and an ability to read speeches convincingly. For these reasons, you cast her as the president and spokesperson in your video. (After all, by the time the video is

Figure 11-2 Video hobbyists are likely candidates for your crew.

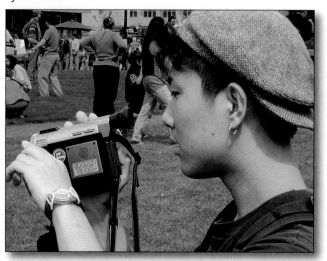

Figure 11-3 Local theater groups are good sources of talent.

(Corel)

Chapter 11 Production Planning 207

Case Study:
The McKinley Booster Club

Throughout this chapter, we illustrate the discussion with a fictional organization called the McKinley Booster Club.

In the small city of McKinley, the Booster Club is an organization of business and professional men and women who meet for lunch weekly to discuss civic beautification and other urban improvements. The Boosters conduct fund-raisers for park benches, playground equipment, and landscaping projects. They work to save historic buildings and return them to productive use. On weekends, they spread out in squads to collect litter, clean up graffiti, and paint houses in low-income areas.

Now, they want a ten-minute video program to explain their activities, in order to recruit more members, raise money, and generally increase civic awareness of what they do. Imagine that, as a proud member of the McKinley Booster Club, you have volunteered to produce this show for them.

President McKinley welcomes visitors to the Plaza.

completed and released, the current president will be out of office too, so it makes no difference whom you cast in the role.)

- Find out in advance whether people have other obligations that might conflict with your shooting schedule. No-show actors can bring your entire production to a halt.

In this book, "actor" refers to both genders.

Conducting Effective Auditions

Though you may have no choice in "casting" a personal family program or a professional job, like a wedding, many video projects require you to find appropriate people to fill on-screen roles. To match these roles to the right performers, you may wish to hold auditions at which people try out for parts in the program. You can simplify the tryout process by using these suggestions:

- Ensure privacy. Never ask people to read for a role in front of other applicants for the same part. Instead, have applicants wait in one room while you hold tryouts in another.
- Videotape all acting candidates to see how they look on screen. People often appear different on video, sometimes surprisingly so. Also, by recording tryouts you can later refresh your memory of each applicant (and they do tend to blur in your mind very quickly). Finally, you can edit the footage of various candidates to see how they look together. Suppose, for instance, that you have chosen your Romeo, but three different actors might make good Juliets. By intercutting audition shots of Romeo with footage of each different Juliet, you can see which combination works best.
- Tell actors not to look at the camcorder—and then see whether they can follow this direction. Some amateurs will spoil take after take by unconsciously looking into the lens.
- Give auditioners simple things to do, such as entering a room and sitting down or pouring and serving a glass of water. This will show you how naturally they can behave on camera.

Handling Talent

Auditions are psychologically punishing to all actors because they involve so much rejection, **Figure 11-4**. Only one applicant can be chosen for each role. So, if ten people try out for a part, you will have to reject *90 percent* of them. In doing so, you appear to be rejecting them not as performers, but as people.

Think of it this way: if you turn down an applicant for a lathe operator position you are judging her metalworking skills. If you do not select a violin player for your orchestra, you are evaluating his skills as an instrumentalist. But,

Figure 11-4 Auditioning can be stressful for performers.

an actor's instrument is not a lathe or a violin; it is his appearance, his voice, his personality—in short, his entire self. By rejecting the actor, you may appear to reject the whole person. For this reason, it is important to support the people who audition for you.

First, give them adequate time. Even if you can see at once that applicants are unsuitable, give them time to show what they can do. It makes them feel that they have been taken seriously, and it often reveals talent that you may want for a later production.

Tell auditioners very simply what you want them to do. An audition is not the place for elaborate explanations of character. For example, if the role is that of an unsuitable job applicant, simply tell the actor something like this: "You want this job so badly that you can't help talking too much." A talented applicant will know what to do with this direction.

Finally, show that you are paying close attention to their efforts. Watch them as you listen to them read. Use their names frequently in conversation. Find something in every performance that you can praise.

By treating auditioners decently, you also enhance your own reputation. Whether a production is cast from a village's amateur theater crowd or the ranks of Hollywood professionals, the performing community is always small and the word gets around. By proving yourself sensitive to actors' feelings, you will make actors wish to work for you—both now and in future productions.

Evaluating Audition Results

When auditions are over, you will need to select your cast. Some video makers systematically list and judge the pros and cons of each applicant (**Figure 11-5**); others rely completely on instinct for placing the best person in each role.

No matter how you cast your program, be sure to base your decisions on the audition videotapes. As noted above, the impressions people make on tape are often quite different from the way they appear in real life. People who are dynamic and forceful in person may seem dull on screen, while others who are not impressive in life are remarkably effective on tape. It is said that "the camera loves" these people; you can evaluate that affection only by studying the audition tapes.

Places: Scouting Locations

In scouting a shooting location, you contact the local people whose help you will need, determine the suitability of the place for shooting, and look for potential problems that must be solved in order to conduct a successful shoot there.

Scouting People

First and foremost, you need to establish productive relationships with key people at any location. If they are on your side, these people can give you substantial help, but if they are uncooperative they can make life very difficult.

Figure 11-5 Keep careful casting notes.

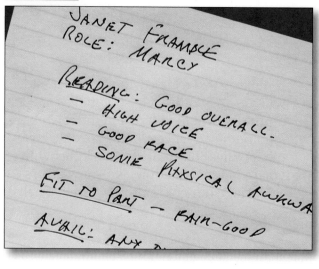

Managing Volunteer Helpers

Unreliable help is one of the biggest problems in producing personal or entry-level professional videos. If you have made videos with volunteers, you have probably experienced behavior like this:

- People arrive late and then announce that they must leave before the day's shooting is over.
- Crew members do not fulfill promises to do preproduction chores, such as finding props and costumes.
- Actors appear for later shooting sessions in clothes that do not match the outfits they wore in previous sessions.
- Crew members work slowly, or refuse to do jobs that do not appeal to them.
- Actors fail to memorize their lines or, sometimes, even to read the script.
- Cast and crew members alike develop last-minute things to do instead, and fail to show up for shooting sessions.

Studio shoot, day one.

Studio shoot, day two—wrong shirt.

Why do volunteers often let the production down like this? Because they are not video professionals and do not pretend to be. Professionals have three powerful motivations for responsible behavior: they are being paid for their work, they will not be rehired if they develop a reputation for irresponsibility, and they enjoy their crafts and take personal pride in practicing them well. Amateurs, however, are motivated only by friendship with you, personal ego, and sometimes an interest in making a successful program.

Obtaining Loyalty

You may not be able to pay your cast and crew, but you can give them rewards that will help motivate them to work reliably and hard. These rewards are importance, attention, praise, credit, and ownership.

- *Importance.* Stress to cast members, in particular, that they are so essential to the production that when they are not present the production literally stops. So, if they do not show up, many other volunteers like themselves are forced to stand idle. (Be careful in using this idea with certain actors, who may secretly enjoy this power to stop and start the whole production.)
- *Attention.* No matter how busy you may be, pay honest attention to the ideas and suggestions made by cast and crew. When people feel that their contributions are ignored, they lose their willingness to help in all areas.
- *Praise.* Notice and acknowledge good work. Praising actors is not too difficult, but your volunteer crew also needs recognition. A simple compliment like, "Great lighting setup—very realistic," may be all the pay that your lighting assistant will get.
- *Credit.* Before production starts, emphasize that cast and crew will receive on-screen credit for their contributions. People love to see their names in the credits.
- *Ownership.* Above all, create the idea that this is a team effort and the production belongs to everybody. In Shakespeare's *Henry V*, the king motivates his troops before a battle by calling them, "We few, we happy few, we band of brothers." That psychology works just as well today.

Cultivate the custodian or building engineer, who knows more about the facility than anyone else, including the locations of the electrical power box, spare furniture, and other potential props and set decorations, **Figure 11-6**. For example, the McKinley Booster Club can avoid high rental costs at their regular meeting room in the McKinley Arms Hotel by borrowing a lectern with the hotel logo on it and using it to stage the president's on-camera remarks at another location outside the hotel. Who knows how to get hold of that lectern? The hotel's maintenance supervisor.

If your location is a public place, you may also need to coordinate with one or two other key people. See the executive in charge for permission to use the facility. For example, the Booster Club might need to ask the hotel manager for permission to shoot establishing shots at a luncheon meeting. Often, management will supply a person to accompany you throughout your shoot at the location, or at least to be on call when needed. For instance, the McKinley Arms Hotel might deputize the Assistant Manager for Banquets and Meetings to accompany the production crew in order to help out.

Consulting key people is important on personal projects, as well. For instance, before moving furniture to get better camera angles on a birthday party, you might want to clear the idea with the party's host.

Scouting Facilities

You scout the facilities to ensure that you will have everything you need for production.

- Check out the space where you will be shooting. Is there enough room? Can you create in it the on-screen environment you need (hotel room, office, dungeon, desert island)?
- Ensure that you will have access to a production area for storing equipment, costumes and props; for recharging camcorder and light batteries (**Figure 11-7**); for changing costumes. Is this area secure?
- Check the electrical service, starting with the power supply. Will the electricity fed to the shooting area power your camcorder, audio equipment, and above all, lighting? If not, can you obtain power by running a cable to the main service box? Verify that enough wall outlets are available for your use.

Tapping an electrical distribution box is strictly a job for a professional lighting person, unless a crew member happens to be a qualified electrician.

- Scout parking for cast and crew. Even a small production can involve a dozen cars that may have to be expensively garaged, or else parked far from the location.

Scouting Video Problems

Perhaps the most important scouting task is checking the location for camera problems,

---

Figure 11-6 Maintenance people know a building's systems.

Figure 11-7 A secure place for recharging batteries is essential.

especially lighting difficulties. In many cases, the shooting area is too dark for good-quality video. By scouting, you can decide whether to bring in movie lights, increase the power of the available lighting, or even recommend moving to a better-lit area.

Another problem is excessive contrast: too much difference between the lightest and darkest parts of the scene. Camcorders cannot record high-quality video in environments where the contrast is extreme. In those conditions, the lightest areas "burn out" to glaring white, or the darkest areas "block up" to a solid black.

Take, for example, the video scene in which the mayor of McKinley praises the civic activities of the Booster Club. This sequence is to be shot

with the mayor sitting at a desk in her office. The problem is that the desk is backed by a window that is very bright, even when the drapery is closed. In this high-contrast environment, the mayor seated at a desk will record as a black silhouette (**Figure 11-8A**). By scouting for camera problems, you realize that you can place your camera where it will exclude the bright windows, as you can see from the office plan in **Figure 11-8B**. Simply direct the mayor to open the sequence facing forward and then swing her chair around to face the camera.

Another potential problem is mismatched light colors. In the mayor's office, light from a big desk lamp has a distinctly orange cast, while the window light is quite bluish. Noting this

Figure 11-8 A—In front of a bright window, the mayor is too dark. B—Repositioning the camera avoids the bright window. C—Moving the desk allows extra camera angles. D—The mayor from the new camera angle.

A

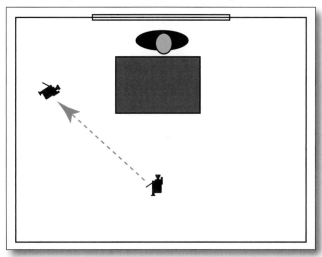

B

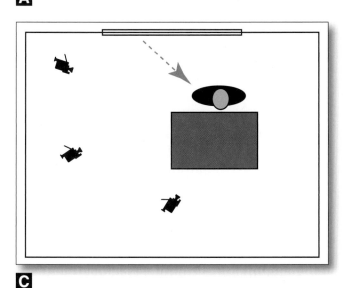

C

D

problem, you can plan to provide blue filtration to convert the lamp output to the color of daylight.

The same scouting trip will also reveal that you have a big problem in getting enough camera angles. Shooting at right angles to the window will give you one setup, but you will be unable to move to the side without running into the wall or else including the bright windows. After inspecting the location, you can solve the problem by planning to "cheat" the mayor's desk forward enough to permit a side angle on her face, **Figure 11-8C**. The shift will not be noticeable to the viewers, **Figure 11-8D**.

Finally, many video problems arise because the camera must be placed too far from the subject. This is often true, for example, in wedding videos, where cameras and their operators are prohibited in certain parts of places of worship. By discovering that you will be far from the action, you can plan to bring a long lens or lens extender, plus a wireless microphone for the wedding couple (**Figure 11-9**).

Scouting Audio Problems

A distant sound source is one of several problems in recording quality audio. Although zoom microphones are available, they have nowhere near a lens' ability to zoom in for more detail and less background. If you catch the problem, you might provide wireless mikes,

as suggested earlier. Or you may be able to feed sound from the facility's own public address system.

Another major audio problem is background noise, and here the McKinley Booster project is in serious trouble. Most of the shooting locations are outdoors at public buildings or in the town plaza. Traffic noise is almost impossible to eliminate, since it reflects from all sides. You can control the problem by staging important scenes with the worst traffic behind the camera and microphone. Alternatively, you could establish the scene in the noisy location and then tape all the close shots of talking in a matching, but quieter background (**Figure 11-10**).

Wireless microphones can sometimes add as many sound problems as they solve, especially if they pick up interference. Auto ignitions and other electrical sources can cause static, and your microphones may share frequencies with other electronic devices. High-quality wireless microphones have fewer interference problems, but they can cost several hundred dollars apiece. So, if you plan to use wireless mikes, take them on your scouting trips and test them at each location.

Things: Equipment and Supplies

After people and places, the most complex components of any production are *things*: cameras and lights, sound equipment, furniture, props, and set decorations. Without planning and careful management, looking after *things* can cause major problems.

Production Equipment

Where simple personal or entry-level professional productions are concerned, preproduction work with video equipment means ensuring that everything is present and in good working order. To do this, make a list of every piece of equipment, even if you have only a very simple video outfit. You would be surprised at how easy it is for small accessories and other items to be overlooked. Easily forgotten items include:

- Batteries and cables, if you use external microphones.

Figure 11-9 Wedding setups often require long lenses and wireless microphones.

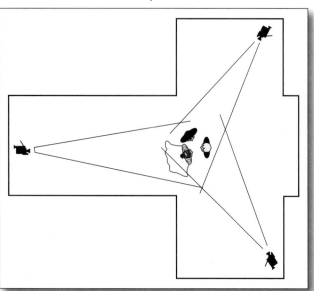

Figure 11-10   Moving the subject away from noisy backgrounds will often improve the audio quality.

Noisy background sounds are picked up by the mike.

Close shot recorded in a matching, quiet environment.

(Sue Stinson)

- Camera quick-release components, **Figure 11-11**. One-half may still be attached to the tripod, but is the other half on the camcorder?
- Spare lamps for movie lights (since these special bulbs cannot be bought at just any supermarket).

Next, make sure that all your equipment is functioning properly by repairing problems as soon as you notice them. If the tripod lock is defective or if the zoom lens control is stiff, get it fixed before you forget it. It is all too common to arrive at a new shoot with equipment that has not been repaired.

Figure 11-11   Quick release plates can become separated from their tripods.

Managing Workflow

Digital media management is now tightly integrated into all phases of production and postproduction. Recording codecs and media, storage of recorded material, and editing protocols vary greatly. In the digital world, just about anything can be transcoded into something else. But, this process is time-consuming and can sometimes lead to losses in image quality.

In preparing a shoot, you need to organize your project's workflow:

- If possible, choose camera and editing equipment that all use the same codec. (Some cameras are now designed to work directly with particular editing programs.)
- Provide storage media. If your camera uses tape or disc, your footage is stored automatically. If you record on an internal hard drive, solid-state drive, or removable flash media, have someone with a laptop computer or portable DVR transferring footage as you shoot.
- Back up your footage. Whatever your storage medium, you should burn copies of all your material. (Of course, if you import footage from camera tapes or CDs, your original media then serve as backups.)
- Provide for immediate review (and, possibly, trial editing). It is now common to make a rough assembly of footage as it is shot, to ensure that coverage is adequate and shots match. Even if you will not edit until later, you should review all your footage on a well set-up monitor.

Pay particular attention to power for the camcorder and for lights, if you use any. Check to see that all camera batteries are fully charged and that the charger is included in your gear for use on location. Have plenty of industrial grade extension cords for movie lights.

Equipment planning also covers special items that you rent or buy for this shoot. Such items might include teleconverter accessories for shooting athletic events or special filters for creating optical effects.

Finally, ensure that you have an ample supply of recording media and, if appropriate, off-line storage, such as a laptop.

A **codec** is a particular computer protocol for encoding visual data to record it and decoding it to display it. **Transcoding** is the process of translating material from one codec into another. For example, transcoding is required when the camera and editing software use different codecs.

Sets, Properties, Costumes, and Makeup

Simple productions rarely involve actual set building, but you will often want to *dress* locations to make them look better on screen. To anticipate set-dressing needs, collect an assortment of potted plants, wall decorations, small furniture pieces, and other items to utilize if needed. It is also a good idea to have a roll of masking tape or double-faced tape.

Properties (usually shortened to "props") are all the items that are called for in the script: the glasses and silverware, the spectacles, the tire iron, the chocolate chip cookies—anything and everything that will appear in your program. To make sure you have every prop, make a list of them and check them off twice: once before you leave for a location and again before you shoot. Nothing is more frustrating than being unable to shoot for an hour while someone runs all over town, frantically searching for a forgotten prop.

In the McKinley Booster Club program, a Certificate of Merit is a key prop because the president will be seen presenting it to a homeowner for an exceptionally beautiful front yard. In this case, the actual certificate will not

be ready in time for shooting, so you will need to create a fake one with a color printer. From a distance, the camcorder will be unable to tell the difference.

In simple productions, costumes are usually provided by the actors (**Figure 11-12**), but you must include them in preplanning by using a costume inventory and checklist. Before every scheduled shoot, call all cast members and remind them of the costume items for which they are responsible. (Perhaps the biggest source of delays in personal and small-scale professional production is actors who do not return in the same outfit they wore during the previous shooting session.)

Like costumes, makeup is usually the responsibility of individual actors. The sidebar *A Basic Makeup Kit* suggests a simple kit that you may want to assemble and have ready to use for touch-ups, if needed.

Releases

Releases are legal documents granting you permission to include people, places, objects, and music in your program. Without releases, you can get into serious difficulties when your program is shown.

Hobbyist video makers rarely bother to obtain releases, and it is true that if you are making a program for *purely personal or family use*, you do not need to worry about permissions.

Figure 11-12 Performers often provide their own costumes.

A Basic Makeup Kit

Very little makeup is used in most personal and entry-level professional videos. Female cast members generally wear their normal street makeup. Men usually wear none at all. Nevertheless, you should have at least the following items to use for touching up your talent:

- Neutral powder for eliminating skin shine.
- Light- and medium-toned powders for minimizing skin imperfections.
- One or two powder brushes.
- Neutral blemish spotter in a variety of skin tones.
- Facial tissues.
- Mirror.
- Hairbrush, comb.
- Hair spray, to keep hair in place.
- Trim scissors.
- Small plastic tool or tackle box to hold everything.

If your programs include performers of different ethnicities, you will want to customize your powder and spotter colors to suit. In most larger cities, you can buy makeup at theatrical supply houses. In smaller cities and towns, try stores that sell dance supplies or party and Halloween costumes. You can obtain simple products at any drug or discount store makeup counter.

The Internet is also a good source for makeup supplies. However, be careful about buying complete "kits" that may include items you do not want, and may exclude some that you do. It is better to select and buy your components individually.

With your supplies assembled, here are some tips for using makeup in your programs:

- Use as little as possible.
- Fold tissues around collars to keep powder off costumes.
- Use makeup only to cover things up (shine, blemishes), not to alter a performer's appearance.
- Let women apply their own makeup, and allow both women and men to fix their own hair. They know themselves better than you do.
- Check the effect of makeup on an external color monitor before taping.

Some basic items for a makeup kit.

Professional video programs are different. If you are making your program for profit and/or public showing, you risk being prosecuted if you fail to get the necessary releases and rights. What does the phrase *"for profit and/or public showing"* mean? It includes one or more of the following uses for your program:

- You were paid for producing it.
- It will be offered for rental or sale.
- It will be transmitted publicly via broadcast, cable, satellite, the Internet, or any other method.
- It will be shown anywhere outside your home and those of family and close friends.

Note that a program qualifies if it is either for profit *or* public. A private corporate video is for profit because you were paid to make it. A showing outside your home is public, even if no one receives compensation for it.

The McKinley Booster Club video falls into the last category. Nobody will make a penny for working on this program, and the club will not sell tapes or charge admission to showings. However, they will screen it all over town in order to develop support for the club's activities. For that reason, the Booster Club must have all necessary releases on file before the program is shown.

Types of Releases

Generally speaking, there are seven common types of releases and permissions: general, talent, minor, materials, location, stock photos and footage, and music.

U. S. copyright law is nationwide, but other forms of permission may be affected by state laws. Release forms sold in local camera stores will be appropriate for your location, but it is wise to have the language of each release form reviewed by an attorney.

General release. A general release is for people who will not be paid to appear in your video. This applies to subjects of documentaries and volunteer actors alike.

Talent release. A talent release is used for people who are paid to appear. (This is sometimes called a "model release" on commercially available forms.)

Minor release. A minor release is used when the subject is under 18 years of age. It includes provision for printed, signed permission granted by the minor's parent and/or legal guardian. (Often, a general release can be modified by adding a permission statement signed by a parent.)

Materials release. A materials release is for objects included in the program, often photographs or graphic materials. If you want to show a group portrait of the McKinley Boosters

Release Forms

You may take either of two opposite approaches to release forms. Some people feel that unpaid amateurs, especially, may be intimidated or made suspicious by overly complicated language. A clear, simple release like the one below is preferred.

Performer Release

My signature below attests that I, _____, have read this release and that I understand and agree to the following:

1. I permit recordings of my voice and likeness to be made in connection with a media project currently titled _____.
2. I authorize _____, the proprietor(s) of this project, in perpetuity, to process, store, reproduce, distribute, and display recordings of my voice and likeness made in connection with this project, in whatever ways, forms, and media, and by whatever methods and technologies they may choose.
3. I waive, in perpetuity, the rights to any and all compensation for these recordings, other than such rights as may be set forth in other written agreements between myself and the proprietor(s) of this project.

_____
Performer signature

_____
Performer printed name

_____
Date

_____
Witness signature

_____
Witness printed name

_____
Date

---

in 1926, you need to find out who owns it and execute a materials release for using it.

Location release. A location release allows you to shoot on private property.

Stock permission. You need permission to include still or moving images obtained from professional stock footage companies. Typically, permission for a specified use of the material is included when you buy the rights to the footage.

Music permission. In the same way, you must obtain permission to use copyrighted music, whether previously recorded or newly recorded for your program. It is not safe to assume that music is in the public domain. *Happy Birthday,* for example, is copyrighted.

See the sidebar *Music Permissions* for more details.

Music Permissions

It is vitally important to get permission to use copyrighted music in your video, particularly because over 25 "Music Rights Societies" worldwide monitor usage diligently and enforce copyrights aggressively. In the U.S., the two music rights clearinghouses are ASCAP (The American Society of Composers, Authors, and Publishers) and BMI (Broadcast Music, Incorporated). For useful links regarding music rights, look at the website of the National Music Publishers' Association (www.nmpa.org).

Other people feel that a release should address every legal contingency. A sample of a comprehensive release appears below.

Release Agreement

Whereas, _____ (the "Producer") is engaged in a project (the "Video"), and

Whereas, I, the undersigned, have agreed to appear in the Video, and

Whereas, I, understand that my voice, name, and image will be recorded by various mechanical and electrical means of all descriptions (such recordings, any piece thereof, the contents therein and all reproductions thereof, along with the utilization of my name, shall be collectively referred to herein as the "Released Subject Matter"),

Therefore, in exchange for $1.00, receipt of which is hereby acknowledged and whose sufficiency as consideration I affirm, I hereby freely and without restraint consent to and give unto the Producer and its agents or assigns or anyone authorized by the Producer, (collectively referred to herein as the "Releasees") the unrestrained right in perpetuity to own, utilize, or alter the Released Subject Matter, in any manner the Releasees may see fit and for any purpose whatsoever, all of the foregoing to be without limitation of any kind. Without limiting the generality of the foregoing, I hereby authorize the Releasees and grant unto them the unrestrained rights to utilize the Released Subject Matter in connection with the Video's advertising, publicity, public displays, and exhibitions. I hereby stipulate that the Released Subject Matter is the property of the Producer to do with as it will.

I hereby waive to the fullest extent that I may lawfully do so, any causes of action in law or equity I may have or may hereafter acquire against the Releasees or any of them for libel, slander, invasion of privacy, copyright or trademark violation, right of publicity, or false light arising out of or in connection with the utilization by the Releasees or another of the Released Subject Matter.

It is my intention that the above-mentioned consideration represents the sole compensation that I am entitled to receive in connection with any and all usages of the Released Subject Matter. I expressly stipulate that the Releasees may utilize the Released Subject Matter or not as they choose in their sole discretion without affecting the validity of this Release. This Release shall be governed by (name of state) law.

I hereby certify that I am over the age of eighteen, and that I have read, understood, and agreed to the foregoing.

_____
Print Name

_____
Address

Phone No (____) _____

_____
Signature   Date

_____
City, State, Zip:

Remember that these are only samples, and may not be appropriate for your area. *Do not use them without checking with an attorney.*

When to Obtain Releases

In many cases, people start out to make a program for purely personal use, so they do not obtain releases. When the video is finished, however, they decide that they want to show it publicly—perhaps by entering it in a contest or screening it for a local club. At this point, it may seem too late (or too much trouble) to secure releases. In this situation, some producers will enter or show the program anyway, thinking that the program's participants will not care (or perhaps even know) that they are being publicly displayed.

This is not only illegal, but risky as well. It is far better to obtain all necessary releases and permissions before and during shooting, even if the program is not intended for public use and/or profit. In short, if in doubt, *get releases*.

The sidebar *Release Forms* includes sample releases.

When Releases Are Not Needed

If you are shooting on public property, you normally do not need permission to show buildings, people, etc. who happen to be in the background. However, you may still need releases. Here are just a few situations that require them.

- You are shooting into private property (such as a residential backyard).
- Your program implies that someone is doing something unlawful or improper. For instance, if you show pedestrians in a night scene and voice-over narration says, "Many people out at this hour are engaged in illegal activities." You are indicating that the narration applies to the pedestrians.
- An exterior or interior location is put to fictitious use. For example, suppose you show a subject entering the street door of a restaurant, after which you cut to a restaurant set. You are saying, in effect, that your set is the interior of the actual restaurant, so you need to have permission to do so.

Examples could be multiplied, but the general principle is clear: *when in doubt, get releases*.

One solution to the permissions problem that is often seen on news and documentary programs is to throw an entire image completely out of focus, either in the camera or digitally in postproduction, *Figure 11-13*. For example, the soft-focus shot may show a street full of pedestrians as a narrator says, "Eighty percent of adults have broken the law at least once." Since all the people are too blurred to identify, no one can complain that the statement applied to her or him.

Plans: Production Logistics

Logistics is the overall task of organizing and supplying the shoot. In an elaborate production, this can be a formidable job. For simpler programs, the biggest tasks are creating a shooting schedule and providing support services.

Scheduling

Scheduling is the process of deciding which scenes to shoot in which order. The task seems straightforward until the production manager runs into problems like these:

- For a scene between them, the Mayor cannot work on Wednesday and the Booster Club President is unavailable on Thursday.

Figure 11-13 Images can be thrown out of focus in postproduction.

Figure 11-14 Night shooting requires careful planning.

- The bulldozer for a dump cleanup sequence is available only on weekends, when it is not in use; but the city personnel who drive it do not work weekends.
- The statue illumination ceremony has to be shot at night, **Figure 11-14**.

Juggling several conflicting demands at once can make scheduling a frustrating business. Fortunately, there is a well-developed system for organizing a shoot. To sort out shooting priorities, the production manager lists certain characteristics for every scene in the script. (A scene is a single piece of action that happens in a single place at a single time.) Key characteristics of scenes include:

- *Places.* For obvious reasons, it is more efficient to shoot every scene that takes place at one location before moving on to the next location. Try to group scenes together by location to minimize moving the production from one place to another, then back again.
- *People.* At each place, different performers may be needed at different times. So, scenes in which the same actors perform together should be scheduled consecutively.
- *Things.* Properties and large items, such as special automobiles, may be available only briefly, so the scenes in which they appear should be grouped together.

- *Conditions.* Daylight or darkness, sunshine or overcast, some natural conditions are called for in the script. You can sometimes shoot night scenes during the day and even vice versa, but the process is complex and expensive. So, for each location, schedule night or sunset scenes together, **Figure 11-15**.

A typical production breakdown will prioritize scenes as follows:

1   All scenes at location X.
1.1   All location X scenes at special times, especially night.
1.1.1   All location X night scenes with performer A.

If key performers can work only at certain times, you may be forced to organize the shoot around their schedules, even if it means returning to the same locations.

Support Services

Even on small productions, support services can make a big difference in the morale and enthusiasm of cast and crew. Parking is especially important, and if the production is big enough to require supply trucks for lighting and other equipment, close and convenient parking is essential.

Figure 11-15 Downtown McKinley. The conditions before sunset are called the "magic hour."

Food is equally important. It is good to have light snacks, coffee, and soft drinks available, especially for early morning or late evening shoots. In addition, meal breaks should be scheduled and adhered to. If public eating facilities are not easily available, meals should be provided.

Finally, attend to everyone's comfort. Since all production involves considerable waiting, especially for the performers, provide a comfortable lounging area (often called a *green room*). At interior locations, this may be just a room or other quiet area out of the way. Outdoors, it may be as simple as an area supplied with folding chairs and shaded by a tarpaulin. Restroom facilities are important too, and, if you are shooting at an outdoor location, may be required by local ordinance. If you are working, say, in the desert, you will need a water supply for washing as well as drinking, along with appropriate portable toilets.

Budgets: Production Costs

All video productions cost money. Anticipating and controlling expenses is as vital to the success of a project as writing, directing, or videography. The formal process of identifying these production costs and allocating funds to pay them is *budgeting*. Typically, a project budget is created early in production planning, and then refined and updated throughout the course of the production. Though budgets can sometimes be complex, the principles that govern them are simple:

- Identify every item that will cost money, including salaries, rentals, purchases, services, and rights.
- Estimate the amount of each item that will be needed.
- Determine the exact unit cost for each item.
- Multiply the item amount by the unit cost to determine the total cost for that item.
- Add the item totals together to determine the overall expected costs.

Figure 11-16 illustrates this process with three types of budget items: a salary, a rental, and a purchase. (Simpler projects will, of course, cost less.)

Standard Production Budgets

Even simple productions can have so many expenses that it is difficult to identify every one in advance. For this reason, most established production houses develop or buy pre-organized budget breakdowns. These are usually computer spreadsheet or database programs developed for film and video budgeting, **Figure 11-17**.

To customize these standard forms, you delete line items that do not apply to your project, add any specialized items that may be unique to your production, and then fill in the amount and rate for every item. Typically, the software calculates totals, sales tax, and similar charges when applicable.

If you have some experience in developing spreadsheets or other databases, you may find it productive to build your own custom system for production planning, scheduling, and budgeting. In addition to spreadsheets and databases, some project management software packages include financial functions.

Contingency Funds

Since video production is always vulnerable to unexpected events and expenses, every budget should include a contingency fund: an

Figure 11-16	Budget line items.		
Item	Number/ Quantity	Unit Cost	Total
Actor	4 days	$300/day	$1,200
Camera dolly	9 days	$120/day	$1,080
Gaffer tape	2 rolls	$12/roll	$24

Figure 11-17 Summary page of a sample budget.

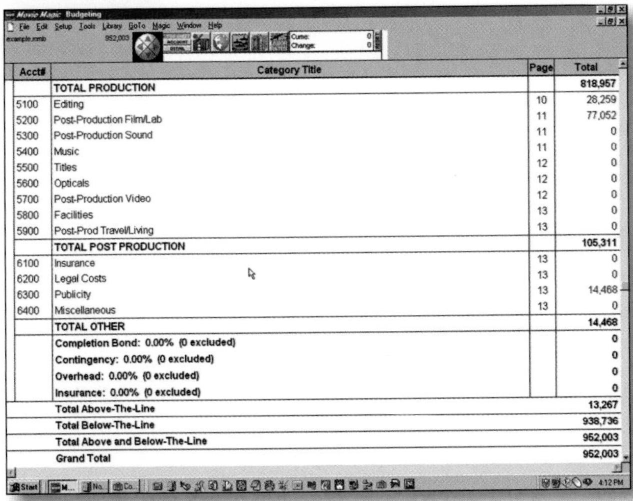

Acct#	Category Title	Page	Total
	TOTAL PRODUCTION		**818,957**
5100	Editing	10	28,259
5200	Post-Production Film/Lab	11	77,052
5300	Post-Production Sound	11	0
5400	Music	11	0
5500	Titles	12	0
5600	Opticals	12	0
5700	Post-Production Video	12	0
5800	Facilities	13	0
5900	Post-Prod Travel/Living	13	0
	TOTAL POST PRODUCTION		**105,311**
6100	Insurance	13	0
6200	Legal Costs	13	0
6300	Publicity	13	14,468
6400	Miscellaneous	13	0
	TOTAL OTHER		**14,468**
	Completion Bond: 0.00% (0 excluded)		**0**
	Contingency: 0.00% (0 excluded)		**0**
	Overhead: 0.00% (0 excluded)		**0**
	Insurance: 0.00% (0 excluded)		**0**
	Total Above-The-Line		**13,267**
	Total Below-The-Line		**938,736**
	Total Above and Below-The-Line		**952,003**
	Grand Total		**952,003**

(Movie Magic Budgeting, Creative Planet)

amount of money set aside to cover unforeseen costs. The amount of a contingency fund is typically calculated as a percentage of the total production budget—anywhere from 1% to perhaps 15%. A 5% contingency in a $10,000 budget would be $500, a relatively modest sum.

To determine how to set your contingency fund, evaluate the uncertainties (like weather) that might prolong production, and adjust your rate proportionally. For example, a training video on how to perform an assembly line procedure might be safe with a very small contingency budget item. But, if you are shooting a nature program on the elusive mountain lion in its natural habitat, provide enough extra money for many additional shooting days.

Ready to Shoot

At this point, you may think that planning even a simple video requires a great amount of work. To put preproduction in perspective, remember that many of the requirements covered in this chapter are unnecessary for short, simple videos—although all programs will benefit from a certain amount of systematic preparation. If you plan to make professional programs of even modest scope, you will find that careful production planning is essential to success.

Summary

- Production planning is critically important for two reasons: time and money.
- Planning for a production of any size includes people, places, things, plans, and budgets.
- The people involved in a production include crew members (behind the camera) and cast members (in front of the camera).
- Scout locations to make sure they are suitable for shooting, and so you can plan accordingly for a successful production.
- The "things" of a production include production equipment, props, costumes and makeup, and set decorations.
- Releases are required when making a program for profit and/or public showing.
- Scheduling a production involves juggling the available people with the available equipment and resources, and the available time.
- Budgeting involves predicting the costs of every aspect of a production, and allocating funds to cover it.

Technical Terms

Budgeting: Predicting the costs of every aspect of a production, and allocating funds to cover it.

Cast: Collective term that refers to all the on-camera performers in a production.

Codec: A particular computer protocol for encoding audio-visual data to record it and decoding it to display it.

Crew: Production staff members who work behind the camera. In larger professional productions, the producer, director, and management staff are not considered "crew."

Dress: To add decorative items to a set or location. Such items are called *set dressing*.

Properties: All the items that are called for in a script; anything that appears in a program. Usually shortened to *props*.

Releases: Legal documents granting permission to include people, places, objects, and music in a program.

Talent: Every production member who performs for the camera.

Transcoding: The process of translating material from one codec into another.

Review Questions

Answer the following questions on a separate piece of paper. Do not write in this book.

1. What is *production planning*? Why is it important?

2. Actors, narrators, and other people who appear in video programs are collectively called the _____.

3. *True or False?* The crew is comprised of production staff members who work behind the camera.

4. Identify some resources for finding crew members for a production.

5. *True or False?* Cast selection is most effectively accomplished by viewing the recorded auditions.

6. What items should you evaluate when scouting the facilities at a shooting location?

7. List some items included in planning for the "things" required for a production.

8. *True or False?* You must obtain releases for any program produced for public showing, even if there is no profit involved.

9. Identify the seven common types of releases and permissions, and state the application of each.

10. What are the key characteristics of scenes that are considered when scheduling a production?

11. *True or False?* Budgeting a production requires that you know the exact cost of every aspect of the production up front.

12. What is a *contingency fund*? How is the contingency amount for a production calculated?

STEM and Academic Activities

>STEM>

1. **Science.** Connectors are made of or contain metal to conduct signals between the cables that they connect. Research the conductivity of various metals. Which metals are the best conductors? Which metals are commonly used in A/V cables and connectors?

2. **Engineering.** Create a flowchart that depicts the video production process from beginning to end. Include the titles of the staff, crew, and talent involved in each part of the process.

3. **Mathematics.** Calculate the contingency fund amount for each of the following productions:
   - A 15% contingency fund for a $250,000 production budget.
   - A 5% contingency fund for a $75,000 production budget.
   - A 10% contingency fund for a $33,000 production budget.
   - A 7% contingency fund for a $350,000 production budget.

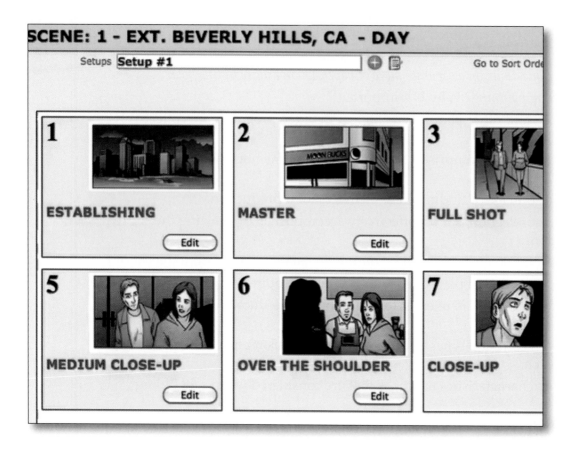

Production planning software may include storyboarding, as well as planning capabilities.

(Jungle Software)

(JVC Corporation)

After studying this chapter, you will be able to:

- Understand the functions of major camera controls.
- Explain the use of major camera support systems.
- Identify steps necessary to prepare camera equipment for shooting.
- Understand the proper setup and cleaning procedures for a camera lens.

About Camera Systems

The camera is, of course, the one indispensable tool for video makers. If you understand how it works and how to use it effectively, you are on your way to professional production. Cameras include many different systems and controls. To make sense of all these features, it is helpful to separate the camera into its two fundamental halves. Video cameras are often called "camcorders"—a combination of "camera" and "recorder"—because the unit contains both components:

- The *camera* section converts incoming light into images, and sound into audio. Both are coded in the form of electrical signals.
- The *recorder* section copies those signals onto recording media, and then plays them back on demand. In units using magnetic tape, the recorder section works exactly like a VCR deck. Some video recording systems use 8 cm discs, internal hard drives, or flash memory cards rather than tape; others use separate recorders, either to back up the in-camera recording or to function as the primary storage.

Separate storage units may be small enough to ride directly on the camera (**Figure 12-1**).

Not only does a camera include two types of machines, it also records two completely different kinds of information. *Video* (picture) information is converted into the video signal by the imaging chip, usually a ***charge-coupled device*** (***CCD***). *Audio* (sound) information is converted into the audio signal by a microphone, either built into or plugged into the camera. Each type of information has its own controls and requires its own procedures, so it will be helpful if you think of video and audio separately.

Audio recording, which is also performed by the camera, is covered separately in Chapter 17, *Recording Audio*.

Camera Functions

The job of the camera half of a video camera is to convert incoming light into coherent optical images (recognizable pictures), and then translate those optical images into an electrical video signal (**Figure 12-2**). Here is how the video signal is created:

- Light passes through the camera lens, which organizes it into a coherent image and projects it onto the surface of the CCD.

The CCD is sometimes referred to simply as "the chip," despite the fact that it is only one of several solid-state chips in the camera.

- The CCD contains hundreds of thousands of microscopic sensors. Each sensor responds differently to the amount and color of light striking it, so each sensor creates a different electrical signal.

Figure 12-1 This portable recorder weighs 12 ounces and records 100 gigabytes of high-definition video. It docks directly with the camera.

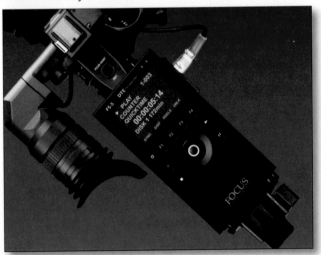

(Focus Enhancements, Inc.)

Figure 12-2 The lens projects an optical image on the imaging chip.

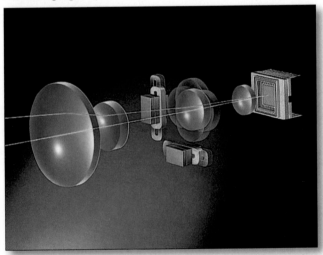

(Canon USA)

- The thousands of separate signals are processed by electronic circuits that combine them into the composite signal for one complete image called a *frame*.
- This process repeats 30 times each second to create a video track coded in electronic form.

Recording Formats

Video recording and TV playback were standard for well over half a century. Then, newly introduced digital recording and display systems allowed better image quality.

Traditional NTSC System

Historically in the North American (NTSC) video display format, each frame was electronically divided into two separate 1/60 second *fields*—one contains all the even numbered picture lines and the other all the odd ones. The two fields are combined to make one frame by a process called *interlacing*. Each frame of the video signal lasts only 1/30 second before it is replaced by the next frame.

All cameras recording standard definition analog signals use this system. Even Mini-DV cameras, which use digital rather than analog recording, lay down 30 frames per second in two interlaced fields.

The changeover from analog to digital broadcast did not replace the 30-fps ("frames-per-second") interlaced system.

High-Definition Systems

Professional and almost all hobbyist cameras now record high-definition images, which offer much better resolution than the older system, **Figure 12-3**.

To compare new and old, look carefully at news and commentary programs on a high-definition display. In split-screen discussions, the reporter in the studio will often be on a high-definition camera, while the interviewee in the field is on a standard-definition camera. The difference in resolution is obvious.

High-definition systems come in several varieties. Some use *progressive scanning* to

Figure 12-3 These images illustrate the difference between a 480i image and 1080p image, as they would appear on a digital, high-definition home television screen.

record and display 30 complete frames per second. Others continue to use the interlace method, displaying 30 frames per second in alternating 1/60 of a second fields. These various systems are not incompatible because modern TV sets automatically convert all digital signals to the sets' native formats—most commonly "1080p" (1080 horizontal lines displayed 30 times per second).

Recorder Functions

To preserve short-lived video images, engineers developed the videotape recorder to record them. Combined in one unit with the camera, the recorder forms the "corder" part of the camcorder package.

One Chip or Three?

Most modern cameras convert optical images into electronic signals by means of microchips called "charge-coupled devices" or "CCDs" or just "chips."

How a Chip Works

The imaging chip is a CCD (charge-coupled device).

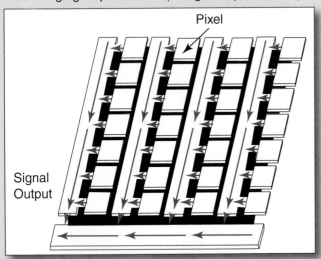

A CCD is a grid of tiny solar cells; so small that 400,000 or more can fit on a rectangle about one-quarter inch wide, measured diagonally. Like all solar cells, each of these minute sensors converts light into electricity—the brighter the light, the stronger the current. Thirty times each second, electronic circuits drain the energy generated in these cells and send the pattern of charges to the recording mechanism. It is this pattern of light and dark that makes the picture.

Recording Colors

Since the strength of the current generated by CCD cells depends on the brightness of the light, rather than its wavelength, these solar collectors cannot distinguish colors. To a cell, a certain amount of blue light "looks" much like the same amount of green or red light.

To record color information, video engineers have devised an ingenious solution. Each full-color picture unit is created by not one, but three sensors working together. Each light collecting surface is covered with a different colored filter: red, green, or blue. The red-filtered sensor responds only to red light, the green-filtered sensor to green light, and the blue-filtered sensor to blue light. As a result, each sensor creates a slightly different amount of current.

Though visible light contains a theoretically infinite number of different colors, they can all be created by combining different intensities of the primary hues: red, green, and blue. By analyzing the currents produced by the three sensors, the camera circuits deduce the color striking them. When you consider that this analysis takes place 30 times per second on hundreds of thousands of cell triplets, you can appreciate the engineering miracle in every camera.

Recording on Tape

Traditionally, video cameras produce the interlaced frames mentioned previously, in which the odd-numbered lines of each frame are recorded as one field and the even number lines as a second field.

Here is how a videotape recorder handles interlaced images:

- The electrically encoded information about each video field (2 per frame) is sent to a recording head spinning on a drum in the recorder.
- The record head preserves the signal by selectively magnetizing metallic particles on the videotape that moves across its path. When set to playback, the recorder passes the tape across the spinning drum again.

This time, separate playback heads recreate the video field from the magnetically coded tape particles.

VCR drums actually have at least two video record and two video playback heads. To create a two-field frame, each head records or plays a single field.

- Sent at a rate of 60 per second to a playback unit, these fields are interlaced line-by-line to create 30 complete images displayed on the surface of a picture tube or LCD display, **Figure 12-4**.

In short, the "corder" section of a camera is a very compact VCR, with all the usual tape controls. Because of their sophisticated recording features, cameras are often used in editing, as either source or assembly decks.

Every color in the spectrum can be created by combining the red, green, and blue primary colors.

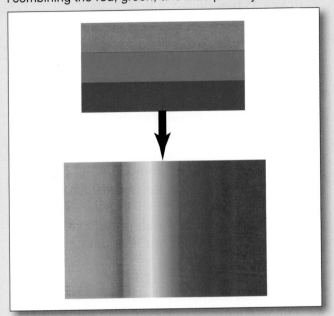

The three-cell design has drawbacks, too. Because it is so difficult to lay these near-microscopic color filters precisely over individual cells, they are not always perfectly positioned, and color inaccuracies result. And, because three cells are needed for every one color element, the chip has only one-third the resolution (sharpness) that would theoretically be possible if every cell were recording a separate part of the image. That is why more sophisticated cameras use not one chip, but three.

Three-Chip Cameras

In a three-chip camera, the entire surface of each chip is covered with a single primary color filter: red, green, or blue. This approach eliminates filter registration problems and increases picture sharpness. Professionals also prefer three-chip models because each chip can be supplied with its own electrical circuitry. In some cameras, that makes it possible to fine-tune the balance among the three chips to obtain the most accurate color rendition.

A three-chip imaging system.

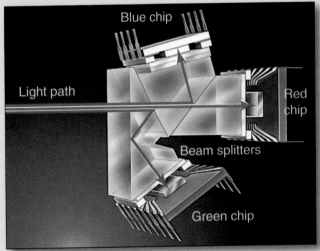

(Canon USA)

Some digital camera systems record every line in the video picture, in order, instead of dividing odd and even lines into separate fields. This single-pass recording system is called "progressive scan."

Digital Tape

Today, most cameras record audio and video signals digitally. This change is vitally important to video postproduction, but it does not greatly alter the basic processes of shooting with the camera. In most ways, digital cameras work much like analog models. The lens, focus, shutter, and aperture admit and process light into images, the CCD converts those light images to coded electrical signals. The recording system preserves the results on videotape.

Figure 12-4 A standard TV image is created by interlacing the lines of fields A and B.

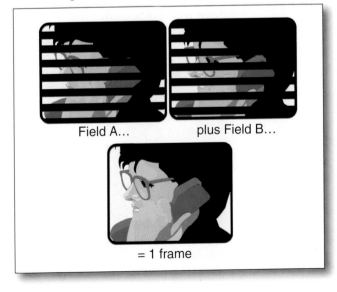

The major difference is inside the camera, between the CCD and the recording mechanism. There, an analog-to-digital (A/D) chip converts the electrical signal from the continuously varying voltage of an analog waveform to the characteristic on-off pulses of digitally coded current. The recorder section captures this signal as digital video.

In some ways, digital recording does affect the way a cinematographer uses the camera. For example, digital signal processing allows extensive control over white balance and color. For this reason, it is sometimes possible to obtain acceptable image quality in working environments with mediocre lighting conditions (**Figure 12-5**).

Digital video also permits the recording of finer detail. The mini DV format that is still popular in amateur cameras captures images superior to similar analog formats.

Today however, even standard-definition cameras that record digitally are giving way to high-definition models.

Tapeless Recording Systems

Eventually, all videotape may well be replaced by one or more tapeless recording systems. Any review of the specifics of these systems will become obsolete almost immediately, due to rapid advances in computer storage capacity, speed, and miniaturization, along with advances in disc recording technology and new types of storage hardware. In general, however, the current trends are likely to continue.

Storage Types

Several types of tapeless storage are available:

- **Hard drive:** As hard drives grow ever smaller in size, larger in capacity, and faster in access time, they can be incorporated directly into cameras (or in some cases, installed in computers, which are then fed real-time data directly from the cameras). Increasingly, cameras are using high-capacity solid-state drives with no moving parts. (These are very similar to the flash media described next.)
- *Flash media*: Very small plug-in chips can store enormous amounts of data, then be plugged into computers for data transfer to a hard drive (**Figure 12-6**).
- **Discs:** Disc recording media (typically 8" in diameter) can be burned directly in recorders built into the camera.
- **Stand-alone DVRs (digital video recorders):** In addition to laptops, special external recording units can be connected to the camera (**Figure 12-1**).

Advantages of Tapeless Recording

Tapeless recording offers many advantages over tape:

- **Random access.** Even digital tape is a linear system: to reach any point in the recordings,

---

Figure 12-5   Even severe color casts can be corrected by a digital camera.

Image with color cast.

Corrected image.

(Pretec)

you must roll through all the footage that precedes it on the tape. Tapeless recording permits instant access to any point in the data.
- **Archival storage.** Once the contents on hard or flash drives are copied onto disc, they form a nearly permanent record—direct disc recording, of course, omits this step. In addition, perfect quality backup copies can be quickly created for insurance.
- **Instant editing.** The more powerful laptop computers become, the more they are used to edit high-quality video in the field. Tapeless recording eliminates the need to slowly transfer tape data to a computer before working on it.
- **Transcoding.** Some DVRs can record in the particular format used by the editing software chosen for the project. This avoids the time-consuming process of transcoding all the data from the camera's system to the editor's system. (Many editing systems accept several formats, so that the editor can work in the camera's native format without the need to transcode the recordings.)
- **Metadata.** Some DVRs (and some computer software, too) allow users or their assistants to add metadata to each shot as it is recorded. Not only scene and take numbers, but other useful information (such as *"wide shot," "actor missed line," "NG: camera shake"*) is

permanently attached to the footage, which greatly simplifies the tedious process of logging and describing shots to identify them for editing.

Instant editing means that news and sports features can be ready for airing almost as soon as shooting finishes. On documentary and fiction program shoots, editors working in parallel with directors can examine footage and determine the need for re-takes or additional coverage.

Disadvantages of Tapeless Recording

Tapes provide instant external storage backups when the footage has been uploaded to computers for editing. Among tapeless alternatives, only disc systems offer this instant backup. Otherwise, editors must burn the footage onto discs, as well as upload it for editing.

Also, like tapes, tapeless storage media have limited capacities. Footage on hard drives in cameras and DVRs must be transferred prior to re-use. Even replaceable flash media cards must eventually be cleared and recycled. This is no problem when an assistant or on-site editor can transfer footage in real time, but for field projects (like nature documentaries), large amounts of storage media must be taken along.

Although flash media devices are more expensive than tapes, they are eventually far more cost effective because they can be reused almost indefinitely.

Camera Support Systems

With the principal camera functions outlined, we can survey the major equipment systems, starting from the ground up, with camera support. Cameras are supported by *tripods*, *stabilizers*, or *dollies*.

Tripods

Tripods are by far the most common camera support (**Figure 12-7**). They have three legs instead of four because a three-point support will not wobble on uneven ground.

To see the problem with four legs, try leveling a card table outdoors.

Figure 12-7 A tripod head designed for a lightweight video camera.

(Manfrotto/Bogen)

Figure 12-8 A ball level mechanism can be added to lightweight tripods between the tripod head and legs.

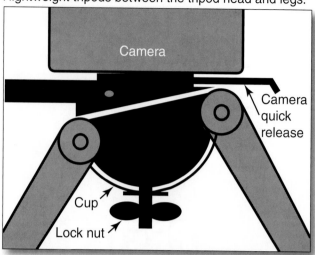

Camera

Camera quick release

Cup

Lock nut

(Manfrotto/Bogen)

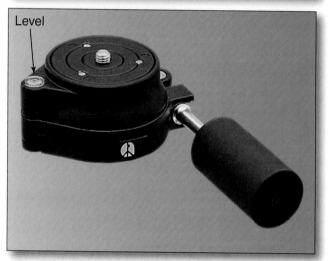

Level

To deploy a camera tripod, set up the feet, the legs, and the head—in that order. Most tripods have two sets of feet: soft, non-skid pads for smooth surfaces, like floors, and pointed pins for softer outdoor surfaces. It is easier to select one set or the other first, while the tripod is still collapsed. Unless you are setting up a low angle shot with the tripod legs unextended, start by extending the legs fully. Different models use different methods for locking the leg sections.

Before you can adjust the tripod head, you need to level it on the legs. Professional tripods let you unlock the head, tilt it until a bubble level shows that the head is level, and then lock it (**Figure 12-8**).

With less expensive units, you must level the tripod head by adjusting the legs. First, use the bubble level to locate the high side of the tripod, then point one leg directly at that side. Shorten the high-side leg until the bubble level indicates that the head is level.

Adjusting the Head

With the tripod set up and leveled, the next task is to set various controls so that the head operates smoothly when you move it and does not move by itself (**Figure 12-9**).

Quick Release Mechanism

If your camera and tripod connect via a quick release system, make sure that the camera

and tripod head are tightly screwed to their docking plates. This helps prevent camera wobble.

Pan Handle

Generally speaking, you want the pan handle installed on whichever side of the tripod head gives you easiest access to camera controls. Normally, the handle is installed on the right side because the viewfinder is on the left. However, if the camera's only record button is on the right-side hand grip, you may need to install the pan handle on the left.

Avoid using the camera's hand grip while working on a tripod. Turning the unit with the pan handle makes for smoother movement and less stress on the camera.

Figure 12-9 The controls on a professional tripod head.

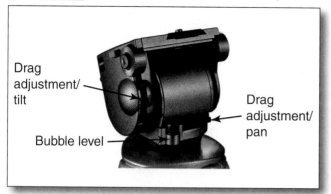

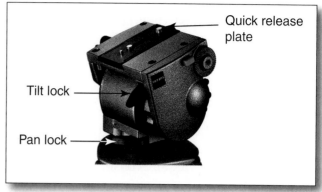

(Gitzo/Bogen)

Camera Balance

Some tripods allow you to balance the weight of the camera, either by sliding the unit forward or backward on the head, or by adjusting a built-in counterbalancing mechanism. Adjust the balance until your camera remains level, even when the tilt lock and tilt drag controls are completely disengaged.

Tilt Lock

A tilt lock prevents an unattended camera from tipping forward or backward. (Some tripod heads lack balancing mechanisms, and even those that do have them may also provide a tilt lock for extra security.) At the start of the shooting session, engage and disengage this lock to ensure that it is working properly.

Tilt and Pan Drag

You execute vertical and horizontal camera swings more smoothly if you are pushing against resistance, called *drag*. Too little drag is ineffective, while too much makes for jerky movements. Adjust the horizontal drag control to your taste, then disengage the tilt lock, and do the same for the vertical drag control. That way, you will not waste time on setup when the equipment is needed.

Stabilizers

If you will be using a *camera stabilizer*, you should pre-assemble, adjust, and test it at the beginning of the shooting session. Camera stabilizers are systems designed to smooth

hand-held shots. The most popular types are available in sizes to fit every camera from consumer mini-cameras to large professional models, **Figure 12-10**.

Unlike internal stabilizers, which steady the image being recorded, camera stabilizers are external hardware designed to steady the entire camera.

Though various camera stabilizers differ in their design and engineering, most of them consist of four parts:
- Head—Supports the camera.
- Counterweight system—Balances the camera.
- Pivot system—Allows the camera and counterweight to move freely.
- View screen—Allows the operator to frame and follow action without touching the camera's internal viewfinder.

Figure 12-10 A camera stabilizer in operation.

(Hollywood Lite)

In some systems, the view screen and/or external battery are part of the counterbalance system.

When well-adjusted and competently used, stabilizers work amazingly well—delivering footage as steady as dolly shots, while allowing the camera operator to follow the action with a flexibility that dollies cannot match (**Figure 12-11**). On the other hand, stabilizers can be time-consuming to set up and counterbalance, and they may need frequent fine-tuning to work properly. Also, they require considerable practice to operate.

In feature film and video productions, camera stabilizer operators are specialists who are hired solely to provide stabilized shots. Usually, they receive screen credit for their work.

Dollies

A *camera dolly* is a wheeled platform used to make smooth moving shots. Dollies are classified by two sets of criteria: design and wheel system. See **Figure 12-12**.

There are two major types of camera dolly designs: pedestal and cart. *Pedestal* dollies consist of a camera head mounted on a telescoping center column. The column, in turn, is supported by a wheeled base. Pedestal dollies can be smoothly moved up or down (this is sometimes called "pedestalling") during a shot. *Cart* dollies are essentially rolling platforms fitted with tripods. Heavier, more expensive types allow booming via a crane arm that moves up and down, but simpler cart dollies generally lack this feature. Some ultra-light dollies consist of tripods supplied with wheels and mounted on rails.

Figure 12-11   Paths of the subject and the camera operator.

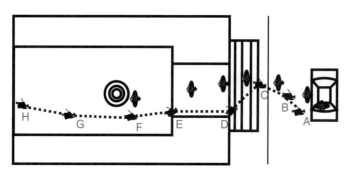

A—Close to the car.

B—Pulls back as the subject emerges.

C—Climbs one step to the sidewalk.

D—Up six steps into the archway.

E—Moves to one side.

F—Begins to draw ahead of the subject.

G—Subject stops at the fountain.

H—Camera continues to the other end of the courtyard.

Do-It-Yourself Stabilizers

Though commercially produced stabilizers are precision equipment and are priced accordingly, you can make simple, inexpensive versions that are remarkably effective. One of the more popular designs was engineered by Johnny Chung Lee, who is shown using his version here.

Johnny Chung Lee using his stabilizer.

(Johnny Chung Lee, http://steadycam.org)

If you Google "do it yourself camcorder stabilizer," you will find links to his and several other ingenious designs. Most of them are easily built with ordinary household tools.

Dollies may be fitted with two different types of wheels: floor and rail. Floor wheels are designed to roll directly on smooth surfaces, like interior floors. Most systems allow the dolly to change the direction in mid-shot by swiveling as they roll. The most sophisticated systems (like the Fisher dolly in **Figure 12-12B**) have four-wheel steering, so that all four wheels can be turned at once and the dolly can "crab" sideways. When used with rails, the Fisher dolly maintains alignment with the small horizontal guide wheels (visible in the illustration). Floor wheel systems are quicker to set up (there are no rails to lay and align) and more flexible (the camera can change direction in mid-shot). On the other hand, rail wheel systems permit dolly shots outside and on floors that are too uneven for wheeled dollies.

Specialized Camera Supports

In addition to tripods, dollies, and stabilizers, many productions utilize special purpose camera supports:

- *Jib arms* allow cameras to range over wide areas, both horizontally and vertically. Operated by remote controls and monitors, they differ from the heavier cranes, which carry the operators, as well as the cameras.
- *Auto mounts* allow stable shooting platforms in moving vehicles. Some mount to seat backs, others attach to windows by clamps or suction cups.
- *Miniature drone helicopters* capture spectacular aerial footage.

The list could be extended by many other even more specialized camera supports.

Figure 12-12   A—Some track and dolly systems are extremely portable. B—A large studio dolly. C—A wheeled spreader can convert a lightweight tripod into a dolly.

A   *(Microdolly Hollywood)*

B   *(J. L. Fisher Corp.)*

C   *(Manfrotto/Bogen)*

Though most of these special supports are out of reach for low-budget shooting, beanbags that fit odd surfaces (such as railings and car roofs) make versatile, inexpensive camera supports. You can buy them from Internet vendors or make your own.

Camera Setup

With camera support ready, the next task is to set up the camera itself, including the power supply, the external monitor (if present), and the recording media.

To avoid repetition, the aperture and shutter systems are covered in Chapter 13, *Camera Operation*, though they too are camera components.

Power Supply

All major camera systems run on electrical power—even the lens needs electricity for its zoom motor. Typically, a wall plug and cord supply 120 volt AC power to a charger/converter. The charger/converter changes the high-power AC to low-voltage DC power, and sends it to the camera to run its systems and charge its battery.

Many countries use 240 volt AC power, rather than the 120 volt that is standard in North America. Professional camera systems can often be set for either voltage.

Though almost all cameras can work directly on house power (often through their battery charging systems), you will operate your camera on battery power much of the time—except in studio-type situations where camera movement is restricted and wall power is conveniently close. Running on DC power, you may be using any of three different battery types: brick batteries, block batteries, or belt batteries.

- Brick batteries fit on or in the camera itself. They are light and convenient, but their capacity is relatively small.

"Brick" batteries were named when they were typically 3" by 6" long, and weighed several ounces. Today's internal batteries are, of course, much more compact.

- Block batteries are carried in cases on shoulder straps. Their capacity is much greater, but they are heavy and relatively clumsy (**Figure 12-13**).
- Belt batteries are as powerful as block batteries, but more convenient to wear because their weight is distributed around the cinematographer's waist.

Block or belt styles are recommended for news, documentary, and event (wedding) cinematography, especially if you also use battery-powered lights. Whichever battery type you select, you should have an external recharger and enough spares to continue shooting while recharging.

Viewfinder

The viewfinder shows you the world as your camera sees it, because all finders display the same image that is recorded on tape. Viewfinders come in two basic styles: internal and external.

An internal finder is essentially a tiny monitor inside a housing on the camera. Internal viewfinders have several advantages. They are fitted with *diopter correction*: eyepiece lenses to make critical focusing easier. These lenses can also be adjusted so that people with vision problems can see the finder clearly without glasses. Safe inside their housings, internal viewfinders are shielded from image-degrading light (like a bright light falling on a TV screen.

Figure 12-13 Professional batteries mount externally.

(Anton Bauer)

Camera Power Checklist

Before shooting, use this checklist to ensure that you will have reliable power.

AC Power

1. Locate power outlets.
2. Supply power cords and extensions, as needed.
3. Hook up and test power supply to camera.

Battery Power

1. Set up battery charger and check condition of all batteries.

2. Mount fully charged battery on camera and test.
3. Recharge second battery for later use.

To decide how many batteries you need for continuous shooting, determine the time required to fully recharge a completely discharged battery. Then, calculate how long a battery lasts in a typical shooting situation, where you are not recording continuously. For example, if a battery typically lasts one hour, but recharging takes twice that long, you can shoot continuously if you have three batteries and two external chargers.

A two-charger system for batteries A, B, and C.

Hour	1	2	3	4	5	6	7	8	9
In camera	A	B	C	A	B	C	A	B	C
In charger 1	B	A	A	C	C	B	B	A	A
In charger 2	C	C	B	B	A	A	C	C	B

External viewfinders, often called *external screens* or *LCD screens*, are small liquid crystal display monitors attached to the camera body (**Figure 12-14**). External finders also have advantages. Their larger viewing areas can be easier to see, except in very bright light. Because they can be pivoted for viewing from a wide range of angles, they permit greater flexibility in positioning the camera. By freeing the cinematographer from pressing the internal finder to one eye, external viewfinders greatly reduce fatigue during lengthy shoots. Because they are obviously TV screens, the user tends to look *at* them rather than *through* them. This makes it easier to evaluate images as two-dimensional compositions, rather than as the real world viewed through a tiny window.

External viewfinders have some problems, as well. Their color, contrast, and sharpness cannot equal those of an external monitor, so they may deliver a less accurate representation of the image being recorded.

Many external viewfinders can be fitted with hoods for outdoor use; and some newer screen designs can be viewed in bright light.

Whether internal or external, all viewfinders display information about what the camera is doing. Different models show different data, but the finder symbols in **Figure 12-15** are typical.

Figure 12-14 An external LCD viewfinder.

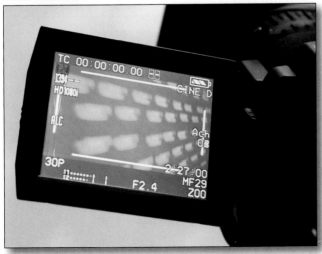

(Source: nikkytok/Shutterstock)

Figure 12-15 Typical viewfinder information.

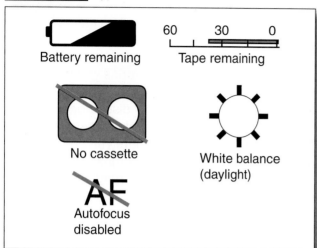

Battery remaining

Tape remaining

No cassette

White balance (daylight)

~~AF~~
Autofocus disabled

The information display is yet another reason for looking at the viewfinder rather than through it. Because it warns you that your battery is out of power or that you have neglected to insert recording media, the viewfinder display can often avert major problems.

Most camera viewfinders are adjustable. At the start of each shooting session:
- Set up the displays to minimize the text overlays that can obscure parts of the image—program your finder to show only the information you want.
- Adjust the brightness and contrast values to match the light in which you are shooting.
- Set the diopter correction. For internal finders, customize the focus to your eyesight by adjusting the diopter control until the finder screen is sharp.

In setting diopter correction, sharpen the text overlaid on the viewfinder rather than the image beneath it. Remember that you are not trying to focus the picture on the screen but the screen itself.

Monitor

If you are shooting at the advanced amateur level, an external color monitor is highly desirable; if your production is professional, it is indispensable. A stand-alone production monitor should have at least a 9″ screen, but a 13″ model is better. Camera-mounted LCD monitors are useful, but they are too small and low-resolution

to use in evaluating video quality. However, monitor-grade LCD screens in several sizes are available, though they can be expensive.

An external monitor performs two vital tasks: it lets you check exposure, lighting, color, and sharpness as shots are being recorded; and it allows other people (like a director or script supervisor) to see exactly what is being recorded (**Figure 12-16**).

Separate monitors may be impractical for programs such as weddings and documentaries. Also, when making long or complex moving shots, you may need to disconnect the monitor cable from the camera.

To ensure accurate color display, the monitor should be set up at the start of each shooting session. All professional and many consumer cameras can output standard color bars for use in adjusting the monitor. If your camera cannot generate bars, set it up as follows:
- Set the white balance on the camera, as discussed below.
- Aim the camera at a standard color model card, obtainable at most larger camera stores.
- Adjust the monitor for most pleasing picture.

High-quality professional production monitors offer both AC and battery operation, and can be calibrated to the same standards as studio monitors. For less elaborate productions, 9″ or 13″ consumer-grade monitors are available at far lower cost, with AC/DC power and the composite and/or S-video and/or HDMI inputs required for a quality picture (**Figure 12-17**).

Figure 12-16 A reference monitor.

(Barco, Inc.)

Figure 12-17   An AC/DC TV set can be used as a location monitor.

Digital cameras that lack standard definition output jacks require high-def monitors. Often, laptop computers do not provide enough image quality.

Recording Media

Set up and test the recording media as carefully as any other component of your cinematographer's kit. If you are shooting tape, verify that the tape has been blacked, that it will record a signal properly, and that it is identified with a unique label. To double-check that a tape has been blacked, play back a few minutes of it from the start of the tape and verify that the screen is black and the time counter or time code readout is functioning.

With flash media and hard drives (both internal and external), check and double-check that all previous recording has been transferred and, preferably, backed up onto disc.

White Balance

With the recording media ready, the next task is to adjust the camera's various exposure systems to suit the shooting conditions. You begin this process by setting the white balance. White balance refers to the overall color cast of the light in which you are shooting. Outdoor light tends to be slightly bluish (**Figure 12-18**), while lightbulbs cast a warmer, redder light. To the human eye, a sheet of paper looks equally white in both kinds of light because the brain

Identifying Media

Many beginners are casual about labeling recording media, until they are taught a painful lesson by an accident like these: a used but unlabeled tape is reused, thereby erasing the original footage, or tape is misplaced because its contents are not evident from its label.

Here are some guidelines for the proper identification of recording media.

Mini-DV and Disc Labels

Small format tapes and miniature CDs offer very little writing space, so give the production a code instead of a title. Such as,

- A program work title: *Prod. 013*
- Tape/disc type and number: *Cam orig #03*

Flash Media and Hard Drives

Since flash media cards and hard drives are reused again and again, it is not practical to identify them, except by numbers. Instead, keep a careful log as footage is recorded.

DVRs

Digital video recorders usually include provision for adding metadata, as described previously. If not, their contents should be logged as they are transferred for editing.

Figure 12-18   Outdoor light is naturally bluish.

automatically compensates for differences in color tint. Unlike the brain, however, a camera can only record colors exactly as its chip and circuits receive them. Without adjustable white balance, outdoor scenes would look too blue and/or indoor scenes too red. The cinematographer's job is to set the camera's white balance to match the color of the light at each new location.

Different levels of cameras offer different degrees of white balance control. Mobile phones, personal media appliances, and snapshot recorders usually set white balance automatically. Circuits in the unit analyze the incoming light, determine its overall color cast, and adjust color sensitivity to produce a tint-free white on playback. With these devices, you do not have to think about white balance at all, but you must accept whatever setting the camera makes.

Most consumer video cameras include a white balance switch for selecting outdoor, indoor, or fully automatic settings. Some cameras include a separate position for fluorescent light. With this type of white balance control, try using the automatic setting only when you are forced to shoot too fast to worry about subtleties of color balance. In many cameras, the individual settings will usually deliver better looking results.

To pick the best setting, try each one in turn, studying the results in your viewfinder.

Still more sophisticated cameras include fully manual white balance. To use it, hold a white card so that its image fills the entire frame (**Figure 12-19**). When you activate the white balance circuitry, the camera reads the light reflected from the card and sets itself to record it as neutral white. A professional cinematographer will typically set white balance to match every shooting location.

The many digital still cameras that also record video will typically have adjustable white balance settings.

Even if you usually use the automatic setting, there are situations in which manual white balance is preferable. When the main light source is an odd color, such as yellowish sodium vapor or bluish mercury lighting, hold the card so that it reflects light only from the colored source. Or,

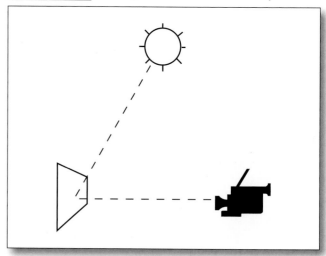

Figure 12-19   Setting white balance manually.

when a colored surface, such as a big, brightly painted wall, tints the light reflected onto the scene, set white balance while aligning the card so that it picks up the colored reflections. This method can reduce or even eliminate the color cast from the strongly colored surface.

The best white balance systems add fully adjustable override controls. Once you have set the white balance, you can gradually make the overall tint warmer or cooler until you obtain exactly the color you prefer. When making adjustments this fine, it helps to use a good external color monitor to check the effect. Whatever your system, it is essential to select (or at least check) the white balance setting each time you move to a new shooting location. Poor white balance can be corrected somewhat in postproduction, but the results are generally not as good as if the camera had been set correctly in the first place.

LED lights and fluorescent lamps intended specifically for media lighting are made with standard color temperatures, which makes white balancing easy.

Lens Setup

The process of recording video images begins at the camera's lens, which gathers light rays reflected from subjects and focuses them on the imaging chip.

Color Temperature

The overall color cast of a light source is described as its *color temperature*, and expressed in degrees on the Kelvin temperature scale. The higher the color temperature, the bluer the light. Here are some typical light sources and their color temperatures ("K" is read as "degrees Kelvin."):

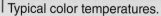
Typical color temperatures.

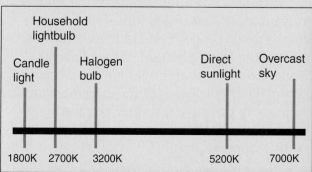

The daylight or outdoor white balance setting usually matches the 5,000–5,500K of sunlight, and the indoor or incandescent setting matches the 3,200K of halogen movie lights. Almost all halogen bulbs, including work lights and home bulbs, are 3,200K.

Many modern cameras have separate white balance settings for fluorescent lights. Fluorescent light differs from both natural and incandescent (lightbulb) illumination. Try both fluorescent and outdoor settings when shooting in fluorescent light and observe the effect on a color monitor. With warm fluorescent tubes and/or yellowed light fixture diffusion panels, you may find that the indoor setting yields the most pleasing color. If you have full manual white balance, use it instead to match the fluorescent lighting precisely.

Most general-purpose video cameras are fitted with zoom lenses that allow you to change image sizes, either between shots or continuously while shooting (**Figure 12-20**).

Lens Mount

On simpler cameras, the lens is permanently attached. More sophisticated models often have lens mounts that permit you to replace one lens with another (**Figure 12-21**). However, changing

Figure 12-20 Cutaway of a zoom lens with optical stabilization.

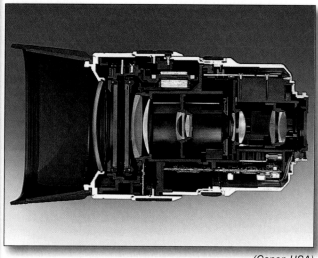

(Canon USA)

lenses during a shooting session is uncommon for two reasons. First, the alignment between a lens and the imaging chips behind it is so critical that it is often safer to mount and calibrate a lens in a studio or lab.

All cameras with interchangeable lenses are three-chip designs. Certain models, notably from Canon, are designed to allow changing lenses in the field. In the movie industry, digital studio cameras have interchangeable lenses.

Second, lenses are usually changed to accommodate the demands of very different shooting situations. For example, you may exchange a general-purpose zoom lens (perhaps

Figure 12-21 A camera with interchangeable lenses.

(Canon USA)

5mm–60mm in some formats) for another zoom with a much longer range (such as 40mm–480mm) in order to shoot wildlife at great distances from the camera.

If your camera does have interchangeable lenses, you should always check to see that the lens is firmly attached to its mount, without any wiggle or "play."

Front Element

The next task is to set up the front end of your lens, including cleaning, installing a filter, and mounting a lens hood.

Cleaning

Visually inspect the lens to ensure that it is free of dust and smudges. Dust can be spritzed away with a blower bulb and brushed off with a lens brush. Smudges should be removed by cleaning, as described in the side bar *Cleaning the Lens*. Any dirt on a lens will degrade its quality. In bright light, the depth of field (focus) of the lens may be so great that the front element (glass) of the lens itself is sharp, including all the dust and smudges on it. Also, a dirty lens increases light flares.

Filter

A *filter* is a piece of optical glass that is usually placed in front of the camera lens. If your lens is not recessed behind a protective glass window (and if it is threaded to receive filters) you should mount a transparent lens cap called a "UV," "1A," or "skylight filter"

to keep the lens clean. In very bright lighting conditions, a *neutral density filter* can replace the clear model. A *polarizing filter* can be installed to enhance the appearance of the sky, **Figure 12-22**. Never install more than one filter at a time; this could degrade image quality and allow the filter edge to show in the picture. A filter should be cleaned exactly like the lens itself.

Lens Hood

If the lens is mounted outside the camera body and the front end is threaded, you may wish to mount a lens hood to help keep direct light off the lens. When selecting a lens hood, check it to verify that it does not appear in the picture when the lens is in the extreme wide angle zoom position. If it does, use the hood only for telephoto applications, like sports and wildlife cinematography.

If you have mounted a filter, screw the lens hood onto the filter's front threads. Be careful: sometimes a filter will displace the hood forward enough so that it is visible in the frame at wide angle settings.

Focus and Aperture

For setup purposes, decide whether to use the auto focus system or focus the lens manually. Most cameras default to auto focus, so you set manual focus by disabling the auto mechanism. In the same way, the camera defaults to auto exposure mode.

If you wish to adjust the aperture manually, disable the auto exposure system (**Figure 12-23**).

Figure 12-22 A polarizing filter enhances clouds in a blue sky.

Polarizing filter.

Image without polarizing filter.

Image enhanced with polarizing filter.

(Tiffen)

Cleaning the Lens

No matter how careful you are, you will have to clean your lens from time to time. So, here are the supplies and techniques you will need: lens fluid, lens cleaning tissues, and a blower bulb tipped with a soft brush. You can often find these items online, sold as a kit.

A combination brush and blower, with lens cleaning fluid.

In lens care, a good rule of thumb is: never do more than the minimum required to get the surface clean. To observe this rule, use the following methods in order, continuing to the next method only if the previous one fails to complete the job.
- Making sure that your mouth is dry, direct short puffs of air *across* the lens to blow away loose dust.
- Wipe the brush in concentric circles around the lens, starting from the center, while repeatedly squeezing the bulb to blow air through the brush.

Brush the lens in an outward spiral pattern.

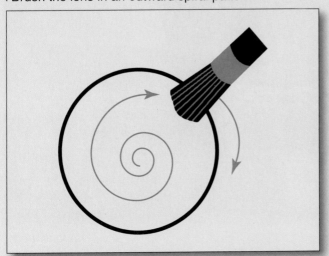

- Dispense a single drop of lens cleaning fluid onto the center of the lens. Crumple a sheet of lens tissue, grasp it between thumb and two fingers, and gently wipe the lens in a circular motion, beginning in the center. When you reach the rim, crumple a second tissue and gently dry the lens, using the same spiral motion.

The lens is the most delicate component of a camcorder.

(JVC Corporation)

Never use facial tissues or paper towels instead of lens tissues, because they can leave lint on the lens or even scratch it. Also, avoid the cloths sold for cleaning eyeglasses, if they have been chemically treated. However, lens cloths intended for binoculars are effective and will not harm the lens coating.

Used gently, microcloths intended for cleaning glasses are safe and effective on lenses.

Figure 12-23 Auto/manual focus controls.

(Canon USA)

Make sure that you are in manual mode by looking for an exposure readout in the viewfinder. Some cameras display actual f-stops (2.0, 2.8, 4.0, 5.6, etc.), while others display increments above or below the automatically determined aperture (+1, +2, +3; –1, –2, –3, etc.).

Typically, each plus or minus number represents 1/3 of a stop, but check your manual to be sure.

Zoom

Always check your zoom lens to verify that it is functioning correctly. If the zoom is motor-driven only, a malfunction means that the camera is essentially inoperative. If the zoom can be operated manually (usually by a ring on the lens) it is less likely to malfunction.

Image Stabilization

Image stabilization is a mechanism for automatically steadying shaky pictures. Stabilization systems use sophisticated circuitry to detect camera shake by comparing each frame with the one that preceded it. (Some systems analyze the actual camera motion, rather than changes in the images.) If just some of the pixels are different from those in the preceding frame, the system concludes that the subject is moving and no compensation is required. In **Figure 12-24A**, the pixels recording the subject have moved with her, but the background remains the same.

If all the pixels have shifted in the same direction, the system identifies camera shake and moves the pixels back to their original locations in the frame for recording. An image stabilizing system would recognize **Figure 12-24B** as camera shake because the entire image has shifted. ("Fuzzy logic" functions can recognize when all pixels are shifting because the camera is panning, rather than just shaking.)

Image stabilization should not be confused with camera stabilizers, which reduce camera shake rather than compensating for its effect on the image.

Figure 12-24 An image stabilization system can distinguish subject movement from camera shake.

Subject movement.

A

(Continued)

Figure 12-24 *(Continued)*

Camera shake.

B

There are two different systems of image stabilization: optical and electronic. Optical stabilization shifts the path of the light on its way through the camera lens, so that the same image strikes the CCD (**Figure 12-25**). Electronic stabilization records on only part of an oversized CCD. As the image shakes, the active recording area shifts in the opposite direction, to compensate.

Both systems work very well, and image stabilization is a highly desirable feature on any camera. If you have image stabilization, by all means use it for hand-held work.

Some manufacturers recommend disabling stabilization when the camera is mounted on a tripod. You should test your particular system and decide whether or not to do this.

Figure 12-25 An optical image stabilizing system.

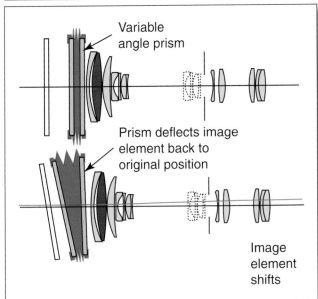

Variable angle prism

Prism deflects image element back to original position

Image element shifts

(Canon USA)

All Systems Ready

With the major camera systems set up and checked out, you are ready to begin the actual process of cinematography. That is the subject of Chapter 13, *Camera Operation*.

Summary

- A video camera converts incoming light into optical images, and then translates the optical images into an electrical video signal.
- Most cameras record audio and video signals digitally.
- Cameras are supported by tripods, stabilizers, or dollies.
- Camera setup tasks include verifying the power supply, setting up the external monitor, preparing the recording media, and white balancing the camera.
- Setting up the front end of a camera lens involves cleaning, installing a filter, and mounting a lens hood.
- Image stabilization systems use sophisticated circuitry to detect camera shake and automatically compensate to create a steady image.

Technical Terms

CCD (charge-coupled device): The camera imaging chip that converts optical images into electronic signals.

Camera dolly: A rolling camera support. A *cart dolly* resembles a wagon; a *pedestal dolly* mounts the camera on a central post.

Camera stabilizer: A mechanical camera support system carried or worn by the operator that permits perfectly steady hand-held shooting.

Color temperature: The overall color cast of nominally white light; expressed in degrees on the Kelvin scale—sunlight color temperature (5,200K) is cooler (more bluish) than halogen light color temperature (3,200K).

Diopter correction: A lens in front of an internal viewfinder that can be adjusted to compensate for the eyesight of the operator.

Drag: The resistance of a tripod head to panning or tilting. On professional tripods, the drag is adjustable.

Field: One half of a single frame of video, consisting either of the odd lines or the even lines. Two fields are interlaced to make a complete frame.

Filter: An optical glass placed (usually) in front of the lens to modify the image being recorded.

Flash media: Data storage on small, thin cards without moving parts.

Image stabilization: Compensation to minimize the effects of camera shake. *Electronic stabilization* shifts the image on the chip to counter movement; *optical stabilization* shifts parts of the lens instead.

Interlace: In video, the process of combining two fields to create one frame of video image.

Neutral density filter: A gray filter that reduces the incoming light without otherwise changing it. Typically supplied in one- to four-stop densities (ND1 through ND4).

Polarizing filter: A rotatable filter that can suppress reflections and specular (very small) highlights, as well as darken blue skies.

Progressive scan: The process of displaying 30 complete frames of video per second.

Tripod: A three-legged camera support that permits leveling and turning the camera.

Review Questions

Answer the following questions on a separate sheet of paper. Do not write in this book.

1. Identify the two components of a camcorder and explain the function of each component.
2. Video information is converted into the video signal by the _____.
3. What is *interlacing*?
4. *True or false?* A three-chip camera yields better images than a one-chip camera.
5. What are the advantages of tapeless recording over videotape recording?
6. Identify some common camera support systems.
7. List the camera power checklist steps for both AC power and battery power.
8. What is the purpose of the diopter correction on a viewfinder?
9. *True or false?* An external monitor can replace the camera's viewfinder during shooting.
10. What is *white balance*?
11. The color temperature of direct sunlight is _____.
12. What is the function of a lens hood?
13. Identify the two different systems of image stabilization.

STEM and Academic Activities

STEM

1. **Science.** The fluid used in a fluid head tripod creates resistance to movement. This resistance property of fluid is called viscosity. How is viscosity measured? Research the viscosity of various common fluids. Give an example of a liquid that is highly viscous and a liquid that is not very viscous at all.
2. **Technology.** Research the changes in battery technology. How have these changes impacted battery life, the materials used to manufacture batteries, and how power is delivered from the battery to a device?
3. **Mathematics.** To convert Kelvin values to degrees Celsius, subtract 273.15 from the Kelvin value, or $°C = K-273.15$. Using this formula, convert the following Kelvin values to Celsius:
   - 2000K
   - 3200K
   - 4300K

   To convert degrees Celsius to degrees Fahrenheit, multiply the Celsius value by 1.8 and add 32 to the product, or $°F = (°C \times 1.8) + 32$. Using this formula, convert the Celsius values you calculated above to degrees Fahrenheit.

(JVC Corporation)

Professional camcorders range from compact to on-the-shoulder full-size models.

(Panasonic Professional Products)

Objectives

After studying this chapter, you will be able to:

- Understand the characteristics and applications of various lens focal lengths.
- Recall the relationship between f-stop settings and aperture.
- Identify the factors that affect depth of field.
- Explain how to set and adjust appropriate lens settings based on shooting conditions.
- Understand how to frame and maintain good video compositions.

About Camera Operation

Just as gourmet foods are more than the recipes on which they are based, quality videography is greater than the sum of its recording techniques. The process is more art than science, and watching skilled videographers at work is like observing chefs in their kitchens. What makes them artists? What is the difference between a true videographer and a casual snapshooter? The answer has to do with the characteristics of the footage recorded—its quality, quantity, and usefulness.

Footage Quality

Professionally videotaped images display high technical and esthetic quality. That quality comes not only from technique, but from attitude. Good videographers are fanatical about achieving the best images possible. This is their area of expertise, and they love to utilize their skills and experience. They are never satisfied—setup position, lighting, focus, and camera movement can always be improved. For a professional videographer there is no such thing as "good enough."

Footage Quantity

Amateurs typically videotape everything only once. They record a view or a piece of action, and then move on to the next one. Videographers are just the opposite, often shooting far more material than will appear in the finished program. They know that when shots are less than adequate (or omitted accidentally), they often cannot be taped at some later date. A wise videographer ensures that everything is recorded and recorded properly, **Figure 13-1**.

Footage Utility

Though good videographers are fanatical about quality and patient about shooting enough footage to achieve it, they never forget that the whole purpose of their art is to enhance the effectiveness of the overall program. Videographers are not working alone, but reinforcing the intentions of the director, the writer, and (in most professional work) the client. Working in teams

Figure 13-1 With disc and flash memory systems replacing videotape, the cost of recording ample footage is almost negligible.

does not mean that videographers cannot be creative, but they exercise their creativity in productive collaboration with their colleagues and in service to the needs of the production.

In short, professional-quality videotaping results not only from skill with the hardware, but from an attitude toward the work. Having said that, we can begin to look at the techniques of videography. We will cover:

- Working with lenses
- Setting focus
- Controlling aperture, shutter, filtration, and gain
- Moving the camera
- Composing the shot

Working with Lenses

Video recording starts at the camcorder lens, which turns light rays into visual images and determines the pictorial qualities that those images will display.

Lens Focal Length

The way in which a lens renders images is determined by its focal length. The *focal length* of a lens is the distance (expressed in millimeters) from the front element of the lens to the focal *point* (the point at which the light rays intersect on their way to the focal *plane*, where the chip is positioned), **Figure 13-2**.

Figure 13-2 The principal dimensions used in describing lenses.

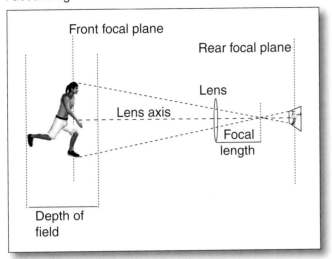

Lenses are named according to focal length, such as a "5mm" lens or a "75mm" lens. Originally, the focal length of every lens was fixed. To change focal lengths, a photographer had to dismount one lens and replace it with another. Today, however, almost all video optics are zoom lenses, whose focal lengths can be varied continuously between their shortest and longest positions. On Mini-DV cameras, for instance, the focal length of a zoom lens might range from 3mm to 48mm. Since the ratio of 48 to 3 is ¹⁶⁄₁, this lens would be described as a "16-to-1 zoom."

As explained in the sidebar *Wide Angle or Telephoto?*, zoom ranges are dependent on the size of the imaging chip. Most of this discussion uses Mini-DV focal lengths as examples. This format is employed in many standard- and high-definition cameras.

Within a zoom range, the number of focal lengths is infinite. With a 3mm to 48mm zoom, you could theoretically set the focal length at, say, 35mm or 35.2mm or 35.205mm (though you would need optical calibration equipment to do so). In actual production, however, professionals rarely refer to lenses by exact focal length. Instead, they divide them into three broad groups of focal lengths: normal, wide angle, and telephoto.

- Normal focal length lenses produce images that closely resemble human vision.
- Wide angle focal lengths include a wider field of view than normal lenses.

- Telephoto focal lengths show a narrower, more restricted field of view.

For proper contrast with "wide angle," long focal length lenses should probably be called "narrow angle" instead of *telephoto.*

Though wide angle, normal, and telephoto settings are actually just focal lengths within the zoom range of a single lens, they are usually discussed as if they were separate lenses. So, when you read "telephoto lens" or "wide angle lens," remember that this typically refers to a setting on a zoom lens. As we discuss the optical qualities of wide, normal, and telephoto lenses, keep in mind that each category includes a whole range of individual focal lengths. In our Mini-DV example, 3mm, 3.5mm, and 4mm are all different degrees of wide angle lens, and 14mm, 26mm, and 45mm are all telephoto focal lengths.

Finally, remember that the pictorial qualities of a particular focal length are dependent on the size of the camera chip on which the image is created. For example, the 8mm focal length that is considered "normal" on a ¼" chip, would be an ultra-wide angle "fisheye" lens on a 35mm still camera. This is explained more fully in the sidebar *Wide Angle or Telephoto?*

Lens Characteristics

All lenses have three characteristics that affect the way the videographer uses them: angle of view, maximum aperture, and depth of field.

Angle of View

Every lens "sees" within a certain arc or angle of view, **Figure 13-3**:
- Wide angle lenses typically cover 70°–140°.
- Normal lenses cover 35°–70°.
- Telephoto lenses cover 18°–35°.

These distinctions are somewhat arbitrary—a "normal" lens that covers 69° is not significantly different from a "wide angle" lens that covers 71°. Because of their broader coverage, wide angle lenses are useful in confined shooting areas, where longer focal lengths would not cover the whole action area.

Wide Angle or Telephoto?

Whether a lens is wide angle, normal, or telephoto depends on the size of the imaging area for which it is intended. The image created by the lens must cover the whole area of the chip that is recording it. Since lens images are circular while video frames are rectangular, the minimum image diameter must slightly exceed the diagonal of the frame, as you can see from the following illustration.

As a rule of thumb, a lens focal length that is slightly longer than the image diagonal produces a "normal" perspective. Shorter focal lengths yield wide angle images, and longer ones create telephoto images. Since the visual perspective of any focal length depends on the area upon which the image is formed, a focal length that is wide angle on a frame of 35mm still film, is telephoto on a ¼" chip. To illustrate this, here are typical focal lengths for three image formats:

- A nominally ¼" chip for a Mini-DV camcorder.
- A ½" chip for a full-size professional camcorder.
- A 24mm × 36mm frame of 35mm still film.

The diameter of the lens must be slightly greater than the diagonal measurement of the frame.

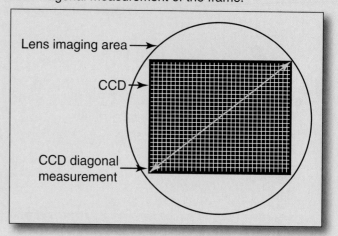

Professional digital single lens reflex still cameras (DSLRs) are now often adapted for capturing full-motion video. The sensors on most of these cameras are essentially the same 24 × 36 size as the classic 35mm still film format, and use the same lenses.

Figure 13-3 Angles of view of typical lenses.

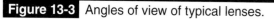
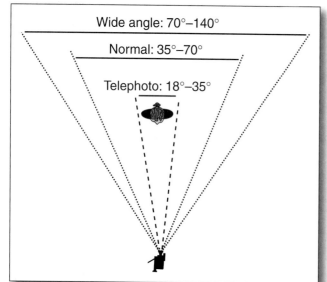

Maximum Aperture

As a rule, normal and wide angle lenses have larger maximum apertures (lens openings) than telephoto designs. That is because the "speed" (light gathering ability) of a lens depends on both the lens diameter and the focal length. For a lens of any diameter, the shorter the focal length, the larger the maximum potential aperture.

In fixed focal length designs, wide angle lenses should theoretically have larger maximum apertures than normal lenses. Because of other constraints in designing wide angle lenses, however, normal focal lengths often have the largest maximum apertures.

Since a zoom lens has a fixed diameter but a variable focal length, its maximum aperture often will become smaller as the lens zooms from wide angle to telephoto. For example, a lens that can open to f/1.4 in the wide angle position may open only to f/2.8 at full telephoto. That is why you will often see identification markings on lenses like "f/1.4–2.8." This indicates f/1.4 is the maximum aperture at full wide angle, while f/2.8 is the maximum at full telephoto.

By comparing 35mm focal lengths to those of different video chips, you can get a relative idea of how different focal lengths behave in different image formats. As you look at the chart, notice three things:

- The 35mm focal lengths vary somewhat from straight multiples (like 17.5mm, 35mm, 70mm, 140mm) because still camera lenses are more typically designed in slightly different focal lengths (18mm, 35mm, 70mm, 135mm).
- The comparisons are not exact because the 3-to-2 proportions of the 35mm frame are different from the 4-to-3 proportions of the classic video frame and the 16-to-9 ratio of high-definition video.
- Because of the mathematics of optical design, the difference (expressed in millimeters) between focal lengths is quite small at the wide angle end of the spectrum, but much larger at the telephoto end. For example, a 4mm wide angle lens produces an image twice

1/4″ Chip Focal Length	1/2″ Chip Focal Length	35mm Focal Length	Lens Characteristics
2mm	4mm	18mm	Extreme wide
3mm	6mm	28mm	Wide angle
4mm	8mm	35mm	Mild wide
6mm	12mm	45mm	Short normal
8mm	16mm	55mm	Normal
12mm	24mm	70mm	Long normal
16mm	32mm	105mm	Short
24mm	48mm	135mm	Telephoto
32mm	64mm	200mm	Long telephoto
48mm	96mm	300mm	Very long

CCD chips for different visual formats.

- 1/4″ chip (3.75mm × 5mm)
- 1/2″ chip (7.5mm × 10mm)
- 35mm film (24mm × 36mm)

as large as a 2mm, though the difference between them is only 2mm. At the telephoto end, a 48mm lens also has twice the magnification of a 24mm lens. But, the increase in focal length required to double the image size is 24mm, rather than just 2mm.

Some professional zoom lenses are designed to maintain the same maximum aperture throughout their zoom ranges.

Understanding how maximum apertures work in zoom lenses will help you use them effectively. In low light levels, for example, you may wish to restrict yourself to normal or wide angle lens settings, because their larger maximum apertures admit more light. They will yield better quality images in dim illumination.

Depth of Field

Depth of field is the distance, in front of and behind the plane on which the lens is focused, in which objects in the image appear sharp (**Figure 13-4**). Like maximum aperture, depth of field grows smaller as the lens focal length grows longer. At any given aperture and distance from the subject, wide angle lenses have the deepest depth of field and telephoto focal lengths have the shallowest.

You need to be aware of these characteristics when following sports or other action in telephoto mode, and when setting focus for zooming. A scene that looks perfectly sharp in wide angle may turn soft when you zoom to telephoto. That is because the wide angle setting's greater depth of field includes the distant subjects, but the telephoto setting's shallower range does not.

The Pictorial Qualities of Lenses

The only true optical difference among lens focal lengths is *angle of view*—the arc within which a lens gathers light to form images. All other seeming differences are merely appearances. Nonetheless, those appearances are quite visible to viewers and videographers use them to shape the images they create. We may call these appearances the *pictorial qualities* of lenses. They include *magnification*, *perspective*, *movement*, and *distortion*.

Apertures and f-stops

How much light gets through a lens depends on two factors:

- Its focal length (such as 10mm or 100mm or 200mm).
- Its maximum aperture—the full diameter of the lens opening.

The light-gathering ability of a lens is called its *speed*; a sensitive design is called a "fast lens." The speed of a lens is determined by a simple formula:

focal length ÷ maximum aperture = lens speed

According to this formula, if a lens has a focal length of 100mm and a diameter of 50mm, then its speed would be 100 divided by 50, or f/2 (100mm ÷ 50mm = f/2).

By convention, apertures are indicated with a lower-case "f" and referred to as *f-stops*.

A 200mm lens with the same 50mm diameter would have a speed of f/4 (200mm ÷ 50mm = f/4).

An iris diaphragm varies the effective aperture.

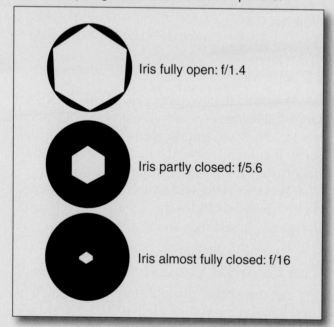

Iris fully open: f/1.4

Iris partly closed: f/5.6

Iris almost fully closed: f/16

Aperture Range

Every lens includes an internal iris diaphragm ("iris" for short) that can be adjusted to progressively reduce the diameter of the lens opening. For this reason, a lens has a whole range of apertures.

Each reduction in aperture changes the result of the equation, so:

100mm ÷ 50mm (iris wide open) = f/2

100mm ÷ 12.5mm (iris partially closed) = f/8

Apertures on most lenses have been standardized at f/1.4, f/2, f/2.8, f/4, f/5.6, f/8, f/11, f/16, and f/22. Though these numbers seem arbitrary, a moment's study will reveal that they are all based on square roots.

Smaller Equals Higher

As you can see from the diagram, the smaller the f-stop number, the larger the aperture. Commonly, f/1.4 is the largest aperture and f/22 is the smallest, though certain specialized lenses can "stop down" to f/64. By another convention, the speed of a lens is expressed as its maximum aperture. Thus, an f/2 lens is said to be "two stops faster" than an f/4 lens, because it includes the two extra stops (f/2.8 and f/2) below f/4.

Apertures grow smaller as their f-stop numbers grow larger.

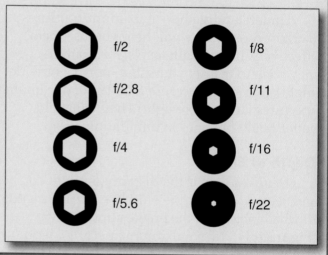

f/2

f/2.8

f/4

f/5.6

f/8

f/11

f/16

f/22

Figure 13-4   In this telephoto shot, depth of field is limited to the area between the red lines.

Depth of field (view from above)

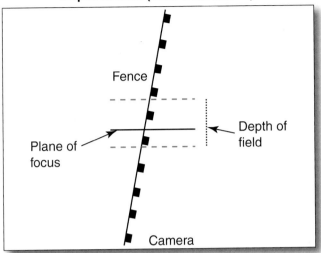

Telephoto lens shot

Magnification

The most obvious difference between wide angle and telephoto lenses is *magnification*. An object that appears tiny in wide angle may loom huge in telephoto. In comparison to human vision:

- *Normal lenses* make objects seem about the size they would appear to the human eye.

- *Telephoto lenses* make objects seem bigger.
- *Wide angle lenses* make objects seem smaller.

Apparent magnification is an effect of angle of view, because every lens must fill the frame from edge to edge. Since the screen size does not change, filling it with a 5° view will make objects appear much larger than filling it with a 90° view. This concept is illustrated in **Figure 13-5**.

Figure 13-5   Wide angle and telephoto lens fields of view. Because the frame remains the same size, filling it with a telephoto view makes the subject larger.

Fields of view

Wide angle

Telephoto

First-time camcorder users quickly discover that zooming from wide angle to telephoto enlarges the subject in the frame.

Perspective

Perspective is the illusion of depth in the picture; the apparent distance from the camcorder to the farthest part of the scene. Wide angle lenses exaggerate this virtual depth, making distances seem greater and faraway objects seem smaller. Telephoto lenses reduce apparent depth, seemingly bringing distant objects closer. **Figure 13-6** shows the same setup photographed with both wide angle and telephoto lenses. Notice that much of the scene is excluded from the telephoto shot because of the lens' narrow angle of view. Objects in the telephoto image are much larger than in the wide angle view, and the buildings appear squeezed together.

In **Figure 13-6**, the lens remains the same distance from the subject at both wide angle and telephoto settings. But, if you move the camera you will achieve a quite different effect. In **Figure 13-7**, the same subjects are photographed using both a wide angle lens and a telephoto lens. For the telephoto lens image, the camera was placed much farther away. As a result, the near subject is exactly the same size as in the previous wide angle setting, but the far subject appears much larger and closer.

Movement

Because wide angle and telephoto lenses affect the apparent depth of an image, they also modify apparent movement toward or away from the lens. In wide angle shots, moving people and objects seem to rush quickly toward the camera or away, **Figure 13-8**. In telephoto shots, however, fore-and-aft movement is minimized, and people do not change size dramatically as they move forward or backward. For these reasons, wide angle lenses are useful for dynamic action scenes (like fights and chases), while telephoto lenses convey a contrasting sense of distance and detachment.

Distortion

Since wide angle and telephoto lenses affect the way objects "shrink" and "grow," depending on their distance from the camera, they also can distort the shapes of those objects. Note how the wide angle lens in **Figure 13-9** makes the vehicle look longer and more powerful, while the telephoto shot seems to squeeze it together.

Be careful in using wide angle lenses on the human body. **Figure 13-10** shows what happens when wide angle perspective stretches arms and legs. For the same reason, try to avoid wide angle lenses in portraiture, since the extra depth can distort chins and noses unpleasantly.

Moderate telephoto settings are generally preferable for portraits of women. However, a mild wide angle lens can make a male portrait look more rugged.

Figure 13-6 Lenses change perspective.

Wide angle lens

Telephoto lens

Figure 13-7 Wide angle and telephoto camera setups. The camera is moved for the telephoto image to maintain the foreground subject size.

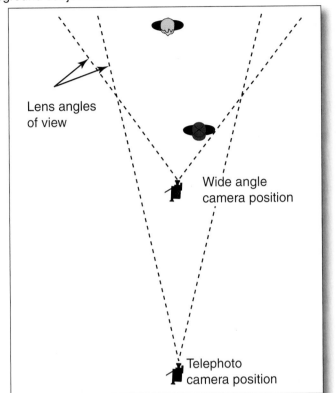

Lens angles of view

Wide angle camera position

Telephoto camera position

Wide angle image

Telephoto image

Focal Length and Composition

To arrange images effectively with wide angle and telephoto lenses, first decide whether you want to create a composition in three dimensions or two. Wide angle lenses help you enhance the illusion of a third dimension in a composition by emphasizing apparent depth. Telephoto lenses let you organize compositions on the screen surface, rather than "behind" it. As you can see from **Figure 13-11**, both approaches can yield very effective compositions.

Finally, keep in mind that the lens focal lengths most like human vision are in the normal range. When you want viewers to subconsciously think that they are looking directly at the action, rather than at a TV screen, normal focal lengths render images without the exaggerations and distortions that announce the presence of a lens.

Operating the Zoom

As a rule, use the zoom sparingly because the seconds required to move between the start and end compositions are sometimes a waste of screen time. When you zoom, try to set the shot up in such a way that the editor can remove the zoom by cutting directly from the start to the finish, **Figure 13-12**. A zoom cannot be easily removed if it contains action. For example, if you shoot someone walking, removing the zoom will result in a jump cut, as you can see

Figure 13-8 In a wide angle shot, both the truck and the sidewalk grow dramatically as they approach the camera.

Figure 13-9 Lens distortion. The wide angle lens makes the vehicle look longer, and the telephoto lens squeezes the vehicle together.

Wide angle lens

Telephoto lens

Figure 13-10 Wide angle lenses can distort human proportions and make unpleasant-looking portraits.

Figure 13-11 A wide angle lens enhances the illusion of depth. A telephoto lens creates the composition on the surface of the screen.

Wide angle lens

Telephoto lens

Optical vs. Pictorial Lens Qualities

Photographers know that wide angle and telephoto lenses render image magnification, perspective, movement, and distortion in markedly different ways. Optical engineers, however, insist that there is only one difference between wide angle and telephoto lenses: angle of view. All other seeming differences are merely illusions.

Which side is right? Both. On the one hand, the many photos in this chapter clearly show the pictorial differences between wide angle and telephoto images. On the other hand, if you enlarge the center of a wide angle image until it matches the angle of view of a telephoto, the perspective of the two shots will be almost identical.

In the photo panel, the first photo is a wide angle scenic and the second is a telephoto shot made from exactly the same spot. The third photo is the very center of the wide angle image, enlarged to match the telephoto picture (hence its relatively poor quality).

Wide angle shot	Telephoto shot	Center portion of the wide angle shot

Studying the building doors and railings, notice that the scale and perspective in the blowup are essentially the same as those in the telephoto shot. From an optical standpoint, there is no difference between wide and long lenses except angle of view.

from **Figure 13-13**. One solution is to insert a cutaway shot in the place of the removed zoom, **Figure 13-14**.

Current viewers are so sophisticated that jump cuts are often used purposely, especially in energetic programs aimed at youthful audiences.

Figure 13-12 With no action in the shot, it is easy to cut out the zoom from wide angle to telephoto.

Wide angle	Telephoto

Figure 13-13 Omitting the zoom from an action shot creates a jump cut.

Zooming Techniques

Skilled videographers often prefer to zoom manually for better control. (If your motorized zoom lacks a manual mode, it probably offers variable speed operation.) Here is a technique for professional looking zooms:

- Begin slowly, accelerate during the first half of the zoom.
- Decelerate during the second half.
- Finish the zoom slowly and come to a stop.

Never correct a zoom after you have initially stopped by "bumping" it a bit closer or farther away. Either accept the end composition where you first halted, or retake the shot.

At one time, zooms were rarely used in feature films. Nowadays, the prejudice against them has lessened.

Digital Zoom

Digital zoom is an electronic feature that can magnify images beyond the ability of the lens to enlarge them. True zoom, called *optical zoom*, adjusts the lens to fill the entire imaging chip with a smaller and smaller angle of view. Because the entire chip is always used to form the image, the picture quality always remains the same. Digital zoom, by contrast, works by selecting a smaller and smaller percentage of the image on the chip and electronically enlarging it to fill the whole frame. Since the number of pixels is fixed by the total number on the chip, using just part of the chip requires filling the frame with fewer pixels. The fewer the pixels, the worse the image quality (**Figure 13-15**).

Pixels (short for "picture elements") are the dots that make up the image.

Figure 13-14 Inserting a cutaway shot can avoid a jump cut.

Shot A (before the zoom)

Cutaway inserted in place of the zoom

Shot A resumes after the zoom

Figure 13-15 When zooming in on a distant subject, optical zoom maintains full image quality and digital zoom degrades image quality.

Distant subject   **Optical zoom**   **Digital zoom**

Special Lens Filters

The front elements of professional lenses are machined with threads to receive screw-on filters. (The lenses of some compact cameras may not have filter threads.) Many filters used in film photography, such as starburst and color shift effects, are not employed as often in video because their effects can be supplied by software during the process of computerized editing.

As a rule, it is better to record the best possible unmodified images and then filter them digitally in postproduction. Any effect that modifies an original camera recording is permanent. Two types of filtration, however, can be achieved only in the camera: *neutral density* and *polarization*.

Neutral Density Filters

Gray-colored neutral density filters are used to cut down the incoming light, either to reduce the intensity of an exceptionally bright scene, or to force the aperture wider open in order to reduce depth of field.

Polarizing Filters

A polarizing filter, or *polarizer*, is a sandwich of two rings. The back one threads onto the lens, and the front one holds the filter and rotates freely. By changing the orientation of the filter with respect to the lens, you can suppress small, bright glints of light (called **specular reflections**) from surfaces like metal, glass, water, and shiny paint. You can also reduce unwanted reflections on window glass, as shown in the accompanying illustration. Polarizers can also darken blue skies to make clouds stand out. A polarizer can do extra duty as a variable-strength neutral density filter. Generally,

A polarizing filter can reduce reflections, such as the glare off this store window.

(though not always) the more you rotate the filter, the less light you admit to the lens. Cinematographers typically choose "circular" polarizers for use on digital imaging systems.

Because image quality is degraded by digital zooming, this technique should be used only when subjects are too far away to be magnified adequately by optical means alone. Common applications for digital zoom include nature videography, shooting sports from the sidelines, and surveillance.

Since digital zooming is also possible in many editing programs, you may wish to record the scene at the limits of optical zoom and then enlarge it further in postproduction.

Setting Focus

Setting focus is the process of adjusting the lens so that subjects at a certain distance from the camera appear sharp on the focal plane. Since the imaging chip (CCD) is located on that plane, objects in focus there will be sharp in the video image. Of course, focus must be changeable because subjects may be videotaped at many different distances: a postcard six inches away from the lens, a group of people six feet away, a distant mountain six miles away.

Like your own eyes, optical systems cannot always keep everything—from a few inches to the horizon—in focus at the same time.

Depth of Field

Mathematically speaking, objects in an image are in perfect focus only if they lie on an imaginary plane surface that is precisely a certain distance from the lens. Objects that are behind or in front of that plane—even by as little as a fraction of an inch—are not in theoretically perfect focus, as shown in the diagram. (Technically, this imaginary plane of perfect focus is called the *front focal plane*, while the plane on which the chip lies is called the *rear focal plane*.)

In reality, however, objects up to a certain distance behind and in front of the plane of "perfect" focus are also acceptably sharp, because they appear that way to viewers. This acceptably focused zone is called the *depth of field*. (Note that the depth of field in front of the plane of focus is shallower than the depth of field behind it, as illustrated. This is always the case.)

The depth of field in an image is determined by three factors working together: lens focal length, lens aperture, and the distance from the lens to the theoretical plane of focus.

Lens Focal Length

The longer the lens focal length, the shallower the depth of field. If you set up two camcorders with a 4mm lens and a 16mm lens, both at the same distance from the subject and using the same aperture (f-stop), the 4mm lens will yield a greater depth of field than the 16mm lens.

In theory, only a single mathematical plane is in perfect focus—the "plane of focus." In fact, the range of acceptable sharpness extends before and behind the plane.

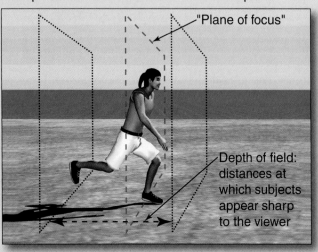

"Plane of focus"

Depth of field: distances at which subjects appear sharp to the viewer

The longer the lens focal length, the shallower the depth of field.

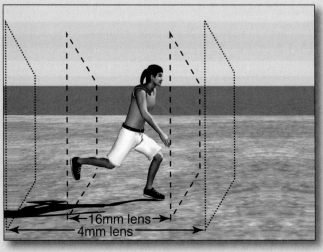

16mm lens
4mm lens

In very bright light and at wide angle lens settings, every object from the postcard to the mountain may be in reasonably sharp focus. On the other hand, when shooting in dim light with a telephoto lens, the depth of field (range of focus) may be as shallow as an inch or two. The reasons for this are explained in the sidebar *Depth of Field*. Almost all camcorders can compute and set focus automatically, and modern auto focus systems work remarkably well. But, they are not perfect.

Auto Focus

Auto focus is a camcorder system that adjusts the lens continuously and automatically to keep objects in the image sharp. Though different camcorders employ different approaches to auto focus, most of them involve the same basic procedure: the system calculates the distance to objects inside a focusing zone within the image—usually the central portion (**Figure 13-16**)—and a servo motor in the lens adjusts focus to the calculated distance. Most camcorders default to the auto focus setting.

Lens Aperture

The wider the lens aperture, the shallower the depth of field. If you set up those same two camcorders, both at the same distance from the subject and both using the same lens focal length, an f/16 aperture will produce the greater depth of field and an f/1.4 stop will produce the lesser.

Subject Distance

The shorter the lens-to-subject distance, the shallower the depth of field. If your two camcorder lenses have both the same focal length and the same aperture, but one camera is set closer to the subject than the other, the more distant camera will realize the greater depth of field and the closer one the lesser.

Depth of field can be confusing to calculate because it is determined by focal length, aperture, and distance, all working at once. For example, some people think that if the depth of field produced by a telephoto setting is too shallow, they can fix the problem by switching to a wide angle setting with its inherently greater depth of field. The problem is that in order to match the view of the telephoto lens, they would have to move the wide angle lens closer to the subject. Since moving closer reduces depth of field, the eventual wide angle focus would be no deeper than the original telephoto focus.

The wider the aperture, the shallower the depth of field.

Aperture f/4
Aperture f/16

The shorter the distance, the shallower the depth of field.

Lens-to-subject distance: 3′

Lens-to-subject distance: 12′

Figure 13-16 Many camcorders use the center of the image for focusing.

(Sue Stinson)

Auto focus has several limitations that can make it unsuitable for professional shooting. If the subject is not within the focusing zone, it may be out of focus. If the camera and/or subject moves, the auto focus may not respond quickly enough to maintain uninterrupted sharpness. If unimportant objects move into the focusing zone, the system may focus on them, rather than on the center of interest. For these reasons, better camcorders offer three alternate ways to maintain focus: *auto focus lock, manual focus,* and *pulling focus.*

Auto Focus Lock

Auto focus lock uses the automatic mechanism to focus on a subject and then prevents that setting from changing, even if the subject moves. To use the auto focus lock system, focus on the subject, then engage the focus lock control on the camcorder.

When the shot is complete, remember to disable the focus-locking mechanism.

Locking the auto focus function is especially useful when the center of interest is outside the focusing area. Simply center the frame on the subject, focus the lens, lock the focus, and recompose the shot. The subject will stay sharp.

You can also use auto focus lock when unimportant foreground objects "steal" the focus from your subject matter. If you lock focus on the subject, the focusing system will ignore the intruding objects. In **Figure 13-17**, for example, the subject is taped as she moves along a sidewalk behind a series of hedges.

The auto focus system does not always work reliably during zooms, because the depth of field narrows as you zoom in. This means that a focus setting made at a wide angle focal length may lose sharpness as you zoom into a distant part of the scene. This is especially true of a zoom in, where the auto focus system has set focus in the wide angle start position (**Figure 13-18**). (In a zoom out, the start focus will be properly set and the system will usually adjust itself satisfactorily as the shot widens.)

Figure 13-17 Comparing auto focus and auto focus lock functions.

The intended subject is sharp and focused in the shot.

The auto focus system shifts focus to the hedges filling the screen in the foreground, and softens the subject.

Locking the auto focus on the intended subject prevents the unwanted shift in focus.

Figure 13-18 A scene focused at a wide angle setting may not stay in focus at the end of a zoom in.

Wide angle

Zoom in

For this reason, you should use the auto focus lock option to set up every zoom in, as shown in **Figure 13-19**. In most situations, this procedure will keep your shot sharp throughout the range of the zoom:

1. Zoom to full telephoto, allowing the auto focus system to focus.
2. Lock the focus.
3. Zoom out to your start position.

Manual Focus

In some situations, the auto focus lock method will not work because the auto focus mechanism fails to get a fix on the subject. This can happen in low light, when shooting through a window,

or when the subject lacks features that the focusing system needs for reference. The procedure for focusing manually is similar to focusing for a zoom:

1. Disable the auto focus system.
2. Zoom to the full telephoto position to obtain the shallowest depth of field.
3. Manually adjust the focus until the subject is sharp.
4. Zoom back out to the desired composition.

Except in news and sports applications, most professional videographers use manual focus whenever practical.

In low light levels, a long zoom may require pulling focus.

Figure 13-19 Use the auto focus lock option to set up every zoom in.

Zoom in   Auto focus lock

Zoom out

Pulling Focus

Pulling focus is the technique of shifting manually from one focal setting to another during the shot.

This procedure is also called *follow focus*, *racking focus*, and other names.

Pulling focus is practical only when the camera has an external zoom lens fitted with a focus ring. It usually requires an assistant to revolve that ring to change focus. To prepare a shot for pulling focus:
1. Wrap a narrow piece of white tape around the lens barrel, next to the focus ring.
2. Focus the lens at the starting position and mark a reference line on the tape.
3. Repeat the process for each additional focus position, **Figure 13-20**.

Focus pulling is often used to keep a moving subject sharp. To do this when the subject approaches the camera, set focus on points at the far, middle, and near points of the movement. As the subject moves toward the camera, shift the focus continuously, using the reference marks on the tape. (In professional camera setups, focus points are marked on a white disc surrounding a focus knob fitted to the lens.)

Controlling Exposure

Setting exposure means literally *exposing* the camcorder circuits to the exact amount of light required to form a quality image. If the imaging chip receives too much light, the picture will be too light (overexposed) or even "burnt out" to white. With too little light, the image will be too dark (underexposed) or "blocked up" to black. Sometimes the camcorder cannot handle the ultra-bright light in an environment, like a beach or a ski slope. On the other hand, when taping at night or in a dark room, there may be too little light to form a good image. To solve the problems of too much or too little light, the camcorder relies on four exposure control systems: *aperture*, *shutter*, *gain*, and *filtration*.

Aperture

As explained in the section on lenses, the *aperture* is the hole through which light enters the camcorder. The bigger this hole, the more light gets in. The aperture is set by an *iris*—a circular opening inside the camera lens that can be expanded to nearly the full diameter of the lens or contracted until it is almost completely closed.

The aperture is usually polygonal rather than truly circular. That is why lens flares (from shooting into the sun or other bright lights) appear as multiple polygons.

Aperture Exposure Control

The auto exposure system built into every camcorder works by adjusting the aperture.

Automatic Aperture

In "auto" mode, your camcorder sets exposure by analyzing the intensity of the incoming light and automatically changing the size of the lens opening, so that the same amount of light reaches the chip, regardless of the light level outside. Auto exposure systems adjust aperture only, leaving the shutter at its default speed. Auto exposure systems work so well that you can use them much of the time, as long as you avoid panning too quickly from, say, a very bright scene to a dark one.

When light levels are too different, try restaging the action to avoid the light change, or get a cutaway shot to replace the badly exposed transitional section when you edit the scene later.

| **Figure 13-20** | White tape marked with interim focus points. |

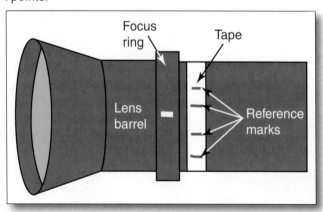

Manual Aperture

Many amateur and all professional camcorders also allow you to set the aperture manually. One way to do this is by locking the exposure:

1. In "auto" mode, frame the scene that you wish to expose for.
2. Lock the aperture.
3. Make the shot.

Figure 13-21 shows a typical exposure problem that can be solved by this method. Notice that it is very similar to the focus problem discussed previously.

You can also set aperture in full manual mode to fine-tune overall exposure. This is especially useful when backlighting is a problem, or when the center of the frame is much brighter than the rest of the image. Excessively bright areas are more irritating, visually, than overly dark ones. For this reason, it is often better to correct backlighting problems by lighting the foreground subject or moving it away from the bright background. However, when it is the subject that is too bright you can expose for it and allow the darker areas to "block up." To do this:

1. Zoom in until the bright subject fills the frame.
2. Lock the auto exposure.
3. Zoom back out to the original composition.
4. Make the shot.

Programmed Aperture

Most camcorders offer programmed aperture settings to address various exposure problems.

Though different models offer different programs, two of the more popular aperture-based options are *backlight* and *spotlight*.

Backlight programs automatically open the aperture a certain amount to increase the exposure for dark subjects in front of bright backgrounds. Unfortunately, many backlight programs have only one setting. If it doesn't match the situation exactly, the results may not be satisfactory. If your camcorder has manual aperture control, it is better to compensate for backlighting by increasing exposure in small steps until you are satisfied with the result.

The spotlight program is the opposite of backlight. It is intended for theater performances and other situations in which the subject is brightly illuminated against a much darker background. The spotlight program reduces the contrast between the two elements, and improves the picture considerably.

Shutter

While the aperture controls the size of the opening through which light is admitted to the camcorder, the **shutter** controls the amount of time allotted to forming each video frame (image). In a film camera, the shutter is an actual window that opens and shuts for each image. In video, however, the shutter is a set of circuits that control how long the incoming light is allowed to build up electrical charges on the chip before they are removed and sent on to be processed and recorded.

Figure 13-21 Comparing auto exposure to manual aperture.

Original exposure.

Auto exposure increases exposure for the column, overexposing the subject.
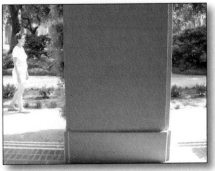

Manual aperture retains original exposure.

Overexposure Zebras

Many professional cameras warn the operator when part or all of the scene is too bright to record properly. Typically, this warning overlays the too-bright portions of the image in the viewfinder with a striped pattern. For this reason, it is called a *zebra*.

A striped overlay identifies the overexposed portion of the image.

The most common form of zebra warning appears when part of the image is brighter than 100% of the camera's ability to record it. Another type is programmed to appear when the center of the image is too bright to record Caucasian flesh tones properly (about 70% of maximum brightness). Most professional camcorders offer both settings.

Some cameras can be set to offer a warning in the viewfinder when skin tones are overexposed.

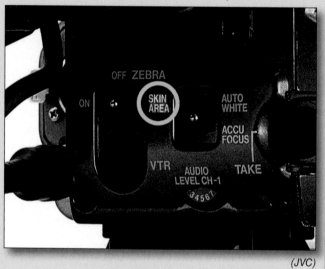

(JVC)

NTSC video is displayed at 30 frames per second, with every frame made up of two interlaced fields. This means that the effective normal shutter speed is ¹⁄₆₀ second. That is, when the shutter circuits are in "normal" mode, they allow the chip to build up electrical charges for each field for ¹⁄₆₀ second before collecting them and starting over.

Some interlaced systems combine the information from adjacent scan lines. This electronic processing can increase the effective exposure from ¹⁄₆₀ to ¹⁄₃₀ second. In terms of recording movement, however, the shutter speed is still ¹⁄₆₀.

Shutter speed usually remains locked at ¹⁄₆₀ second so that exposure control is handled entirely by the aperture. There are special situations, however, in which you may want to change shutter speed. For this purpose, many camcorders offer shutter control programs.

Programmed Shutter

Like exposure programs, shutter programs adjust one or more camcorder settings for you automatically. All you do is select the effect you want and the circuitry does the rest. Shutter programs have three purposes: forcing open the aperture, freezing rapid motion, and compensating for extreme light levels.

Portrait

The portrait program increases shutter speed moderately in order to force the aperture wider open. The wider aperture reduces depth of field so you can keep your portrait or still life subject sharp in the foreground, while throwing a distracting background out of focus (**Figure 13-22**).

As explained in the sidebar *Apertures and f-stops*, each time you double shutter speed, the aperture opens by one f-stop to compensate.

Figure 13-22 The portrait program opens the aperture to throw backgrounds out of focus.

(Sue Stinson)

Figure 13-23 The high light level program reduces overexposure.

(Sue Stinson)

Sports

The sports program increases shutter speed more than the portrait program in order to sharpen fast action that might be blurred at the normal shutter speed. Sharpening the image improves detail, but can make movement look jerky.

High Light Levels

The high light level program (often called "sand and snow," or some other proprietary term) increases shutter speed when outside conditions are so bright that the smallest aperture still admits too much light. Since each doubling of the shutter speed reduces incoming light by one full f-stop, a moderate shutter increase (say, to $\frac{1}{125}$ second) is usually enough to compensate for the brightest scene (**Figure 13-23**).

Most professional camcorders control excessive brightness by reducing the gain setting, enabling built-in neutral density filtration, or both.

Low Light Levels

Some shutters can be set slower than normal, typically at $\frac{1}{15}$ of a second. This allows the CCD four times as long to form an image in very low light. The drawback is an obvious blurring of the action; and since the circuitry must record each frame twice for normal playback speed, the resulting movement is somewhat jerky. In general, a slow shutter

should be used only when the light is otherwise too dim to record an image. A slow shutter can also create an eerie, dreamlike effect for music videos and other special-purpose applications.

Manual Shutter

Some camcorders allow you to change shutter speed manually. For instance, you can manually control depth of field in a portrait shot more precisely than with an automatic portrait program. Very high shutter speeds are often useful for scientific analysis and similar purposes.

As you manually adjust shutter speed, keep two important points in mind. First, increasing shutter speed always widens the aperture proportionately (in order to maintain the same exposure). This can seriously reduce depth of field, making it hard to hold focus at long telephoto settings. Second, shutter speeds faster than about ¹⁄₁₂₅ second jerk and strobe too much to yield normal-looking movement. Because of these motion problems, exposure is adjusted less often by manual and programmed *shutter* settings, than by *aperture* controls.

Gain

Gain is the electronic amplification of the video signal. In most camcorders, this "video volume" can be turned up to compensate for low light levels. Unlike aperture and shutter, gain does not affect the amount of light used to form the image. Instead, it operates *after* the image has been coded as an electronic signal by increasing the strength of that signal. Gain control can operate in several different ways, depending on the sophistication of the camcorder:

- In simple units, the gain turns on automatically whenever light levels drop too low.
- Better models allow you to turn the gain on and off manually.
- Still more sophisticated cameras let you select different amounts of gain compensation, to suit lighting conditions precisely.

Professional video cameras are often designed to operate in normal light with a certain amount of gain enabled. That way, the videographer can compensate for excessively bright conditions by reducing the gain instead of using filtration.

Gain control is useful where the situation is otherwise too dark for recording. But, it cannot improve the poor quality of the image, it can only brighten it. The result is often washed out color, coarse detail, and video noise that appears on the screen as "snow," **Figure 13-24**.

Filtration

The final way to regulate exposure is by placing a neutral density filter over the lens. This is a gray filter that reduces the brightness of the light without changing its color. Popular neutral density filters are offered in shades labeled ND3, ND6, ND9, and ND12, which reduce incoming light by 1, 2, 3, and 4 f-stops, respectively. In unusually bright shooting conditions, neutral density filters enable the camera to capture a good quality image. Used

Figure 13-24 Gain control circuits can lighten the image, but the quality is often poor.

at normal light levels, they can reduce depth of field by widening the aperture without increasing the shutter speed.

Filters for Bright Conditions

Very bright sunshine in reflective locations can overpower your camcorder's chip and circuitry. The result is flares of white and smeary colors that "bloom" beyond their proper borders. In these conditions, an appropriate neutral density filter will reduce the incoming light to a level that the electronic circuitry can handle, **Figure 13-25**.

Because video recording tends to lose picture detail in very bright areas, it is often better to "expose for the highlights"—that is, reduce exposure until the lightest areas still retain details. Neutral density filters offer an easy way to do this.

Figure 13-25 Result of neutral density filters.

Overexposure (no filtration) **ND3 (one-stop) filter** **ND6 (two-stop) filter**

(Sue Stinson)

Filters for Aperture Control

Even when brightness is not excessive, you may want to reduce the incoming light in order to force open the aperture. You could do this by increasing shutter speed, as explained previously, but fast shutters often result in undesirable stroboscopic effects. To widen the aperture without changing shutter speed, use a neutral density filter instead.

There are two reasons for working at a wider aperture. First, the reduced depth of field lets you separate your subject from the background by keeping it sharp, while the area behind it is soft.

Some high-definition cameras are unsuitable for shooting theatrical movies because their small imaging chips result in depth of field too deep for techniques such as shifting focus within a shot. To avoid the problem, the best cameras now use chips as big as 35mm movie film frames, or even 35mm still camera size, which is twice as large.

Second, almost all photographic lenses record their best images at apertures in the middle of their range. If, for example, the apertures of a lens range from f/2 (widest) to f/22 (narrowest), the lens will perform best between about f/5.6 and f/11. Since even average sunshine may require a very small aperture, a neutral density filter can improve picture quality and help control depth of field.

If, for instance, the auto exposure system sets the lens aperture to f/22, an ND9 neutral density filter over the lens will reduce the aperture three f-stops, to f/8, where the depth of field is extensive but not excessive, and the optical quality of the lens is at its best.

Since the gain control is enabled only in low light levels and most filters reduce light levels, the two controls are rarely employed at the same time.

Operating the Camera

Managing the camcorder capably is so important that fine camera operators are highly regarded. No matter how good the acting, costumes, sets, or lighting, a shot can be no better than the competence with which it is recorded. Basically, a camera operator's job is to frame (compose) the shot well, then to move the camera as needed to maintain an effective composition, **Figure 13-26**.

Framing the Shot

Framing a shot means creating a pleasing composition within the frame. The key to success is your ability to see the viewfinder not as a window, but as a picture within the frame.

Visual composition is covered in Chapter 6, *Video Composition*.

Correcting

To compensate for movement within the image, the camera must move as well. Professional camera operators rarely lock the tripod's pan and tilt mechanisms. Even when the shot is supposed to be fixed, they are constantly moving the camera slightly, to maintain a good composition. In a skilled operator's hands, this camera movement is so smooth and unobtrusive that it is invisible

Factors Affecting Exposure

Like focus, exposure is the result of several factors working together. In order of importance, these factors are aperture (f-stop), shutter speed, filtration, and gain. Filtration and gain operate only when you enable them, but aperture and shutter are always working.

A Fixed Amount of Light

The key to understanding exposure is the idea that to make a good quality image, the camcorder must always receive exactly the same amount of light. Whether the original light is on a beach lit by the sun at noon or by a fire at midnight, the light that eventually reaches the imaging chip must be at the same brightness level—even if the noon sunlight is 250 times as bright as the firelight.

The most common problem is to reduce excessively bright light so that it does not overwhelm the sensitive chip. To do this, the camcorder uses first the aperture and then, if necessary, the shutter.

Aperture

At its widest aperture (such as f/1.4) the lens admits nearly all the light that strikes it. If the raw light is too bright for the chip, the aperture grows smaller and smaller. Each full stop reduction in the aperture size is called one f-stop (or just "stop"), and each smaller stop admits one-half as much light as the next wider one. The usual f-stops are 1.4, 2, 2.8, 4, 5.6, 8, 11, 16, and 22 (to control very bright light, some camcorder lenses will "stop down" to f/32). Most aperture control systems can change in partial stops, closing, say, from f/2 to f/2.4, which is half way to the next full stop, f/2.8.

More sophisticated models divide stops into one-third stop increments, rather than one-half stop increments.

Shutter

In situations where the aperture alone cannot fully control exposure, the shutter is brought into play. Typical shutter speeds are 1/60 second, 1/125, 1/250, 1/500, 1/1000, 1/2000, 1/4000, and 1/8000. Each time you double the shutter speed, you reduce the light by half, because you allow the chip half as much time to form an image before it is removed and processed.

Aperture Plus Shutter Equals Exposure

The amount of light used to form an image depends first on how much is allowed to reach the chip (aperture) and how long the chip is allowed to soak it up (shutter speed). A one-stop decrease in aperture reduces the light by half and so does each doubling of shutter speed.

To see how aperture and shutter work together, imagine a sort of equation:

Aperture + Shutter = Correct Exposure

- Pretend that "Correct Exposure" has a value of 10 (Aperture + Shutter = 10).
- Give the aperture f/1.4 a value of 1, f/2 a value of 2, f/2.8 a value of 3, and so on, increasing the value by 1 for each full stop.
- Give shutter speed 1/60 second a value of 1, 1/125 a value of 2, and so on, also increasing the value of each doubled speed by a value of 1.

to the audience. To achieve this fluid movement, you aim the lens much as a hunter aims at a constantly moving target. But instead of following the central subject, you "aim" at the entire composition within the frame.

Compensation

Frequently, you correct the composition to compensate for added or subtracted elements.

Suppose you have a composition in which a second subject joins a first subject. **Figure 13-27** shows the effect without camera compensation. As you can see, failure to reframe for the second subject ruins the composition. **Figure 13-28** shows correction for the second subject.

You can value both f-stops and shutter speeds in multiples of 1 because each change represents the same reduction in light.

With these values, you can see that 1 + 9 = 10, 2 + 8 =10, 3 + 7 = 10, 4 + 6 = 10, 5 + 5 = 10. If you translate these numerical values back into apertures and shutter speeds, the result would look like this for bright sunlight:

Aperture		Shutter		Total Light
1 (f/1.4)	+	9 (1/16000)	=	10
2 (f/2.0)	+	8 (1/8000)	=	10
3 (f/2.8)	+	7 (1/4000)	=	10
4 (f/4.0)	+	6 (1/2000)	=	10
5 (f/5.6)	+	5 (1/1000)	=	10
6 (f/8.0)	+	4 (1/500)	=	10
7 (f/11.0)	+	3 (1/250)	=	10
8 (f/16.0)	+	2 (1/125)	=	10
9 (f/22.0)	+	1 (1/60)	=	10

In a darker environment, like a living room at night, the available exposure combinations might resemble this:

Aperture		Shutter		Total Light
1 (f/1.4)	+	9 (1/250)	=	10
2 (f/2.0)	+	8 (1/125)	=	10
3 (f/2.8)	+	7 (1/60)	=	10

In this darker location, f-stops smaller than f/2.8 and shutter speeds faster than 1/250 are not available because there is not enough light to allow them.

As you review these examples, keep in mind that "aperture + shutter speed = exposure" is not a true mathematical equation, but only a rough analogy.

Look Room

When framing people, allow more room on the side toward which they are facing. This avoids the claustrophobic feeling that they are boxed in by the frame. **Figure 13-29** compares two different compositions of a subject. Since people move as they speak, a common application for the floating frame is maintaining adequate look room.

Figure 13-26 A professional videographer at work.

Figure 13-27 Without camera compensation, the new subject spoils the composition.

Figure 13-28 Camera compensation corrects the shot.

Lead Room

When subjects move so that the camera too must move to hold them in frame, look room is called *lead* room. The idea is exactly the same: allow extra room on the side of the frame toward which people or objects are moving, to prevent them from crowding the frame line (**Figure 13-30**).

Framing Moving Shots

Maintaining lead room will keep a moving subject properly framed, but it will not guarantee a good composition. To do that, you must consider the background as well as the subject. To ensure good composition overall, set up an effective start composition that accommodates both subject and background. Pre-frame an equal or better ending composition with the subject blocked to complete the move at the designated spot.

In video, as in theater, a subject's predesigned and rehearsed movements are called **blocking**.

Moving the Camera

Composing moving shots naturally requires moving the camcorder, and this is not always as simple as it sounds. The operator must keep

Figure 13-29 Comparing look room in a composition.

Inadequate look room

Ample look room

Figure 13-30 Maintaining lead room in a composition.

Subject is "crowding the frame."

Centering the subject improves the composition.

Ideally, the subject should be slightly behind the center line of the frame.

the move smooth, while maintaining a good frame and operating camera features, like the zoom control. Whether pivoting the camera on its tripod or rolling on a dolly, the key to a professional-looking move is variable speed. Start the move slowly and then accelerate to the desired maximum speed. Then, slow smoothly and gradually to a stop.

Panning and Tilting

When rotating the camera horizontally (panning) or vertically (tilting), follow these suggestions:

- Set the drag (resistance) on the tripod head so that you are pushing against it hard enough to smooth the move, but not hard enough to cause jerks in the motion.
- Position your body at the midpoint of the movement. By twisting first toward the start of the movement and then toward the end, you can make smooth pans of 180°. See **Figure 13-31**.
- Use an external monitor, either camera-mounted or stand-alone. By not looking through the built-in viewfinder, you can position yourself far enough away from the camera to avoid impeding its free movement.

Dollying

These suggestions apply equally to moving the camera on a dolly. The main problem with dolly shots is keeping perfect pace with the moving subject so that the lead room remains constant. It is very distracting to have a moving subject shifting forward and backward in relation to the frame. To make dollying effective, remember that even the fastest-moving subject is standing still in relation to the frame. To give the shot a feeling of dynamic movement, be sure to include a distinctive background. On screen, it is the background that will fly by, not the subject.

Hand-Holding

All the suggestions for smooth panning and dollying apply as well to hand-held moving shots. The challenge with these shots is to keep them smooth. Here are some tips for professional hand-holding:

- Use a wide angle lens setting to minimize the appearance of camera shake.
- Use an external view screen. This keeps your head from transmitting bumps to the camera, and it lets you use your peripheral vision to watch where you are going.
- Keep your arms away from your sides so that your bent elbows can act as shock absorbers.
- Walk with knees slightly flexed, and try to glide along rather than striding.
- Carry the camcorder as if it were a very full and very hot cup of coffee.

If you combine these techniques with a good lens stabilization system, you can get hand-held shots that rival professional dolly work.

Figure 13-31 Begin by facing the center of the pan. Twist your upper body to frame the start of the pan and follow through with your body to the end of the pan (up to 180°).

Center of the pan

Starting point of the pan

Ending point of the pan

Hand-Holding Appliance Cameras

Many cell phones, personal players, still cameras, and tablets can record video, but they can be hard to control because of their small size and because some types are not designed for comfortable hand-held shooting. These tips can help you get better results:

- Hold the unit lightly in both hands, for greater steadiness.
- When practical, prop or brace the unit on any handy surface, such as a tabletop, a railing, or the side of a doorway. The external screen will still let you view what you are shooting.
- Stay close to your subjects for better recording of details.
- Avoid "fire-hosing"—sweeping the unit around constantly without framing and holding compositions. Instead, frame a subject, shoot at least three seconds of material, then move the camera smoothly to a new composition and repeat the process.
- Do not move yourself unless you must. These tiny units are difficult enough to hold steady while remaining still.

Roll Camera

At this point, you have worked your way through the basics of professional videography, and now you are ready to shoot. Just one final reminder: always roll at least five seconds of tape before signaling to begin the action; and keep on rolling at least five seconds after the action has ended or moved off-screen. The editor will thank you.

Summary

- The focal length of a lens is the distance from the front element of the lens to the focal point.
- Within a zoom range, the number of focal lengths is infinite. However, focal lengths are commonly divided into three broad groups: normal, wide angle, and telephoto.
- The pictorial qualities of any focal length are dependent on the size of the camera chip on which the image is created.
- All lenses have three characteristics that affect the way the videographer uses them: angle of view, maximum aperture, and depth of field.
- For a lens of any diameter, the shorter the focal length, the larger the maximum potential aperture.
- The smaller the f-stop number, the larger the aperture.
- The depth of field in an image is determined by three factors working together: lens focal length, lens aperture, and the distance from the lens to the theoretical plane of focus.
- The aperture is the hole through which light enters the camcorder. The bigger this hole, the more light gets in. The aperture is set by an iris.
- The shutter controls the amount of time allotted to forming each video frame. NTSC video is displayed at 30 frames per second.
- Framing a shot means creating a pleasing composition within the frame.

Technical Terms

Angle of view: The breadth of a lens' field of coverage; expressed as an arc of a circle, such as 10°.

Aperture: The opening in the lens that admits light. The iris diaphragm can vary the aperture from fully open to almost or completely closed.

Auto focus: The camcorder system that automatically adjusts the lens to keep subjects sharp.

Blocking: The predetermined movements that a subject makes during a shot. Camera movement is also blocked.

Depth of field: The distance range, near-to-far, within which subjects appear sharp in the image.

Digital zooming: Increasing the subject size by filling the frame with only the central part of the image.

f-stop: A particular aperture setting. Most lenses are designed with preset f-stops of f/1.4, f/2, f/2.8, f/4, f/5.6, f/8, f/11, f/16, and f/22.

Focal length: Technically, one design parameter of a lens, expressed in millimeters (4mm, 40mm). Informally, the name of a particular lens, such as a 40mm lens.

Gain: The electronic amplification of the signal made from an image, in order to increase its brightness.

Iris (iris diaphragm): A mechanism inside a lens (usually a ring of overlapping blades) that varies the size of the lens opening (aperture).

Magnification: The apparent increase or decrease of subject size in an image, compared to the same subject as seen by the human eye. Telephoto lenses magnify subjects; wide angle lenses reduce them.

Optical zoom: Changing a lens' angle of view (wide angle to telephoto) continuously by moving internal parts of the lens.

Pulling focus: Changing the lens focus during a shot to keep a moving subject sharp. Also called *following focus* and *racking focus*.

Setting focus: Adjusting the lens to make the subject appear clear and sharp.

Shutter: The electronic circuitry in a camera that determines how long each frame of picture will accumulate on the imaging chip before processing. The standard shutter speed is 1/60 per second for NTSC format video.

Specular reflections: Hard, bright reflections from surfaces, such as water, glass, metal, and automobile paint, that create points of light on the image. Often controllable by a polarizer.

Speed: The light-gathering ability of a lens; expressed as its maximum aperture. Thus, an f/1.4 lens is "two stops faster" (more light sensitive) than an f/2.8 lens.

Review Questions

Answer the following questions on a separate piece of paper. Do not write in this book.

1. What is the *focal length* of a lens?
2. *True or False?* Telephoto focal length lenses show a narrower, more restricted field of view.
3. The shorter the lens' focal length, the larger the maximum potential _____.
4. What is the speed of a lens? How is the speed determined?
5. What is *depth of field*?
6. The only true optical difference among lens focal lengths is _____.
7. Describe the difference in perspective (depth) created by a wide angle lens and a telephoto lens.
8. Which range of lens focal lengths is most like human vision?
9. How can a zoom that contains action be removed from a shot?
10. Explain the difference between *digital zoom* and *optical zoom*.
11. A(n) _____ filter can be used to suppress specular reflections.
12. What is the benefit of using *auto focus lock*?
13. What is the *iris* in a camera lens?
14. The _____ controls the amount of time allotted to form each video frame.
15. What is *gain*? How is gain different from aperture and shutter?
16. What is the effect of using a neutral density filter at normal light levels?
17. List the typical f-stop values found on lenses.
18. The space on the side of the frame toward which people or objects are moving is called _____.
19. What is *blocking*?

STEM and Academic Activities

1. **Science.** This chapter compares the capabilities of the camera lens to the field of view of human vision. What is the field of view of human vision? What are the limits of the field of view of human vision?

2. **Technology.** The most commonly used CCD image sensors are available in three different designs (or architectures). What are the three designs? How do they function differently?

Camera stabilizers are widely used in video documentary projects.

CHAPTER **14** Lighting Tools

(Litepanels Inc.)

Objectives

After studying this chapter, you will be able to:

- Identify the tools available for controlling the quantity, direction, and color of video lighting.
- Recall the characteristics of different lighting instruments.
- Summarize the differences among the types of lamps used in video lighting.
- Explain the function of various lighting accessories.

281

About Lighting Tools

Considered from one perspective, video is about *light*. You may think you are recording beautiful backgrounds, exciting action, and expressive performers, but you are not. What you are actually recording is only the light that bounces off your subjects, streams through the lens, and strikes the imaging chip. Since that light is all that your camera sees, you can shape video images by controlling the light that illuminates them.

This discussion of lighting equipment is the first of three chapters on video lighting. The next chapter covers the principles of lighting design and the final chapter brings tools and designs together in a survey of typical lighting applications.

Because camcorders can capture a satisfactory image in almost any quantity and quality of illumination, it is possible to make videos without paying much attention to lighting. But if you ignore lighting, you neglect one of your most powerful tools. One difference between a casual shooter and a video artist is that the beginner merely *records* images while the videographer *creates* them—in part, by managing light.

Video lighting is the art of creating images with light. To do this, of course, you need to understand the special tools used to create lighting designs:
- Tools for controlling available light
- Tools for adding light
- Lamps
- Accessories

Since it is difficult to discuss lighting equipment (**Figure 14-1**) without mentioning its uses, this chapter anticipates some of the content in the next two chapters.

Tools for Available Light

In thinking about video lighting, it helps to distinguish the light you may provide from the illumination that is already present at the scene. In many circumstances, indoors as well as out, available light will provide some or all of your video lighting.

Available light is whatever illumination already exists at the location you are preparing to light for video.

Controlling available light means modifying and redirecting it to improve the quality of the image. In working with available light, you control its *quantity*, *direction*, and *color*. In doing these things, of course, you also determine the light's *character*, especially how relatively direct or diffuse it is.

Controlling Quantity

Controlling the quantity of available light means reducing it when it is too intense. (By definition, you cannot *increase* the amount of light that is naturally there.) In reducing light quantity, your principal tools are *flags*, *screens*, *silks*, and *filters*.

Flags are flat, opaque cards, usually constructed of metal, thin plywood, or fabric and mounted on a metal frame. Most are rectangular, ranging from squares to long, thin

---

Figure 14-1 Lighting equipment and accessories.

Reflectors

Spotlights

Reference monitors

(Bogen Manfrotto)   *(Lowel Light Inc.)*   *(JVC Corporation)*

paddles. (Flags are also used extensively to control spill from video lights.)

Reflectors can serve as flags when set to block light rather than reflect it.

Foam board, used mainly for reflectors, also works well as a flag. Flags may be held by crew members or clamped onto stands. Flags completely block the path of directional light, **Figure 14-2**.

Screens are large expanses of plastic mesh stretched on frames. Small screens can be held by stands or by crew members, but most are mounted on support structures (**Figure 14-3**). Unlike flags, screens block only part of the light; and unlike silks, they do not change the light's directionality or character. The screening material can be obtained in different densities, depending on how much light reduction is needed.

Figure 14-3 A screen reduces light intensity without altering its character. Part of the light falling on the subject is blocked by a screen.

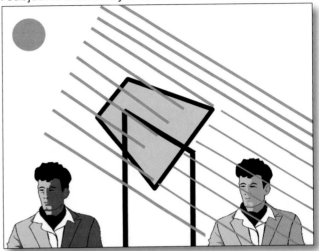

Figure 14-2 A flag blocks light completely, placing the subject in the flag's shadow.

Silks look and work like screens, except that they are sheets of thin white synthetic fabric. Like screens, silks reduce the intensity of light on the subject. But, as you can see from **Figure 14-4**, silks also diffuse the light, changing it from a single, directional beam to an overall glow. Because silks diffuse the light as well as reduce it, they change the light source's quality as well as its quantity.

Actual *silk* fabric is not used on "silks" anymore.

Neutral density filters are sheets of gray-tinted plastic large enough to cover entire windows. Daylight is typically much brighter than interior lighting. To prevent on-screen windows from "burning out" to blank white, neutral density filter plastic is applied over the windows, **Figure 14-5**.

Figure 14-4 A silk diffuses light and reduces light intensity. The light falling on the subject is softened and reduced by the silk.

Neutral density filtration material is supplied in several strengths, reducing light intensity by one, two, three, or four f-stops—coded "ND3," "ND6," "ND9," and "ND12."

Controlling Direction

If you have enough direct light to cast shadows, you can multiply its effect by bouncing it back on your subjects.

Reflectors are large boards with bright surfaces that are positioned to receive light from directional sources and aim it back at subjects. Large studio-style reflectors are made of stiff materials and mounted on yoked stands, **Figure 14-6**. Because of their weight, they are stable in winds, but awkward to handle.

Figure 14-5 A neutral density filter reduces incoming light, but does not change its bluish outdoor color.

Unfiltered window

Filtered window

Hand-held reflectors have just the opposite advantages and disadvantages: light and easy to handle, they tend to wobble in breezes, making the light on the subject waver unsuitably. The lightest models are made of fabric stretched on wire hoops, **Figure 14-7**. These can be folded and stored in very small spaces.

Reflectors come in four main types:

- **Hard metallic:** Highly directional, these silver reflectors throw a light beam a long distance and/or fill a small subject area. They can be painfully bright when directed at an actor's face, **Figure 14-8**.

Figure 14-6 A heavy-duty studio-style reflector.

(Lowel Light Inc.)

Figure 14-7 Hoop and fabric reflectors are versatile tools. This one has been folded to one-quarter of its working size.

(Bogen Manfrotto)

- **Soft metallic:** Less directional, these silver reflectors still throw light a long distance, but disperse it over a wider area (**Figure 14-9**).
- **White:** Extremely diffuse, these reflectors provide a soft, even fill light that does not hurt actors' eyes (**Figure 14-10**). The most common white reflector is made of foam board—a sheet of rigid plastic foam laminated between two sheets of white paper. Stiff, light, and disposable, foam board is an indispensable lighting tool.

Figure 14-8 Hard reflectors reflect light over long distances, but can be hard on subjects' eyes.

Hard metallic reflector

Figure 14-9 Soft reflectors are a compromise between hard metallic and white.

Soft metallic reflector

Figure 14-10 White reflectors deliver the softest light.

White reflector

- **Tinted:** Metal, cloth, or foam board reflectors can be obtained in colors—notably gold tone for metallic surfaces and pale amber or pale blue for paper. By changing the color of the reflected light, these reflectors can warm up flesh tones, simulate moonlight, and deliver other specialized tints (**Figure 14-11**).

Controlling Color

Tinted reflectors offer only one way to change the color of available light. More commonly, the light is tinted by large plastic *sheet filters* (also called filtering gels) placed over windows.

Figure 14-11 Gold-tinted reflectors are useful for warming up cool light.

Gold-tone metallic reflector

(Bogen Manfrotto)

Filters placed directly on the camera lens change *all* of the light entering the camera. Sheet filters, however, affect only the light sources that they are covering.

Window filtering gels are used to change the color temperature of light to match halogen movie lights. If your interior lights and camcorder are balanced for incandescent color temperature (3200K), the daylight streaming through visible windows will look too blue. Orange-tinted color correction filters positioned outside these windows will warm up the cool daylight to match the interior lights. These filters are often referred to by their code designation: "85."

The color temperature 3200K is read as "3,200 degrees Kelvin." The Kelvin scale was named for its inventor, Sir William Thomson, Lord Kelvin.

By itself, a color correction filter can seldom reduce the outdoor light *intensity* to the level of the indoor lighting. For this reason, color-correcting filters are used alone when the window light contributes to the illumination, but the actual windows will not appear on-camera, **Figure 14-12**.

If the windows *will* appear in the frame, both light intensity and color must be controlled. So, color correction tinting is added to neutral density filtration, **Figure 14-13**. These combination filters are designated 85ND-3, -6, -9, or -12, depending on how much they reduce the light.

Figure 14-12 A color correction filter converts daylight color temperature to indoor light, but the window is still too bright to show on-screen.

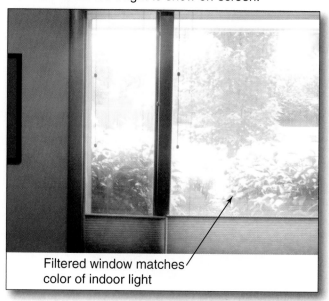

Filtered window matches color of indoor light

You can also change the color of available artificial light (lamps and fluorescents) already in use at a location. Sheet filters can be placed over fluorescent fixtures to match the warmer color of movie lights, or tube-shaped filters can be slipped over the fluorescent lamps themselves. Sometimes, the existing fluorescent tubes can be replaced with special lamps that supply matching color temperature.

Figure 14-13 Combined in a single filter sheet, neutral density and color correction match the interior light color and darken the window enough to be included in the image.

Filter with color correction

As fluorescent video lights become more versatile, color-filtering of daylight and/or ambient fluorescent lighting is required less often. Instead, the camcorder white balance is set for fluorescent color temperature or halogen units that supplement available light may be gelled to match, **Figure 14-14**.

Tools for Adding Light

Tools for adding light are, of course, lighting instruments. Sometimes, you will use them to supplement available light. In other situations, you will exclude ambient illumination and design with video lights exclusively.

Technically, a video light is called a lighting **instrument** (**Figure 14-15**), and the light source inside is the **lamp**. Several units are collectively called "lights."

Professional lighting instruments can seem confusing because there are so many to choose from. The lights of one manufacturer are designed differently from those of others, and each manufacturer offers several competing families of lighting instruments with different characteristics engineered for different applications, **Figure 14-16**.

In practice, many lighting designers eventually standardize on a particular make and family of lights, simply because those are they ones they happened to have mastered.

Figure 14-14 Blue filters, mounted in frames change the color temperature of these halogen lights to daylight.

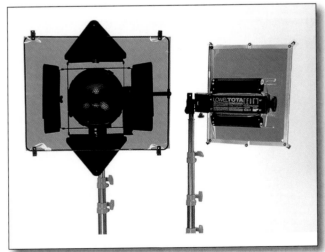

(Lowel Light Inc.)

Figure 14-15 The parts of a typical lighting instrument.

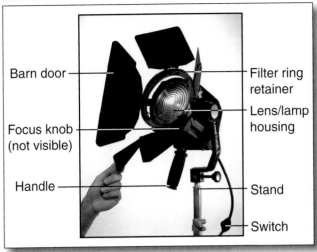

Barn door — Filter ring retainer — Lens/lamp housing — Focus knob (not visible) — Handle — Stand — Switch

(Lowel Light Inc.)

For all their individual differences in features and accessories, most lighting instruments fall into one of the following categories:
- Spotlights
- Floodlights (broads and scoops)
- Softlights (umbrellas, soft boxes, pans)
- On-camera lights
- Practicals

Spotlights

Spotlights ("spots" for short) have small, intense light sources that produce a highly directional light pattern. This means that the light illuminates relatively small areas, casts distinct, hard-edged shadows, and is easily masked by barn doors or flags, **Figure 14-17**. Small source spotlights are often used as key (main) lights on subjects, or to simulate the light from sources ranging from the sun to a table lamp.

The most effective spotlights are fitted with Fresnel lenses for even better light control, but these units are often somewhat larger and heavier.

A spotlight can be focused by moving its lamp forward or backward in relationship to the reflector behind it and, if present, the lens in front. Focusing changes the beam from narrow and intense to wider and less intense. In the "spot" position the light is bright, concentrated, and relatively hard-edged. In the "flood" position, the light beam is larger and less intense, with more gradual falloff at the edges (**Figure 14-18**).

Figure 14-16 Three different lighting kits offered by one manufacturer.

Soft box light, focusable light, wide angle light, umbrella, and stands.

3 accent lights and stands.

2 fluorescent fixtures and stands.

(Lowel Light Inc.)

Figure 14-17 Spotlights are available in different sizes.

(Lowel Light Inc.)

The advantage of small-source lights is that they are very easy to control, **Figure 14-19**. On the other hand, small-source light beams are relatively harsh and several used together throw unrealistic multiple shadows.

Floodlights

Floodlights are larger lights that are commonly used to fill in shadows (typically created by spotlights) and to light backgrounds. Compared with spotlights, floodlights have the opposite advantages and drawbacks. On the positive side, their light is somewhat softer and less directional, their shadows are fainter, and their beam pattern is broad and even. On

Figure 14-18 A spotlight in full spot focus has a narrow beam and defined edge. In flood focus, the beam has a larger diameter, lower intensity, and softer edge.

"Spot" position

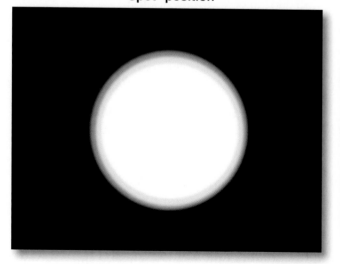

"Flood" position

Fresnel Lenses

Fresnel lenses are named for their inventor, the French engineer Augustin-Jean Fresnel ("Freñ-**nell**").

In a simple lens, called a plano-convex (p-c) or flat-curved lens, light enters the flat surface and exits the curved surface—refracted to create a beam of parallel light rays.

In 1822, a huge lens was needed for a new lighthouse called Cardovan Tower on the Gironde River. But, a p-c lens massive enough for a lighthouse would be impossibly heavy because of all the glass between its flat and curved surfaces.

A p-c lens (A) compared to a Fresnel lens (B) with the same size and optical qualities.

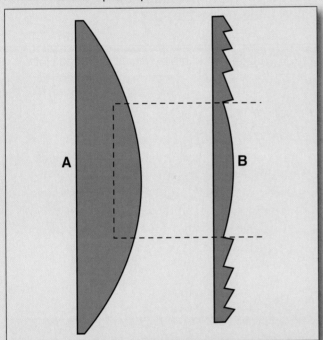

Fresnel realized that most of that interior glass was not needed because all the refraction was being done by the curved surface. So, he designed a lens with that surface divided into concentric rings and pancaked to reduce the amount of interior glass. Because the curvature of each ring is identical to the same area on a p-c lens, the light refraction is also identical (though the result is not quite as sharp). Lighthouses to this day employ Fresnel's innovation, usually in multiple lenses.

A multi-faced Fresnel lens in the lighthouse at McKinleystone.

(Sue Stinson)

The "F" is always capitalized to commemorate him.

Facing a similar problem, the theatrical lighting industry long ago adopted the concentric ring design and the Fresnel spotlight was born.

A Fresnel lens in a spotlight (behind a wire grid filter holder).

(Lowel Light Inc.)

Figure 14-19   Focusing a small-source light.

(Lowel Light Inc.)

the negative side, their intensity is less at any given part of the beam pattern and they are more difficult to control by masking. The most common floodlights are *broads* and *scoops*.

Broads are shallow rectangular pans that are small and portable, **Figure 14-20**. Because their light is moderately directional, they can be fitted with barn doors and frames to hold diffusion or filters.

Scoops are large bowl-shaped reflectors, used mainly in TV studios.

Softlights

Softlights include umbrellas, softboxes, and pans. Softlights are very large light

Figure 14-20   A "broad" floodlight.

(Lowel Light Inc.)

sources—ranging from one- to four-feet square—that deliver almost shadowless, directionless illumination. Softlights are easy to use because they do not throw distinct shadows. On the other hand, their light output is so soft-edged that it is difficult or nearly impossible to mask.

Umbrellas are really umbrella *frames* that can be fitted with different types of cloth, **Figure 14-21**. Some covers contain silver threads for greater reflectivity, and others are translucent white "silk." Typically, a light is mounted at the base of the umbrella, which is pointed at the subject. For additional diffusion, the white umbrella can be turned around so that the spotlight is shining through it at the subject, **Figure 14-22**.

Figure 14-21   An umbrella with silver-thread cloth.

(Lowel Light Inc.)

Softboxes are assemblies of fabric over wire frames, **Figure 14-23**. Some types enclose small lights, and others use proprietary lamps. Softboxes are about the size of umbrellas and throw similar light patterns. Some softboxes are hybrids: fabric boxes supported by umbrella structures, **Figure 14-24**.

Pans are very large light sources fitted with fluorescent tubes, **Figure 14-25**. When covered with diffusion material, pans are—at least theoretically—the softest lights available, because of their exceptionally large size. Also, their fluorescent lamps use less power and emit less heat than halogens.

Figure 14-22 A silk fabric umbrella can be reversed for even softer lighting.

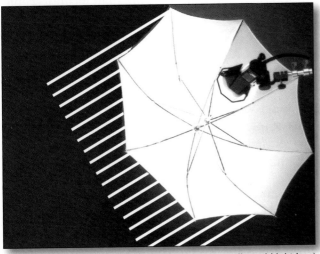

(Lowel Light Inc.)

On-Camera Lights

On-camera lights are miniature spotlights that clip onto camcorders, **Figure 14-26**. Some models are fitted with a diffusion disc and barn doors. Other types of on-camera lights have multiple heads and lamps, so that you can vary both their light intensity and beam spread.

Practicals

Practicals are lights that will be seen by viewers, like table lamps or wall sconces (**Figure 14-27**). In set lighting, you will sometimes replace the lamp in a working practical to adjust

Figure 14-23 A softbox.

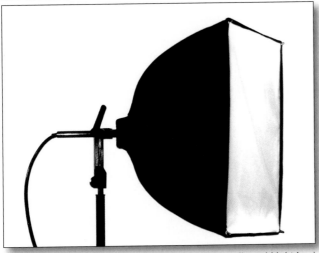

(Lowel Light Inc.)

Figure 14-24 An umbrella-framed softlight.

(Photoflex)

Figure 14-25 A fluorescent pan, with its external ballast functioning as counterweight.

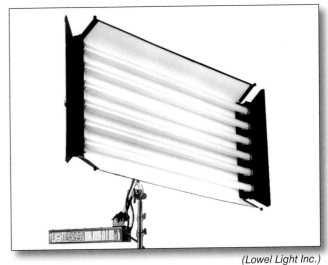

(Lowel Light Inc.)

Figure 14-26 A camera light with a barn door and diffusion disc that can be swung into the light path.

(Lowel Light Inc.)

Figure 14-27 The reading light is a practical.

its intensity or color temperature, or both. Practicals are easy to color-balance because 3200K halogen bulbs can be obtained with the medium screw-type bases used for household lighting.

The term "practical" comes from the theater, where on-stage lights that actors can actually turn on and off require special wiring that must be placed or removed with each scene change.

Lamps

The lightbulbs used in video lighting are called *lamps*, and there are four basic types: incandescent, halogen, fluorescent, and LED.

Incandescent Lamps

Incandescent lamps contain metal filaments glowing in a near-vacuum. They are universally found in household lighting. Ordinary incandescents have a color temperature (relative whiteness) of 2700K-2800K, while the special models sold for photographic lighting burn at 3200K—the white balance of a camcorder's "indoor" or "incandescent" setting. Older lighting instruments take incandescent lamps, but they have been largely replaced by halogen units.

Halogen Lamps

Halogen lamps contain metal filaments in a special mixture of gasses. All halogen lamps burn at 3200K. Unlike incandescent light, halogen

illumination does not grow more yellow as the lamps age. Halogen lamps are also smaller and emit more light, watt-for-watt, than incandescents, **Figure 14-28**. They are generally less expensive than 3200K incandescents, though more expensive than household lamps.

Halogen lamps have one potential safety problem: a tendency to shatter. For this reason, always follow precautions in working with them. Never use a halogen instrument that does not place the lamp behind a safety glass. Never touch a lamp with your fingers. Your natural skin oil can etch the quartz envelope of the lamp. When the lamp heats up, it can shatter at the etched area.

To change a halogen lamp, wait until it has cooled before removing it. Then, wrap the replacement lamp in a facial tissue or use the foam sheet that is usually packed with it. Some types of halogen lamps are enclosed in a second outer envelope that can be touched. Others are mounted within integral reflectors, which can also be handled safely (**Figure 14-29**).

Fluorescent Lamps

Most camcorders have white balance settings optimized for standard *fluorescent lamps*, and you can often obtain satisfactory video colors this way. But if you use fluorescents as movie lights, you will want to obtain specialized tubes that produce more pleasing color. When mixing fluorescent and halogen lights, use tubes rated at 3000K. If your whole lighting setup

Figure 14-28 A professional halogen lamp.

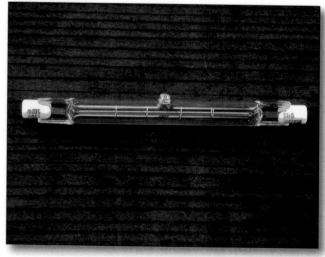

Figure 14-29 A household reflector-style halogen lamp.

is fluorescent, you may get better results with "sunshine" fluorescent lights, whose 5000K color temperature closely matches that of direct sunlight (**Figure 14-30**).

When choosing fluorescent tubes, note that some manufacturers provide a "Color Rendering Index" (CRI), which is a scale from 1–100 that rates the ability of a fluorescent tube to deliver natural-looking color. For example, inexpensive "shop light" tubes may have an index rating of 67. Better quality lamps have index ratings in the 70's. The best sunlight tubes have a CRI rating over 90. For more consistent output, make sure that all your tubes are the same make and model, and all are replaced at the same time.

Figure 14-30 When the available light is fluorescent, it is often better to supplement it with fluorescent video lights.

(Lowel Light Inc.)

Small spiral-design ("curly") fluorescent lamps intended for video lighting are also available. Fitted with medium screw bases like household bulbs, they can be used as practicals or ganged in special lighting instruments (**Figure 14-31**).

LEDs

LED (light-emitting diode) lamps are very small, specialized semiconductors that are common in flashlights. Because single LEDs are not bright enough for video lighting, they are typically ganged together in "arrays" of up to several hundred units. These LED arrays are exceptionally light and compact, **Figure 14-32**, and they consume relatively little power.

Figure 14-31 This compact scoop light uses three 26-watt fluorescent lamps.

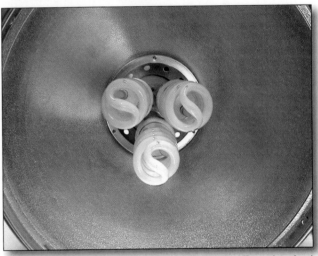

(Equipment Emporium, Inc.)

Figure 14-32 An LED camcorder light.

(Litepanels Inc.)

LED Lighting

LED (light-emitting diode) floodlights have been widely adopted for video lighting. Like broads and scoops, their output is somewhat directional, and because of their large size, they can also work as soft lights when fitted with plastic diffusion. In fact, LEDs can replace every type of lighting instrument except spotlights. Though they cannot serve every lighting need, LED panels rival fluorescent and incandescent lights in versatility and convenience. As for energy consumption, they are the most efficient lights available by far.

A battery-powered LED array lights an interview.

(Litepanels Inc.)

Power Consumption

A one foot-square panel draws 40 watts, with light output as great as a 500 watt incandescent. That is, the LED unit uses less than 13% of the energy required by a comparable halogen lamp. Even a fluorescent array needs 160 watts—four times as much as the LED.

The snap-on battery pack on the back of this one foot panel provides power for 1.75 hours.

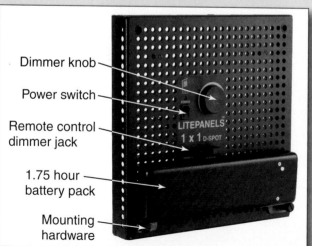

- Dimmer knob
- Power switch
- Remote control dimmer jack
- 1.75 hour battery pack
- Mounting hardware

LITEPANELS
1 x 1 D-SPOT

(Litepanels Inc.)

Versatility

Units of this type can be dimmed continuously from 100% to 0%, with minimal shift in color temperature. (Halogen lights grow more yellow as they dim, and dimming fluorescent units is complicated at best.)

LED arrays are available in either daylight or indoor versions. Daylight models can be converted to indoor use with orange filters. One model is switchable between indoor and outdoor color temperatures. Another version can be focused like a spotlight (though its more diffuse light is not as controllable).

Some models are small enough to mount directly on cameras. These are fitted with flip-down frames that can hold diffusion or orange filters.

The small LED light units mounted on cameras include flip-down frames for filters.

(Litepanels Inc.)

The foot-square panels can be clipped together into 2 × 2 or 4 × 4 arrays, to provide even larger lights sources that are useful for green screen and other applications that need extremely soft, diffuse light. Foot-square units can be fitted with grids to help control light scatter.

These 16 foot-square panels are mounted on a roll-around stand. The light output from 640 watts of power equals that of an 8,000 watt incandescent unit.

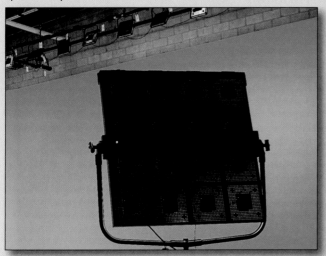

(Litepanels Inc.)

Convenience

Unlike halogen lights, LEDs produce almost no heat, so they are safer to handle and more comfortable for subjects in interviews and similar situations. They are also quite light—the smallest unit weighs four ounces and the foot-square models weigh just three pounds. This means that the two three-pound key lights and two 10-ounce back lights required for an interview would weigh less than eight pounds, exclusive of stands and accessories. (They would also use less power than a 100-watt lightbulb, and could run on clip-on battery packs for 90 minutes.)

Drawbacks

There is no one type of light that excels in every category. As noted, LEDs are no substitute for spotlights (especially models with Fresnel lenses). They are among the more costly professional lights available. On the other hand, their working life (50,000 hours) and ultra low-power consumption can more than recoup the higher initial cost of LEDs.

Chapter 16, *Lighting Applications* covers some of the situations in which LED units are particularly useful.

Lighting Accessories

Lighting accessories, such as flags and filters, have already been mentioned in passing. Following is a more extensive survey of the items that commonly complete a lighting director's kit.

Accessories for Lighting Instruments

The beams from small source lights (spotlights and narrow floods) can be modified by accessories that fit on the front of the instrument to mask, reduce, or diffuse the light.

Masking

The highly directional beams of spotlights can be masked to produce hard-edged cutoff patterns. Masking accessories include barn doors and flags.

Barn doors are pairs of metal flaps hinged to the sides and/or top and bottom edges of the lamp. Their front edges are moved into and out of the light path to control the beam edge, **Figure 14-33**.

Flags, as noted earlier, are opaque cards—usually mounted on stands—that block part of the light beam, **Figure 14-34**.

Figure 14-33 All four barn doors on this spotlight can be shaped for better control.

(Lowel Light Inc.)

Figure 14-34 Small flags on movable arms can be positioned more flexibly than barn doors.

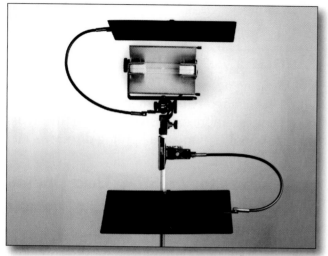

(Lowel Light Inc.)

As a rule, the farther a masking device is placed from the light, the harder-edged its shadow will be. Barn doors attached to the light throw softer shadows than flags mounted on century stands placed closer to the subjects.

Light Reduction

Spotlights can be fitted with circular frames covered with nets (actually made of screen wire) to cut down the light intensity, **Figure 14-35**. The most common nets are:

- **Singles:** One-layer nets for moderate light reduction.

- **Doubles:** Two-layer nets for stronger light reduction.
- **Halfs:** Frames with one-half netted and the other half clear. By rotating a half net, you can selectively reduce the illumination on just part of the area lit by the beam.

Half nets are also supplied in one- and two-layer versions. So, a "half double" is a two-layered half-circle net.

Diffusion

All light beams can be softened by placing *diffusion* material in front of them. Like filters, circles of milk-white plastic can be installed in metal rings and placed in holders at the front of spotlights. For a more diffuse effect, white "spun glass" sheeting can be secured to the edges of barn doors or hung in front of lights, **Figure 14-36**.

Umbrellas and softboxes seldom benefit from additional diffusion, but large pans are often covered with spun glass. White cloth sheeting is especially effective on fluorescent softlights, since the lights emit very little heat.

Color

Lighting instruments are sometimes gelled with colored plastic to simulate firelight, moonlight, or other effects, or simply to warm or cool the light on an area. Earlier, we noted

Figure 14-35 This graduated screen uses 2, 1, and 0 layers of net.

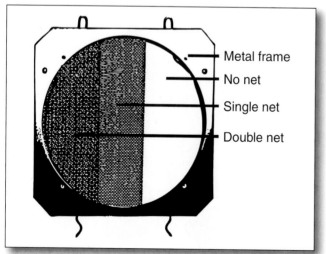

Metal frame

No net

Single net

Double net

(Lowel Light Inc.)

Figure 14-36 Clamps or clothespins can be used to clip filters or diffusion to barn doors.

(Lowel Light Inc.)

the use of orange filtration to convert daylight to incandescent. Similarly, blue filters are often placed in front of halogen lights to convert their color temperature to daylight, **Figure 14-37**. (Usually, these blue filters are coded number 80.)

Other Lighting Accessories

In addition to accessories placed on or in front of the lighting instruments, a wide variety of other lighting tools are available. The following is a summary of the more common accessories.

Century Stands

A *century stand* ("C-stand" for short) has folding legs, a telescoping central column, and a universal clamp that will grip flat items or pipe arms (**Figure 14-38**). Each is supplied with one pipe arm, ending in a second universal clamp. Century stands will hold lighting instruments, flags, reflectors, diffusion, cloth drapes, or just about anything else. They are, perhaps, the most versatile accessory in the lighting director's kit, and large productions may carry dozens of them.

C-stands are widely obtainable from Internet vendors.

Cookies

Specially cut sheets called *cookies* are placed in front of spotlights to create patterns on walls and other surfaces. Cookies can simulate leaves, bare tree branches, Venetian blinds, or any other

Figure 14-38 A century stand will support any lightweight object, in any position.

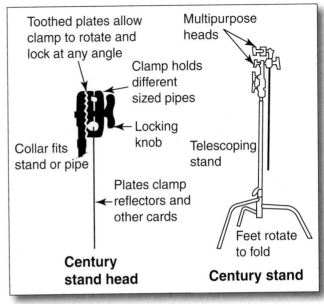

patterned required, **Figure 14-39**. Cookies are covered in detail in Chapter 15, *Lighting Design*.

Mounts

Light stands are the most common accessories for supporting lights. To illuminate larger areas, horizontal overhead lighting pipes spanning the center of a set may be supported by heavy stands and serve as mounts for hanging several instruments. Lightweight instruments can be attached to walls with gaffer tape, and small units can be clamped almost anywhere, **Figure 14-40**.

Figure 14-37 A number 80 blue filter converts halogen color temperature to daylight.

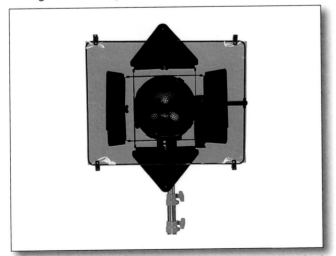

(Lowel Light Inc.)

Figure 14-39 For demonstration purposes, this spotlight carries a variety of special-purpose cookies.

(Lowel Light Inc.)

Figure 14-40 Mounting lights.

Overhead and stand mounts

Taped to a wall

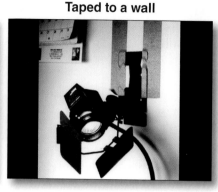

Clamped to a stable object

(Lowel Light Inc.)

Electrical Accessories

Certain accessories are considered essential in professional lighting.

Cables

For location lighting, extension cables are essential, **Figure 14-41**. Make sure that your cables are safe:

- Wire gauge should be no thinner than no. 14; no. 12, or even no. 10 is better. (The lower the number, the thicker the wire.)
- Extensions should not exceed 25′ in length.

Figure 14-41 An appropriate cable for location lighting.

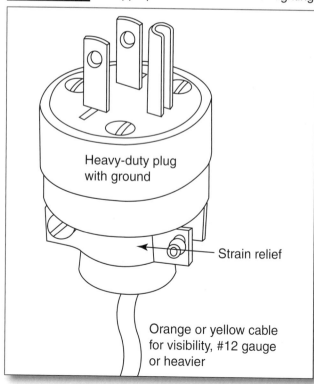

Heavy-duty plug with ground

Strain relief

Orange or yellow cable for visibility, #12 gauge or heavier

- Use single plugs rather than multiple outlet boxes to avoid overloading circuits.
- Use high-visibility yellow or orange cables for safety.
- Tape loose cables to the floor for additional safety. Tangled in a foot, a cable can topple a light on a stand, with disagreeable results.

Dimmers

A dimmer can be used to reduce the output of a light. Dimming an incandescent light warms up its color temperature, so dimming can be used to adjust color as well as output.

Gloves

Leather gloves are a must around hot lights. Find the heaviest industrial grade gloves you can, because incandescent lighting instruments can heat up to hundreds of degrees.

Gaffer Tape

Gaffer tape looks just like ordinary duct tape, but costs five times as much and works five times as well. It is sticky enough to tape lightweight instruments to walls and other surfaces. At the end of a shoot, it can be peeled from the wall (with patience) without lifting paint. Gaffer tape is available from video supply companies. Because these vendors are uncommon outside large production centers, gaffer tape and other specialized accessories are most easily found online.

Gaffer tape is a special high-performance version of duct tape. It was originally offered for lighting use in 1959 by Ross Lowel, the founder of Lowel-Light Manufacturing, Inc.

Lighting Evaluation Tools

In addition to an educated eye, two tools are useful for assessing a lighting design: a monitor and a light meter. Meters can be indispensable in certain situations, and a good monitor is essential for almost *all* types of shooting.

Monitors

No professional videographer likes to work without a reference monitor—a high-quality display that is calibrated to show exactly what the camcorder is recording, **Figure 14-42**.

A good monitor delivers "what you see is what you get" information about the lighting. With it, you can light a set using basic common sense: simply check the effect on the screen as you work and adjust the lighting until you are satisfied with the image.

Light Meters

Even with a reference monitor, a light meter is a useful measurement tool. You can use it to regulate contrast by checking the relative brightness of different image areas and adjusting light levels until the meter indicates the desired contrast ratio.

You can also use a light meter for checking overall brightness levels in order to calculate the f-stop at which your lens is working. This helps you determine whether to use neutral density filters (to reduce light) or gain control (to amplify low-light signals).

Incident meters, like the studio model in **Figure 14-43**, are aimed at the lights and measure illumination directly. *Reflective meters* are aimed at the subject and measure the light that is reflected from the subject to the camera lens. *Color temperature meters* precisely measure the relative blueness or redness of "white" light at any point in a scene. This information is useful if you are working with a mixture of light sources, so that you can decide which ones to adjust for color, and how much adjustment to make.

Some lighting directors like the precision afforded by light meters, while others seldom use them, except in specialized and/or very difficult lighting situations.

Inexpensive Lighting Equipment

Although professional lighting equipment is generally worth the money it costs, you can sometimes buy (or build) less expensive alternatives. This section covers the more common items in a "guerrilla lighting" kit.

Lighting Instruments

Video lighting benefits from several facts:
- All halogen lighting has a color temperature of 3200K, whether intended for video use or not.
- Better quality fluorescent tubes can simulate outdoor light with up to 90% accuracy.

Figure 14-42 A professional reference monitor.

(JVC Corporation)

Figure 14-43 This incident light meter is calibrated in "foot candle" units.

(Sekonic Corporation)

- Both halogen and fluorescent lamps are available with standard medium screw bases for use in any household fixture.

In effect, this means that your choice of lights is limited only by your imagination. The following are a few suggestions.

Halogen Work Lights

Halogen work lights are available at hardware and building supply stores at reasonable prices, **Figure 14-44**. Intended for light industrial use, these fixtures meet all code requirements for lamp safety, cord protection, and switches. Depending on your needs, you can select wattages from 150 to 1,000, single or twin heads, heads fitted with separately switchable lamps, and lights on telescoping floor stands, short legs, or clamps.

When shopping for work lights, look for these desirable features:

- Long support columns—some units stand only about five feet high.
- Handles for moving the heads (to avoid touching the hot units).
- Heads that swivel up and down, as well as back and forth.

For convenience, you may wish to install a switch on the power cord, so you can turn the lights on and off without reaching up to the heads.

Single work lights on spring clamps are useful for backlighting. Wrap the clamp jaws with duct tape (or dip them in liquid plastic) to avoid marring surfaces.

| **Figure 14-44** | A halogen work light. |

Reflector Scoops

Lightweight scoops fitted with clamps make useful instruments for screw-based halogen or fluorescent lamps. Though relatively flimsy, their inexpensive cost allows for easy replacement, and their very light weight makes them ideal for clamping to door moldings and ceiling grids.

Household Reflector Lamps

Halogen lamps with built-in reflectors (designated "R" or "PAR" lamps) are available in a wide variety of diameters and wattages. Where beam edge control is not important, they can be used as crude, but effective, spotlights.

Light Pans

If you are handy at construction, you can mount several individually switched fluorescent fixtures in shallow plywood trays to construct very effective pan lights that use less power and emit far less heat than halogens. Be sure to buy units with better quality ballast, since "shop light" fixtures may not adequately power the high-accuracy fluorescent tubes required for good color temperature.

Reflectors

Reflectors are easily constructed by taping aluminum foil over foam board sheets or light plywood rectangles (24" × 36" sheets is a good size). To create the hardest surface with the longest throw and brightest light, use spray adhesive to glue the foil, shiny side up, to the board. For a slightly softer surface, place the matte side of the foil up instead. For a still more diffuse light, ball up the foil lightly and then spread it flat again before gluing. This breaks up the surface into thousands of tiny facets that refract the light and diffuse the overall reflection, **Figure 14-45**.

For soft fill effects, any white cardboard will do, but white foam board is more rigid. For greater permanence, invest in one-inch board, rather than the thinner type. If you occasionally need to warm up the reflected light, choose a board with gold paper on one side.

The most common type of reflector combines a hard surface (shiny or matte aluminum) on one side with a soft surface (faceted aluminum or white) on the other side.

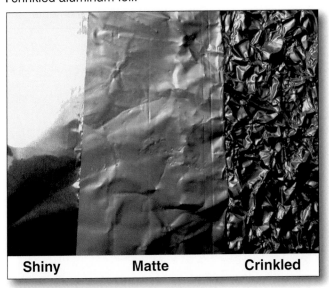

Figure 14-45 The surfaces of shiny, matte, and crinkled aluminum foil.

Shiny	Matte	Crinkled

Folding auto windshield sunshades made of fabric-covered hoops are inexpensive. They are available in both silver and white.

Flags

Foam board can be used for flags, and plywood reflectors can also be used to mask off unwanted light. However, the most versatile flag shape is a long, slim rectangle of thin plywood. (Plywood holds up better than foam board when clamped in a century stand.) Flags should be painted black, to minimize light bouncing off them, **Figure 14-46**.

Figure 14-46 A typical flag.

![A typical flag diagram showing dimensions 15", 30", 1/4" plywood painted black, Post for clamping in century stand]

Screens and Silks

Screens reduce the intensity of light, and silks diffuse light. Hand-held models are inexpensive enough to buy, and you can find them at most larger camera stores. Stretched on wire hoops, these models fold into small areas for storage and transport. For larger models, you will need to construct a mounting frame. **Figure 14-47** diagrams a frame of plastic pipe that disassembles for storage.

Old bedsheets make excellent silks. For screens, plant nurseries and garden supply stores carry black plastic netting in large rolls. You can select the density of the screening, or double-layer lighter screens to block more light.

Weights

The great enemy of outdoor lighting control is wind, especially with screens and large reflectors. To hold down C-stands or screening frames, weights are essential accessories. Sandbags are cheap and effective weights, but they are heavy, dirty, and tend to leak. For this reason, some production companies use weights filled with water instead. By filling and emptying the bags at the location, they avoid transporting most of the weight.

Commercial water weights are available from production supply houses, but the collapsible plastic water bags sold for camping

Figure 14-47 A frame made of inexpensive plastic pipe.

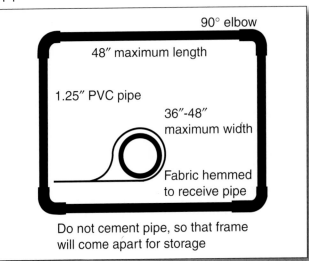

90° elbow

48″ maximum length

1.25″ PVC pipe

36″-48″ maximum width

Fabric hemmed to receive pipe

Do not cement pipe, so that frame will come apart for storage

are inexpensive and work just as well. They are typically supplied with handles and hooks, to help hang them on light and reflector stands. If you need to improvise, one-gallon drinking water bottles are available at any supermarket at very low cost. They can be secured to stand bases with duct tape or bungee cords, **Figure 14-48**.

Reference Monitor

Professional reference monitors are expensive (**Figure 14-49**), but very good 13″ color TV sets are widely available for much less. Many will operate on wall power, 12-volt auto plug, or batteries. Make sure the set accepts composite video input (identifiable as a yellow jack for an RCA plug).

Designing with Lighting Tools

This chapter has covered most of the lighting equipment used for location and small studio shooting. The following chapter explains how this equipment is used to create lighting designs, and the final lighting chapter shows how to employ these tools in a variety of real-world lighting situations.

Figure 14-48 Water bags and plastic drinking water bottles make excellent weights.

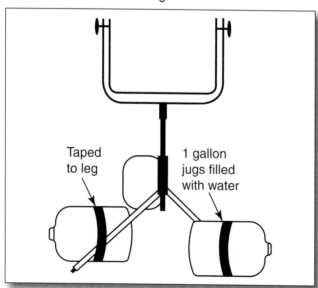

Taped to leg

1 gallon jugs filled with water

Figure 14-49 Professional quality flat panel reference monitors are easily carried, and snap-on batteries can supply their low power requirements.

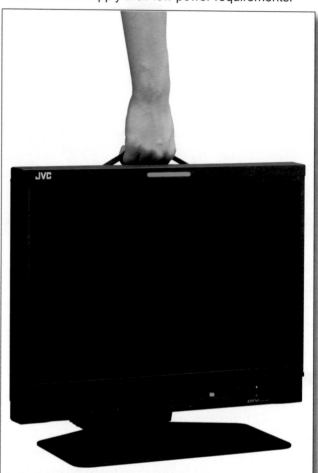

Snap-on battery pack

(JVC Corporation)

Summary

- Video lighting is the art of creating images with light.
- Available light is whatever illumination already exists at a location.
- The principal tools used to reduce light quantity are flags, screens, silks, and filters.
- Reflectors provide control of the direction of light.
- Filters placed directly on the camera lens change the color of all the light entering the camera. Sheet filters affect only the light sources that they are covering.
- A video light is called a lighting instrument, and the light source inside is the lamp.
- Most lighting instruments can be categorized as spotlights, floodlights, softlights, on-camera lights, and practicals.
- The different types of lamps used in video lighting produce light in different ways with different light characteristics.
- Various accessories are necessary with lighting instruments to adjust the light, support the instrument, power the equipment, and evaluate the light and resulting image.

Technical Terms

Available light: The natural and/or artificial light that already exists at a location.

Barn doors: Metal flaps in sets of two or four, attached to the front of a spotlight to control the edges of the light beam.

Broad: Small, portable floodlight in the form of a shallow rectangular pan.

Century stand: A telescoping floor stand fitted with a clamp and usually an adjustable arm for supporting lights and accessories. Commonly referred to as a *C-stand*.

Color temperature meter: A light meter that measures the relative blueness or redness of nominally "white" light.

Cookie: A sheet cut into a specific pattern and placed in the beam of a light to throw distinctive shadows, such as leaves or blinds.

Diffusion: White spun glass or plastic sheeting placed in the light path to soften and disperse it.

Flag: A flat piece of opaque metal, wood, or foam board that is placed to mask off part of a light beam.

Floodlight: A large-source instrument that lacks a lens, but is still moderately directional; used for lighting wide areas.

Fluorescent lamp: A lamp that emits light from the electrically charged gasses it contains.

Halogen lamp: A lamp with a filament and halogen gas enclosed in an envelope of transparent quartz.

Incandescent lamp: A lamp with a filament enclosed, in a near-vacuum, by a glass envelope. Also called a *bulb*.

Incident meter: A light meter that measures illumination as it comes from the light sources.

Instrument: A unit of lighting hardware, such as a spotlight or floodlight.

Lamp: The actual bulb in a lighting instrument.

LED (light-emitting diode): A cool, low-power light source arranged in panels that are used like flood or soft lights.

On-camera light: A small light mounted on the camera to provide foreground fill.

Practical: A lighting instrument that is included in shots and may be operated by the actors.

Reflective meter: A light meter that measures illumination as it bounces off the subjects and into the camera lens.

Reflector: A large silver, white, or colored flat surface used to bounce light onto a subject or scene.

Scoop: A large bowl-shaped floodlight used mainly in TV studios.

Screen: A mesh material that reduces light intensity without markedly changing its character.

Sheet filter: In lighting, a sheet of colored or gray-tinted plastic placed over lights or windows to modify their light. Also called a *filtering gel*.

Silk: A fabric material that reduces both light intensity and directionality, producing a soft, directionless illumination.

Softlight: A lamp or small light enclosed in a large fabric box, which greatly diffuses the light.

Spotlight: A small-source lighting instrument that produces a narrow, hard-edged light pattern. Also called a *spot*.

Umbrella: A fabric-covered umbrella frame. Metallic cloth umbrellas are used to reflect light onto subjects; white cloth models can also filter lights placed in back of them.

Review Questions

Answer the following questions on a separate piece of paper. Do not write in this book.

1. Illumination that already exists at a shooting location is _____.
2. *True or False?* Flags are tools used to reduce light quantity.
3. What are *silks*? How are they used to control light?
4. Identify the four main types of reflectors.
5. How are filters used to control light?
6. Explain the relationship between an *instrument* and a *lamp*.
7. Broads and scoops are types of _____.
8. What are *softlights*?
9. What are the four basic types of lightbulbs used in video lighting? Explain the unique characteristics of each.
10. *True or False?* The farther a masking device is placed from the light, the harder-edged is its shadow.
11. _____ material is placed in the light path to soften and disperse the light.
12. What are C-stands used for?

13. *True or False?* Dimming a light cools down its color temperature.
14. Which type of light meter is aimed at the lights to measure illumination directly?
15. How are weights used on a production site?

STEM and Academic Activities

1. **Science.** What is the *frequency* of color? What is the unit of measurement used to indicate color frequency?
2. **Technology.** The use of LED lights has evolved from indicator lights to general lighting applications to video display. How do LEDs create light? What are OLEDs?
3. **Social Science.** Fresnel lenses provided a great advancement in the construction of lighthouses, but what is the first known lighthouse? Where did it operate? How did it project its beam of light?

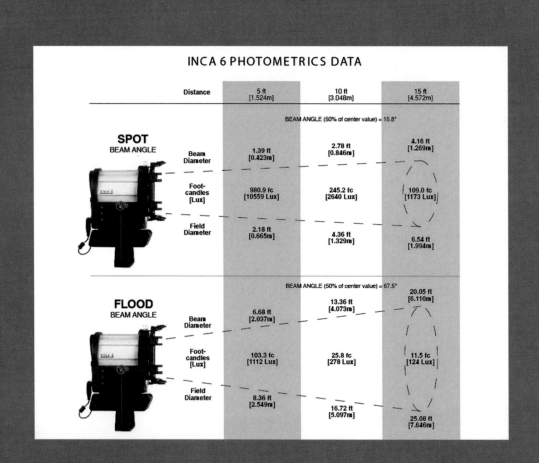

Litepanels Inca 6 LED Fresnel light specifications

Size: 11 x 10 x 15 inches.
Weight: 9 pounds (approx).
Maximum power draw: 104 watts.
Power requirements: 14-18VDC (battery)
Or
100-240VAC (cabled).

(LitePanels Inc.)

(Photoflex)

Objectives

After studying this chapter, you will be able to:

- Recall the techniques for adjusting the quantity, contrast, and color of video lighting.
- Explain the unique characteristics of the basic lighting styles.
- Understand how various lighting strategies are used to meet the goals of the lighting design.
- Recall the procedures and technique for working with lighting equipment.

The Development of Video Lighting

Historically, film and TV scenes have been lit quite differently. Since video is a hybrid of both older technologies, it takes something from each and adds techniques of its own.

Film Lighting

Despite its many advantages, film has historically suffered from two major lighting problems: its relative insensitivity to light, and its need to be processed and printed for viewing. The relatively slow speed (light sensitivity) of film meant that sets needed to be lit with many lighting instruments that consumed considerable power. Although today's color film is more sensitive than it was when film lighting techniques were developed, these techniques are still widely practiced, so that film lighting tends to be a relatively elaborate procedure.

Because film must be processed and printed before it can be viewed, it is impossible to see the precise effect of film lighting during shooting.

On most film productions today, the camera images are viewed on video monitors. Unlike reference monitors for video shooting, however, these displays cannot show the lighting exactly as it will look on film.

To overcome this inconvenience, professional cinematographers and lighting gaffers become experts at predicting how their lighting designs will look on film. To help in this process, they use specialized light meters to measure incident and reflected light levels at important points in the scene. Nonetheless, even with these sophisticated metering instruments and today's sensitive film stocks, film lighting is an art that can take years of practice to master.

The **gaffer** is the head of the lighting crew on a shoot. The term is derived from an old British word meaning "grandfather," or simply "old man."

Television Lighting

Studio-based television also presents lighting problems. For many years, the tube-type cameras employed could not handle *contrast* competently. This meant that the darkest parts of the image had to be no less than about one-third as bright as the lightest parts. Moreover, television programs are recorded more or less continuously by multiple cameras shooting from different directions and often covering different parts of the set. For this reason, every subject, place, action, and camera setup must be lit at the same time. Although technical progress has allowed television studio lighting to improve substantially, the demands of low-contrast lighting and full-set coverage can still make television lighting more flat and bland than film lighting.

TV studios use low-contrast lighting.

(Corel)

Video Lighting

Video is a true hybrid—a medium that combines the shooting techniques of film with the recording technology of television. Cameras fitted with modern imaging chips are more sensitive to light than many standard film stocks, and accept much wider contrast ratios than classic television cameras. Like TV lighting directors, cinematographers can see exactly what their images look like as they are setting them up and shooting them. For this reason, learning to light for video is easier than mastering film lighting. In effect, if it looks good on a quality reference monitor, it *is* good.

Perhaps the fairest assessment might be that today, all three visual media can support sophisticated lighting. While film and television continue using traditional methods with technical improvements, video lighting has evolved its own approach.

About Lighting Design

The previous chapter shows what lighting tools can do, and the next chapter puts them to work. Between them, this chapter explains the ideas behind video lighting design—the basic *styles*, *principles*, and *procedures* that apply to all lighting assignments. With these design fundamentals reviewed and summarized, we will be ready to move on to the final lighting chapter, and apply them to typical real-world lighting applications.

Lighting Standards

Whatever else it may add to a production, professional video lighting begins by delivering high-quality images. Without this technical quality, the subtler contributions of style and mood would be beside the point. To achieve high technical standards, images must be created with light of the proper *quantity*, *contrast*, and *color* (**Figure 15-1**).

Light Quantity

The camera's imaging chip (or chips) must receive exactly the right quantity of light in order to form a high-quality image. Too much light makes colors smear and "bloom," and forces the aperture to very small f-stops (openings). Too little light makes a murky picture with dull colors and little detail, and may push the electronic gain compensation up to a level that creates grainy, noisy images. The first task, then, is to decrease or increase the quantity of light to suit the needs of the camera.

The cinematography chapters (Chapter 12, *Camera Systems* and Chapter 13, *Camera Operation*) show how to adjust the camera to suit the light. This chapter shows just the opposite: how to adjust the light to suit the camera.

Decreasing the Quantity

Outdoors, you often need to control the light by reducing its quantity (intensity). To do this, you can either block or screen it, **Figure 15-2**.

Figure 15-1 To make a good image, the light must have the right…

…quantity…   **…contrast…**   **…and color.**

(Adobe)

Figure 15-2 Decreasing light quantity.

Direct sunlight   **Blocked by a flag**   **Filtered by a screen**

To *block* light entirely, you place an opaque shield, such as a lighting flag, between the light source and the subject. This technique works when there is enough light from other sources to illuminate the blocked area of the subject.

To *screen* the light instead of blocking it, you place a large framed screen of mesh or cloth between the direct sunlight and the subject.

For consistency, some illustrations are repeated in these three lighting chapters.

Increasing the Quantity

Often, the problem is not too much light but too little, so your job is to increase the quantity of illumination. You have two options:

- Outdoors, you generally redirect the light, making double use of the illumination available.
- Indoors, you typically add lighting to the available illumination.

To *redirect* light that is already present, you use reflectors. Reflectors are quick, simple solutions for outdoor lighting especially (**Figure 15-3**), and they can be used indoors as well.

To *add* light to the available illumination means using one or more lighting instruments. Video lighting can be as simple as an on-camera fill light, or as elaborate as a set lit with 20 instruments. In professional video work, almost all interior locations are lit for shooting, sometimes by supplementing the light already available at the location.

Light Contrast

After quantity, the biggest lighting problem is contrast. Contrast is the difference in brightness between the lightest and darkest areas in the image.

Lighting contrast is expressed as a ratio. A "four-to-one" ratio indicates that the lightest areas are four times as bright as the darkest ones.

Too little contrast makes subjects appear flat and bland. Too much contrast makes light areas appear stark white, or dark areas solid black, or (in extreme cases) both at once, **Figure 15-4**. Regardless of the lighting mood or style, a well-lit image has an even range of brightness levels:

- True black in the darkest areas.
- Shadows with details visible in them.
- Several gradations of brightness in the middle range.
- Highlights that retain some detail.
- True white in the lightest areas.

Light Color

To control the quality of light, you need to manage not only its brightness and contrast, but also its color. Commonly, this means setting and maintaining white balance. There are special situations, however, in which you will want to exercise highly selective control over light color.

White balance is a function of color temperature, which is discussed in Chapter 12, *Camera Systems*.

 Figure 15-3 Using a reflector.

Direct sunlight

Shadow filled with a reflector

Estimating Contrast

The human eye is not usually reliable in judging contrast for video because we can distinguish brightness levels exceeding a ratio of 100-to-1. No video (or film) recording is anywhere near that discriminating. For this reason, it is important to check contrast on a reference monitor.

If your production level is too basic to include a monitor, you can use your camera's external screen, well-shielded from ambient light. As noted elsewhere, you can obtain inexpensive cloth hoods for these small monitors.

If you are checking a location even before bringing a camera, you can still guesstimate the contrast produced by the available light by using an old photographer's trick:
1. Look at the scene to be recorded.
2. Close one eye completely.
3. Squint the other eye until it is almost shut and your eyelashes form a sort of filter.

4. Look at the darkest parts of the scene. If you cannot make out details in the shadow areas, the contrast is too great.

Though this improvisation is no substitute for an accurate monitor, it can give you a rough approximation of the scene contrast as the camera will record it.

As described previously, you control contrast by adjusting the light quantity—either lightening the dark areas by filling them in with more light, or darkening the bright areas by screening them. You can also reduce contrast by diffusing the light—scattering the straight-line rays of a light source so that they bounce around the subject and soften or eliminate the shadows. You can diffuse direct light by dispersing it through cloth or plastic sheeting. Reflected light can be diffused by using white reflectors instead of metallic ones. Lighting instruments can be fitted with plastic, cloth, or spun glass diffusion material to scatter the light.

For example, many locations require mixing light sources of different color temperatures, whether combining incandescent and fluorescent lights or managing daylight coming through a window (**Figure 15-5**). In other situations, you may want to suggest the warmth of a sunset or fireplace. Or, you may need to simulate cool moonlight to make daytime footage look like night. Whether creating special effects, balancing multiple color tints, or simply achieving good white balance, you are always dealing with the color of light.

Recording a Standard Image

In managing the quantity, contrast, and color of light, your goal is to record a "standard" video image: an image in which both shadows and highlights retain full detail, and apparent "white" remains the same from shot to shot. To put it another way, you want to record an image that does not need correcting in editing.

Before digital postproduction, many effects were best created in the camera. For example, outdoor "night" scenes shot in daylight were underexposed, with the white balance purposely

Figure 15-4 Lighting contrast.

Contrast is too great

Contrast is too small

Contrast is just right

(Sue Stinson)

Figure 15-5 The window light coming from the right is bluer than the supplementary movie lights.

set to incandescent in order to give all the footage a bluish, "moonlight" cast.

Today, however, you can control image characteristics in the edit bay, so it is best to start with a neutral, high-quality image. If you do not like an effect added in editing, you can easily return to the original footage and try again. If you created that effect in the camera, however, you are stuck with whatever you recorded, like it or not. Moonlit nights, sunsets, and other overall color tints are easy to add during editing. Contrast and brightness can be adjusted to change day shots to night.

This rule is even more important with certain camera effects that cannot be undone by digital postproduction.

On the other hand, if you have colored light on just *one part* of your scene, it may well be easier to create the effect while shooting than to add it in postproduction. For example, the flashing red light of an off-screen "police car" colors only the left side of the subject in **Figure 15-6**.

Lighting Character

If you master the technical aspects of good lighting, your images may be well-illuminated, attractively colored, and pleasing to look at, but they may not contribute much to the program. To pull its weight artistically, lighting must also have character. Lighting character is the set of traits that individualizes the look of a scene by giving it a distinctive *style*.

Lighting Styles

Over the centuries, painters, stage designers, photographers, and cinematographers have evolved many approaches to lighting. Though there is no standard classification, we may distinguish several basic lighting styles and label them *naturalism*, *realism*, *pictorial realism*, *magic realism*, and *expressionism*.

Figure 15-6 Simulating a police car's red light. The intermittent red glow on the subject's face appears to be coming from an off-screen police car. In fact, a grip is holding a revolving light.

A

B

(Continued)

Figure 15-6 *(Continued)*

C

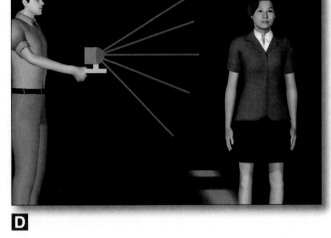

D

Although these terms are not universally agreed upon, they are common in critical writings about art, theater, film, and photography.

Classifying lighting styles is not just a theoretical exercise. By choosing a basic approach to lighting and then customizing it to your personal taste, you can give your cinematography a consistent "look and feel," a distinctive quality that enhances its effectiveness and increases its professionalism.

Naturalism

Naturalism is the lighting style that most closely imitates real-world conditions. A naturalistic scene does not look "lit." Instead, a naturalistic image makes you feel that you are looking directly at the scene rather than at an image of it on a screen. Naturalistic lighting uses available light as much as possible, and does not light anything to achieve an obvious effect (**Figure 15-7**). Naturalism is a useful style if you want your program to have the look of a straightforward documentary.

Realism

Realism is naturalism heightened and intensified somewhat to add dramatic effect. Where naturalism records actual lighting conditions as far as possible, realism *recreates*

actual-looking conditions through lighting. The average viewer may not notice it as lighting, but it is visible to an informed eye (**Figure 15-8**). Realism is the most common style for TV programs, whether police dramas shot on location or comedies recorded in a studio. If you make program videos, you will probably work most often in the realistic style.

Pictorial Realism

Pictorial realism is a "painterly" style, in which the lighting effects are frankly visible, although they still imitate real-world lighting, such as lamps, streetlights, headlights, sunlight,

Figure 15-7 Naturalistic lighting.

(Photoflex)

Figure 15-8 Although this room seems lit by natural daylight, the "hot spots" in the upper right reveal the lights above the frame border.

(Sue Stinson)

and moonlight (**Figure 15-9**). Through most of Hollywood's history, until the 1960s, pictorial realism was the most popular lighting style. The simpler alternative of plain realism was largely confined to exterior shooting, where the lighting equipment and techniques needed for pictorial realism were often impractical.

At the height of the silent film period, "Rembrandt lighting" was highly regarded. During the era of classic studio sound films, an influential textbook on cinematography was titled *Painting with Light*.

Although pictorial realism sometimes looks artificial to modern viewers, it was accepted

Figure 15-9 The dramatic modeling and strong rim light are characteristic of pictorial realism.

(Corel)

as realistic by the original audiences of classic Hollywood films. Today, pictorial realism is often used in commercials and music videos, where a purposely theatrical feel is required.

Magic Realism

Digital video image processing has led to a hybrid style that combines elements of realism, pictorial realism, and expressionism. Because the overall look is superficially realistic, but with a heightened, intensified quality, we may borrow a literary term and name it *magic realism*.

Though not always manipulated by computer, images in magic realism typically look processed: colors may appear slightly more intense or more uniform than normal.

The hotel interiors in Stanley Kubrick's film *The Shining* are lit in the magic realism style.

Different picture components may be sharp or blurred in ways that do not match the focusing characteristics of lenses. Perspective may be skewed. Overall, the images seem to be combinations of photography and illustration, **Figure 15-10**.

Though magic realism is often created in postproduction, it can also be achieved through lighting, often by using extra rim lighting for an ethereal, dreamlike look.

Expressionism

Expressionism is the most frankly theatrical of the major lighting styles, and does not pretend to reflect reality at all. It is lighting for its own sake, in which any effect is acceptable as long as it *expresses* and intensifies the feeling of the scene. Often, though not always, expressionistic lighting is hyper-dramatic, even extreme.

Expressionism began as a movement in painting and theater in Europe, and then was adopted by film makers, notably in Germany. German directors brought expressionistic lighting and camera work to Hollywood in the 1920s and 1930s. Universal Studios' horror films, such as *Frankenstein*, exploited expressionistic lighting to good effect. Today, expressionism, like pictorial realism and magic realism, is often used for commercials (especially humorous ones) and music videos.

Figure 15-10 Examples of magic realism. A cemetery processed in a magic realism style. The dockside image gets its eerie quality from the combination of twilight available light and the harsh work light on the pier.

Figure 15-10 Examples of magic realism. A cemetery processed in a magic realism style. The dockside image gets its eerie quality from the combination of twilight available light and the harsh work light on the pier.

Cemetery

Dockside

(Sue Stinson)

The most influential early expressionistic film is perhaps *The Cabinet of Dr. Caligari*, which, unlike most European movies, was widely shown in the United States (**Figure 15-11**).

Lighting Strategies

Technical standards and graphic style are relevant to all video lighting, but the *strategies* you employ to achieve them will vary from one application to the next. As you approach each lighting situation, you need to decide which

Figure 15-11 *The Cabinet of Dr. Caligari*. This black and white film print was hand-tinted.

common lighting objectives are most important, which approach will do the job most effectively, and whether the program demands a "high-key" (mostly light) or "low-key" (mostly dark) appearance. Once you have made these decisions, you are ready to start creating your design.

Identifying Basic Objectives

In building your designs, you usually want to achieve three objectives:
- Motivate the lighting.
- Light the subject (not the space).
- Enhance the feeling of depth in the image.

Motivate the Lighting

The real world is full of light sources: the sun or the overcast sky, the lighting in a room, the special light of neon signs, computer monitors, headlights, candles, and numberless others. To make your lighting look "real," you generally try to imitate the light sources that would normally be present—*motivated lighting*.

In practical terms, this means, for example, that you would not create scary cross lighting in an office washed by ceiling fluorescents.

So, your first task is usually to decide what sources would naturally light the scene and then imitate them with your video lights.

An Anti-Style

In addition to the other styles, there is a visual look that might be called an "anti-style," in which the images constantly remind the audience that they are looking at a video or film. We have placed this anti-style by itself because it involves not only lighting, but every other aspect of cinematography as well.

This anti-style exploits the characteristics and limitations of the video medium:

- Shots are purposely set up so that light sources will strike the lens and create intentional flares.
- Dark scenes are shot in low light levels with the gain circuits enabled for a murky, grainy look.
- Exteriors are shot in available light. If light levels are too high, colors smear. If the contrast is too great, faces turn into silhouettes.
- Light levels are not equalized, so the auto exposure system is visibly adjusting exposure as the camera moves.

Ultra low-light shot.

(Sue Stinson)

This anti-style is similar to naturalism, in that it strives to convince viewers that they are looking at unprocessed reality. The difference is that naturalism says, "here is reality," while anti-style insists, "here is a video *recording* of reality."

Some viewers think that this anti-style is an intrinsically honest approach to recording video images. Others find it pretentious and irritating. Either way, it can be powerfully effective in the right type of program.

Intentional sun flare.

(Sue Stinson)

Light the Subject

Images are created by light reflected into the lens from subjects in the frame. For this reason, you do not have to light every cubic inch of your location, but only those areas containing subject matter. To light just the areas that will show on screen, follow this procedure:

- Study the performers' positions and movement (blocking).

- Light the areas that they will use, paying special attention to places where they will stay, or at least pause. In other words, keep performers adequately illuminated throughout their movement, and carefully lit at spots where important action happens.
- Light parts of the background (including furniture and any floor or ceiling areas) that will appear in the image.

In **Figure 15-12**, the subject is lit by key, fill, and rim lights at the rear entrance, the center, and the forward stopping point. Two fill lights provide general area lighting.

The rear key light also helps wash the background. It is very common for lights to do double duty.

In this example, the left key light has been mounted overhead because its light stand would be seen in a wide shot. If there is no place to conceal lights, you may want to provide basic lighting for the entire area, adding or adjusting instruments as needed later to optimize the lighting for setups that focus on specific places within the lit area.

This example is covered in greater detail in the next chapter.

Light to Enhance Depth

Where cinematographers use composition to enhance the illusion of depth, lighting directors use light accents. To suggest the third dimension, you will often create highlights deep in the real-world space that you are lighting. These accents guide viewers' eyes past foreground subjects and "into" the scene.

Though you usually want to achieve all three of these objectives, there are special cases in which you do not. For example:

- In the expressionistic lighting style, motivated lighting is not required or necessarily desirable.

- In some wide shots, so much of the environment is on-screen that your lighting must cover all of it.
- To create compositions on the picture plane, you may purposely avoid background light accents that call attention to depth.

Cinematographers contribute to these purposely flat images by using telephoto lens settings and minimizing the usual visual cues of perspective.

Selecting a Lighting Approach

With your objectives for lighting a scene in mind, you are ready to choose a specific approach to the task. In any lighting environment (except in a studio), you have three different options in lighting a scene: work with the light available, start with the available light and supplement it with video lighting, or neutralize the available light and design exclusively with video lights. Your decision will depend on the lighting existing at the scene, the style you have adopted, the equipment at your disposal, and the amount of electrical power available.

Working with Available Light

Except for news and documentaries, most available light shooting takes place outdoors. If the day is overcast, the main concern is color temperature, which is addressed by adjusting camera white balance. In direct sunlight, however, each camera setup offers you three tactics that you can use alone or in combination:

- Select the light.
- Reflect the light.
- Filter the light.

In outdoor shooting, you will typically find yourself using two or all three of these tactics at once.

Select the Light

When the action permits, the easiest method is to find backgrounds and subject positions that naturally yield high-quality images, **Figure 15-13**. This may mean placing subjects in existing shade, or else in sunlight with a darker background behind them to prevent severe backlighting. If the ground is sand, light soil, or concrete, the light

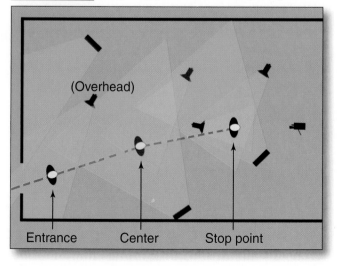

Figure 15-12 A lighting plan for a moving subject.

(Overhead)

Entrance   Center   Stop point

Key, Fill, Rim, and Background

Most lighting designs are combinations of four basic functions—tasks so fundamental that they are part of every lighting style and present in most lighting applications. We cover these functions in detail in a later section; but because we also refer to them here, you need to know what they are.

Key Light

The **key light** is the main illumination on the subject. Typically, it is the brightest light and mimics real-world light sources—usually the actual lights in the room or at the scene.

The key light provides the main illumination.

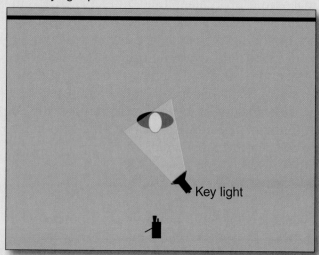

Fill Light

The **fill light** is the secondary illumination on the subject. Usually placed opposite the key light, it fills in shadows to reduce contrast created by the key light.

The fill light fills in shadows.

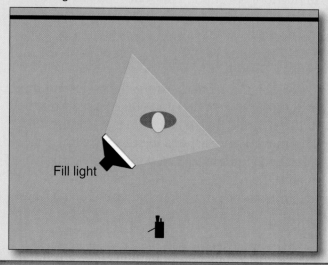

Rim Light

The **rim light** is placed behind and well above the subject. The rim light separates the subject from the background by splashing a rim of light on the subject's head and shoulders.

The rim light separates subject from background.

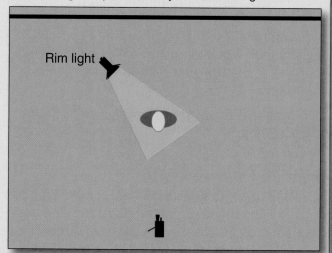

Rim light is also commonly called "back light." This term is not used here, to avoid confusion with *background* light.

Background Light

The **background light** (or lights) raises the light level of the walls or other backing, adds visual interest, and enhances apparent depth by highlighting background features.

The background light illuminates the background.

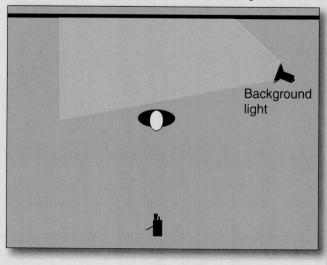

Figure 15-13 The dark, textured wall makes a good background. Note the white reflector for additional fill.

(Sue Stinson)

reflected from it may provide natural subject fill light. Sometimes, the ambient bounce light is so strong that you can use the sun as a rim light and the bounce light as a soft key and fill.

Be careful on grass, which can reflect its green color onto your subjects.

Reflect the Light

The workhorses of outdoor lighting are reflectors, which can be used for key, fill, or backlighting—or in any combination. Reflectors are versatile and easy to use. On the downside, the light from hard reflectors can look harsh, and soft reflectors must be used close to subjects, which makes wide shots difficult to light (**Figure 15-14**).

Filter the Light

Screens and silks are able to control light in larger areas, but big ones require complicated frames and large crews to manage them. Units on small frames can only screen one or perhaps two subjects at once. Like soft reflectors, screens and silks must be used close to subjects, so they cannot be employed in wide shots (**Figure 15-15**).

Adding to Available Light

The second possible approach is to start with the available light, then supplement it with video lights. Outdoors, camera lights can be used to fill facial shadows or simply to enhance under-lit subjects. You can call the viewer's

Figure 15-14 A large silver reflector on a frame.

(Photoflex)

attention to your center of interest by making it just a bit brighter than objects around it, **Figure 15-16**. In **Figure 15-16**, notice how the extra front light slightly emphasizes the subject, without seeming to throw an obvious spotlight on him. (You can, however, see the subject's shadow on the wall.)

The existing-plus-video lighting approach comes into its own indoors, where electrical power is available. Since all cameras can record a good-quality image in well-lit interiors, the easiest approach is often to start with available light and enhance it with video lighting to complete your design.

Figure 15-15 A medium-size silk.

(Photoflex)

Figure 15-17 Natural key light plus rim light (practical).

Lighting design

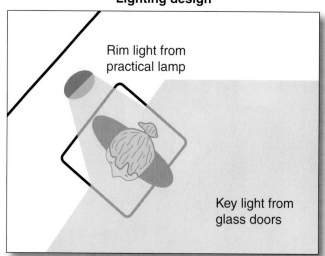

Rim light from practical lamp

Key light from glass doors

Lighted subject

Typically, available light indoors comes from windows, ceiling fixtures or grids, and lamps. In planning your tactics, you need to assign roles to each of the light sources that are available. Ceiling light is usually used for fill and backgrounds, but window light can be either key, fill, or background (depending on its intensity, quality, and position). Once you have given an available light source its task to perform, you complete the design with video lighting.

Complementing an Available Key Light

In **Figure 15-17**, the woman is lit primarily ("keyed") by soft light from an off-screen sliding glass door. General ambient light provides the fill. To enhance visual interest, while preserving the natural-looking design, an added practical (on-screen) light provides rim light.

Figure 15-16 With available light alone, the lighting is murky. An on-camera light adds front fill on the subject.

Available light only

On-camera light

Front fill added

(Lowel Light Inc.)

Complementing an Available Fill Light

In **Figure 15-18**, by contrast, the subject is lit primarily by soft fill from overhead fluorescent lights. In this setup, a soft key light has been added (placed to the side and at eye-level), and small overhead lights provide modeling for the background—note the hot spot on the flowers.

Using Video Lights Only

Sometimes the existing light is difficult to manage, or is simply wrong for your needs. The simplest solution in such cases is to neutralize the available light entirely and create your lighting design strictly with video lights.

You neutralize available light by masking it (as with curtains) or simply turning it off.

Figure 15-18 Natural fill light plus key and rim lights.

Lighting design

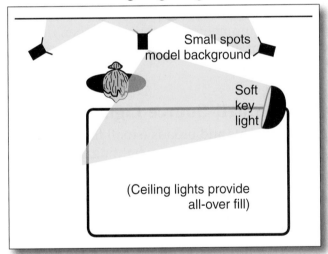

Lighted subject

This strategy offers the tightest control over the lighting, but demands the most resources. Before going this route, make sure you have enough equipment to execute your design, enough power to drive it, and enough time and personnel to rig it and later dismantle and pack it. See **Figure 15-19**.

It is difficult to generalize, but possibly safe to say that a typical four-unit lighting kit can handle up to two subjects at once, in a room up to perhaps 15 feet square.

Choosing a Lighting "Key"

When you have settled upon your objectives and your overall approach, your final strategic option is to pick a "key" for the lighting's overall look, **Figure 15-20**. A *high-key* image is composed mainly of lighter values, punctuated with occasional dark areas. A *low-key* image is the opposite: mainly dark, but usually with a lighter center of interest. A *medium-key* image (to coin a new name for the purpose) has highlights and shadows that are more or less evenly distributed.

High-key and low-key are not styles themselves. It is perfectly possible to use either approach in naturalism, realism, pictorial realism, magic realism, and expressionism. However, it is easier to create high- or low-key designs with spotlights, because they allow greater control over light intensity, direction, and masking. Most

---

Figure 15-19 Lighting this large interior required six soft lights.

(Photoflex)

High-key lighting

Low-key lighting

Medium-key lighting

(Corel)

training, promotional, and other nonfiction programs lean toward the more cheerful look of high-key lighting, without formally adopting it as a design approach.

High-Key/Low-Key Examples

In classic Hollywood films, high-key lighting was used for comedies, romantic movies, and action pictures. Low-key lighting was employed for crime and mystery films, intense dramas, and horror. The "film noir" style of the 1930s and 1940s executed low-key lighting in the pictorial realism style, with great effectiveness.

High- and low-key approaches are easier to study in black and white films. To select two examples, the comedy *Some Like It Hot* is generally high-key and the drama *Dr. Strangelove* is low-key. (These classic films, among the best in Hollywood history, are rewarding to view just as movies.)

In video, high- and low-key lighting has been less common because, until recent years, TV did not offer enough range of contrast to permit it. (Prints of films intended for broadcast were often made with purposely reduced contrast, to accommodate the demands of the medium.)

With today's improved video technology, look for low-key lighting in TV dramas and high-key designs in situation comedies. Studio talk shows and news programs are generally "medium-key," while commercials and music videos often carry high-key/low-key differences to extremes not seen since classic Hollywood.

Lighting Procedures

Moving from the general toward the specific, through lighting *standards*, *styles*, and *strategies*, we come now to lighting *procedures*—the methods you use to deploy your lighting equipment. Since each lighting tool is handled essentially the same way in most applications, it is easier to discuss equipment procedures together, rather than repeat information for every separate lighting situation.

Using Small-Source Lights

Spotlights and broads (small floodlights) are the backbones of many lighting kits. These *small-source lights* are popular because of their portability, versatility, and small size.

Both types grow very hot when in use, so be sure to have leather work gloves handy.

Spotlights

Spotlights are used when you need precise control over the light path. To shape the beam, it helps to follow a regular sequence.

- Raise and aim the light. Always extend the upper sections of the light stand first, to keep as much weight as possible down low.
- Adjust the beam intensity by moving the light and stand toward or away from the subject (as long as they remain out of frame).
- At the desired position, fine-tune light intensity by operating the lamp focus control. Flood the light for lower intensity; use the spot focus for higher intensity.

- If edge masking will not be an issue, soften the beam by installing a milky plastic diffusion filter in front of the lamp, or clipping spun glass diffusion to the barn doors.

The hard-edged spot focus is easier to mask than the flood position. If edge control is wanted:

- Try setting the focus to full spot and adjusting beam intensity with screens.
- Mask the beam edges by adjusting the barn doors. If, as often happens, they will not manage all four sides, supplement the barn doors with clip-on or freestanding flags. (Move the flags toward the light for softer edges or away for harder ones.)
- Check the subject for overly bright areas (often a problem when the light is at an oblique angle to a surface, like a wall). Use half-screens to reduce hot spots, **Figure 15-21**.

If a somewhat warmer color temperature is acceptable, you can also reduce intensity by putting the light on a dimmer.

Though this sequence works well in general, you will often have to go back and forth among focus, barn doors, and screens and diffusion, to get exactly the right effect.

Broads

Small and rectangular, broad floodlights emit light that is directional enough to cast shadows, **Figure 15-22**.

Figure 15-21 Some spotlights use proprietary diffusion screens.

(Lowel Light Inc.)

Figure 15-22 A lightweight broad for location use.

(Lowel Light Inc.)

The procedure for using them is simple:
- Set up the stand and unit.
- Adjust the barn doors (but do not expect precise edge control).
- Add framed or clipped-on diffusion, if desired.
- Adjust intensity by moving the unit (there is no focus control).

Broads are compromise units—not as soft as large source lights, but very small and handy. Also, they are useful for background lighting, where true large-source lights lack a long enough throw.

A light's **throw** is either a) its maximum effective light-to-subject distance, or b) that distance in any particular setup.

Using Large-Source Lights

Umbrellas, soft boxes, pans, and LED arrays are *large-source lights*, which are easy to use but hard to control. With all four types of units,
- Set them up and aim them at the subject.
- Move them forward or back to adjust light intensity.
- Observe their effect on the background and adjust accordingly.

It is difficult to keep large-source light from spilling onto nearby backgrounds. This can be a plus if you are working with very few units and need them to cover background as well as subject. Otherwise, try moving your subjects forward from the background and resetting the lights to reduce spill.

Umbrellas

With umbrellas, you can adjust the light quality by changing the cloth cover: silver threaded for a harder effect, plain white for soft, or translucent light if you wish to aim the umbrella at the subject and shoot the light through it. Umbrellas can be used with small spotlights, but their extra control is wasted in this application. Small broads work well because they do not form a large shadow in the center of the light path, **Figure 15-23**.

Softboxes

The more sophisticated softboxes have internal baffles and/or layers of fabric for fine-tuning the light quality, **Figure 15-24**. Some models also have shallow black grids that can be applied to the front of the unit to limit side spill. Some softboxes accept general purpose lights, but others use proprietary lamps and housings.

Pans

Fluorescent *pans* are desirable where electrical power is limited and/or where heat buildup is an issue. They also work well with available fluorescent light or daylight. Otherwise, their bulk and clumsiness fit them best for studio use, **Figure 15-25**. In fact, all large-source lights are bulky and hard to work with in tight locations. Where a soft light is wanted, it is sometimes better to use diffusion with small, maneuverable spots and broads.

Figure 15-24 This softbox has internal baffles for smoother light distribution.

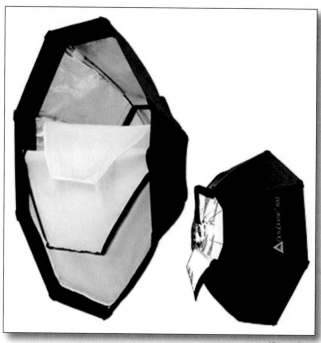

(Photoflex)

Using Cookies

A cookie (a short form of the obscure word "cukaloris") is usually a sheet of opaque material with cutouts, mounted in front of a light to throw patterns on the background. Some cookies throw recognizable shadows, while others simply vary the light on background walls. With an artist's knife and a sheet of foam board, you can create any pattern you need, **Figure 15-26**.

Figure 15-23 Some umbrellas are fitted to specially designed lights.

(Photoflex)

Figure 15-25 Some light pans are designed to include their carrying cases.

(Lowel Light Inc.)

Figure 15-26 A leaf cookie adds character to a blank background wall.

Blank wall   **Leaf cookie added**

Here are a few of the more popular cookie patterns:

- **Bare branch.** A common wall pattern suggests the outline of a leafless tree, **Figure 15-27**.
- **Leaves.** A leaf pattern cookie (**Figure 15-26**) can throw the shadow of leaves on subjects, which makes it useful outdoors as well as inside. Keep the leaf cutouts small, for best results.

If you have permission to cut an actual tree branch, you can clamp it horizontally with a century stand to make a very effective cookie, whose leaves can actually move in the "wind" created by gently shaking the branch.

- **Window blinds.** Venetian blind patterns are easy to make. They are typically cut on the diagonal to throw slanting patterns on backgrounds, **Figure 15-28**.

Some lighting directors carry an actual small blind, which they suspend from a century stand in front of a spotlight and adjust to obtain the desired effect.

Cookies can be used in TV studios to add interest to the plain curtains often used as backings. Simple variegated patterns break up the surface, and cutouts can suggest other backgrounds, **Figure 15-29**.

Figure 15-28 A blind cookie.

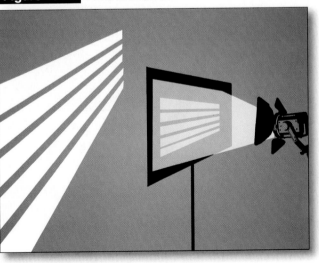

Figure 15-27 A bare branch cookie.

Figure 15-29 A city skyline cookie.

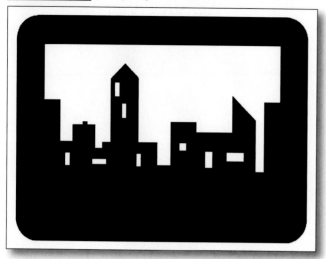

Pattern projections in TV studios are sometimes handled by "gobos," miniature cookies placed inside theatrical lighting instruments called *ellipsoidal spotlights.*

You can cut a cookie to throw any shadow you need for a particular one-time effect. A gallows is a simple (if melodramatic) example, **Figure 15-30**. Cookies with actual images are most believable in the styles of pictorial realism, magic realism, and expressionism. In naturalistic lighting, they tend to look excessively theatrical. Abstract pattern cookies, however, work in any style to add subtle variety to backgrounds.

Figure 15-30 A gallows cookie.

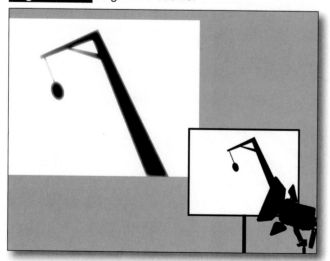

Cookie Procedures

To use a cookie, mount the sheet on the arm clamp of a century stand and focus a spotlight to throw a sharp, distinct shadow through it. (Floodlights do not yield good results and softboxes do not work.) Control the character of the shadow by adjusting the cookie's distance from the light. The closer the cookie is to the light, the bigger the shadow and the softer its edges. For smaller, sharp-edged patterns, place the cookie as far in front of the light source as practical. This setup sequence usually works well:

- Set up both the cookie and the light behind it.
- Watching the background, change the distance between light and cookie to obtain the desired edge sharpness.
- Moving the cookie and the light together (to preserve the distance set between them) adjust the size of the shadow by changing the throw (distance) to the background.

Wide opaque borders around cookie patterns allow the sheets to mask the beam edges when placed farther from the light.

Using Reflectors

Reflectors are so versatile that they are used both indoors and out. They are usually managed in the same way as lights, regardless of the particular lighting situation.

Outdoors, reflectors on stands should be weighted down because they are essentially sails just waiting to blow away. Use rigid reflectors, if practical, because flexible cloth models ripple in the breeze and throw moving reflections on subjects.

Focusing Reflectors

You can adjust the intensity of any reflector by moving it toward or away from the subject. (Be very careful to keep hard reflectors well back from subjects to avoid hurting their eyes with the hard light beam.) With some units, you can also reduce the light level by aiming the beam slightly away from the subject, so that only the edge of the beam is effective. Other units can be focused by flexing the panel, **Figure 15-31**.

Figure 15-31 This reflector is a flexible arc whose curve can be changed to "spot" or "flood" the unit.

(Lowel Light Inc.)

As you adjust fill levels by moving reflectors toward or away from their subjects, watch your monitor carefully to avoid overfilling.

Reflector Lighting Functions

You can use reflectors exactly as you would lights—for key, fill, rim, and background lighting.

Key light. Typically, you key with a reflector when you wish to use the sun as a rim (back) light. To do this, it is usually best to place the reflector on the side opposite the sun, **Figure 15-32**.

In these diagrams, direct sunlight is red and reflected light is yellow.

Figure 15-32 In this arrangement, the reflector is the key light and the sun is the rim light.

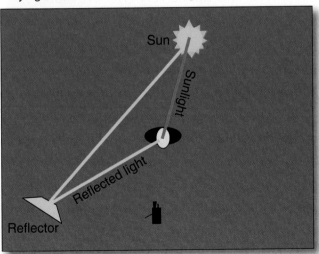

Fill light. Reflectors are used most often as fill lights. In this setup, the sun provides the key light and the reflector, on the opposite side of the subject, adds the fill (**Figure 15-33**).

For closeups, subjects can sometimes hold their own reflectors below the frame line.

Rim light. Rim light is difficult to add with a reflector because the light source should be much higher than the subject. If you can get a crew member up on a ladder that is off-camera, you can make it work. A hard aluminum reflector works best here, and there is no danger of hurting the subject's eyes (**Figure 15-34**).

Figure 15-33 With the reflector as fill and the sun as key light, both are in front of the subject.

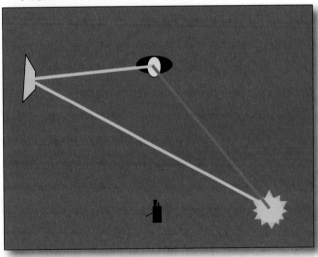

Figure 15-34 If the unit can be placed high enough, the reflector can add rim light to the sun's key light.

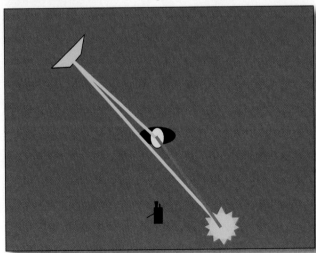

Depending on the height of the camera angle, nearly horizontal rim lighting can sometimes be effective.

Background light. As explained earlier, the easiest way to prevent backlight exposure problems is to set the subject in sunshine, in front of a shaded background. If this makes the image seem dull and lacking in depth, use a reflector to rake the background with light, **Figure 15-35**.

Reflectors Indoors

Indoors, reflectors are used mainly for fill. They can be so effective that you can light a single subject with just one spotlight and a large reflector opposite for fill, **Figure 15-36**. If you have two lights, you can use the second one as a rim light or a background light.

Using Screens and Silks

Screens and silks have more limited uses because they must be placed close to subjects, making them unusable in wide shots.

Figure 15-35 With the sun as key light, the reflector splashes light on the wall behind the subject.

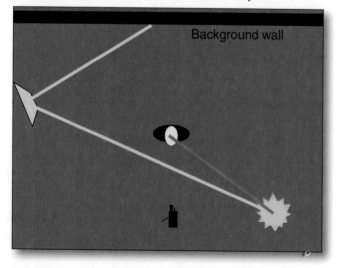

Figure 15-36 The softlight (foreground, right) and the gold reflector provide all the subject lighting.

(Photoflex)

You can solve this problem by ending wide shots with subjects still far enough away so that the lack of light filtering is not obvious, and the closeups seem to match the lighting of the wide shots.

Screens are useful for moderating directional sunlight to reduce excess contrast, but they have limitations. Large, framed models are clumsy and time-consuming to place. They are most useful in static situations, like stand-up narration or interviews, in which subjects remain in the same spot for extended periods.

Silks are less useful in this application because they soften and diffuse the light, as well as reduce it (**Figure 15-37**). That will change the key light quality so that the closeup lighting will not match that of the longer shots. However, silks can be very effective indoors, when an extremely large light source is needed for absolutely shadowless light.

Figure 15-37 The diffusion screens are out of frame in the video images.

Shooting area

Final image

(Photoflex)

From Theory to Practice

Now that you have surveyed lighting tools and reviewed the basics of lighting design, you are ready to apply your knowledge to practical lighting applications. These lighting applications are the subject of the next chapter.

Summary

- To achieve high technical standards, images must be created with light of the proper quantity, contrast, and color.
- Choosing a lighting style (naturalism, realism, pictorial realism, magic realism, and expressionism) can give your cinematography a distinctive quality that enhances its effectiveness and increases its professionalism.
- In building a lighting design, you usually want to achieve three objectives: motivate the lighting, light the subject (not the space), and enhance the feeling of depth in the image.
- In any lighting environment (except in a studio), you have three different options in lighting a scene: work with the light available, start with the available light and supplement it with video lighting, or neutralize the available light and design exclusively with video lights.
- Lighting procedures are the methods you use to deploy your lighting equipment.
- Small-source lights, large-source lights, cookies, reflectors, and silks and screens are standard tools in a lighting designer's toolkit.

Technical Terms

Background light: A light splashed on a wall or other backing to lighten it and add visual interest.

Expressionism: A lighting style that adds a heightened emotional effect, without regard for lighting motivation.

Fill light: The light that lightens shadows created by the main (key) light.

Gaffer: The chief lighting technician on a shoot.

High-key: Lighting in which much of the image is light, with darker accents.

Key light: The principal light on a subject.

Large-source light: A lighting instrument, such as a scoop or other floodlight, with a big front area from which light is emitted. Light from large-source instruments is relatively soft and diffuse.

Low-key: Lighting in which much of the image is dark, with lighter accents.

Magic realism: A lighting style that creates a dreamy or unearthly effect, often enhanced digitally in postproduction.

Medium key: Lighting in which neither light nor dark tones dominate the image (a term used only in this book).

Motivated lighting: Lighting that imitates real-world light sources at a location.

Naturalism: A lighting style that imitates real-world lighting so closely that it is invisible to most viewers.

Pan: In lighting, a large, flat instrument fitted with fluorescent or other long tube lamps.

Pictorial realism: A lighting style in which lighting, though motivated, is exaggerated for a somewhat theatrical effect.

Realism: A lighting style that looks like real-world lighting, but it is slightly enhanced for pictorial effect.

Rim light: A light placed high and behind a subject to create a rim of light on the head and shoulders to help separate subject and background.

Small-source light: A lighting instrument with a small front area from which light is emitted, such as a spotlight. Light from small-source instruments is generally hard-edged and tightly focused.

Throw: A lighting instrument's maximum effective light-to-subject distance. Also, that distance in any particular setup. (The term "throw" is also used to describe camera-to-subject distance.)

Review Questions

Answer the following questions on a separate piece of paper. Do not write in this book.

1. The _____ is the head of the lighting crew on a shoot.
2. What techniques are used to reduce the quantity of light?
3. What effect does too little light have when recording a video image?
4. *True or False?* Contrast is the difference in brightness between the lightest and darkest areas in an image.
5. What are the attributes of a "standard" video image?
6. Which lighting style most closely imitates real-world conditions?
7. The purpose of the _____ lighting style is to express and intensify the feeling of a scene.
8. List the three objectives of building lighting designs.
9. A(n) _____ light separates the subject from the background.
10. What is the function of a background light?
11. *True or False?* Screens and silks must be placed close to the subjects in a shot to effectively filter the light.
12. Describe the characteristics of a low-key image.
13. List two examples of small-source lights.
14. Light from _____ -source lights is relatively soft and diffuse.
15. Cookies with actual images are most believable when used with which lighting styles?
16. What are the effects of moving a cookie closer to and farther away from the light source?
17. In an outdoor shooting environment, the sun provides the _____ light and a reflector can be positioned to add fill light.
18. *True or False?* A screen softens and diffuses light, and reduces its brightness, as well.

STEM and Academic Activities

1. **Science.** To create a video image, light must be applied. The same is true of human vision. Explain how the human eye processes light. How is eye function impaired in someone who has nyctalopia?

2. **Mathematics.** The contrast in a video image is expressed as a ratio. For example, a 300:1 ratio means that the brightness of the white areas in an image is 300 times brighter than the black areas. What methods are used to measure the contrast of various video display devices? How is the contrast ratio determined?

(Litepanels, Inc.)

Objectives

After studying this chapter, you will be able to:

- Recall the techniques and characteristics of both "classic" and "natural" lighting.
- Explain different methods and applications for lighting backgrounds.
- Identify effective solutions for common lighting problems.
- Recall effective methods to light interior and exterior night scenes.
- Identify effective techniques for lighting frequently encountered assignments.

About Lighting Applications

This chapter takes the lighting tools and design principles covered in the two preceding chapters and puts them to work in real-world situations. We will see how to light subjects, locations, and night scenes, how to solve common lighting problems, and how to approach several types of frequently encountered lighting assignments.

There is some necessary content overlap among the three lighting chapters.

Lighting on a "Clock"

For convenience, the horizontal placement of lights is often described in terms of a clock face:

- The subject is at the center, facing the six o'clock position.
- The camcorder is at six o'clock, facing the center.
- The lights are at various "hours" around the clock face.
- The background, if shown, is at the twelve o'clock position.

(View from above.) A four-light setup diagrammed on a clock face, with the back light at 2:30, the key light at 5:00, the camera at 6:00, the fill light at 8:00, and the rim light at 11:00.

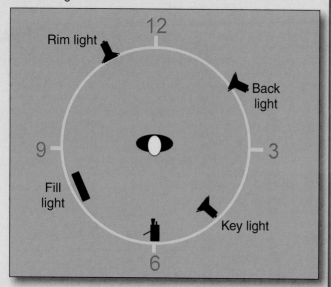

Although this diagram shows the key on the right and the fill on the left, their positions are just as often reversed.

Lighting Subjects

Most of your time, of course, will be spent lighting people. From the simplest production to the most elaborate, there are only two basic approaches to this task: classic studio lighting and soft "natural" lighting.

Classic Studio Lighting

Classic studio lighting uses three lights on the subject and usually one or more on the background, **Figure 16-1**. This is often called *three-point lighting*, despite the frequent use of additional instruments.

Key Light

The key light provides the main illumination, typically mimicking an actual light source like a lamp or ceiling fixture. It is often placed at about 4:30 and 15°–30° higher than the subject's face, **Figure 16-2**.

The key light is typically a spotlight, so the hard-edged beam is often softened with a sheet of spun glass clipped to the barn doors. Even so, it throws distinct shadows on the subject's cheek, upper lip, and neck.

Fill Light

The fill light literally fills in the shadows created by the key light, **Figure 16-3**. Placed opposite the key light, the fill is often farther to the side and not as high as the key, which helps reduce the cheek, lip, and neck shadows.

Figure 16-1 A classic lighting setup, including key, fill, rim, and background lights.

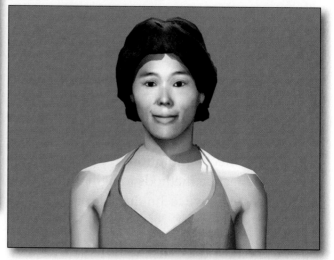

Figure 16-2 Key light placement.

Horizontal placement	Vertical placement	Effect of the key light

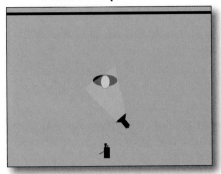 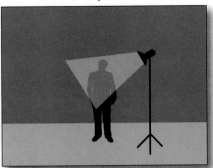 

Figure 16-3 Fill light placement.

Horizontal placement	Vertical placement	Effect of the fill light

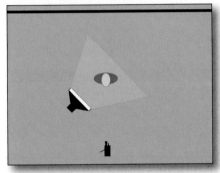 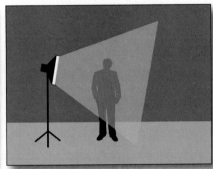 

How completely the fill light moderates these shadows depends on the setting and mood of the scene. In a cheerful interior, the shadows might be slight; in an atmospheric night scene, they might be so deep as to obscure details within them. In any case, the fill light should not be bright enough to make the subject lose the "modeling" that creates the illusion of depth.

Rim (Back) Light

The rim light is typically behind the subject and placed quite high, **Figure 16-4**. If however, its light stand appears in the shot, you can move the rim light aside until it clears the frame.

Rim lights are frequently mounted overhead on clamps or on stands with lateral arms. The brightness of the rim light depends mainly on the lighting style—pronounced for pictorial realism and moderate for realism. For naturalism,

Figure 16-4 Rim light placement.

Horizontal placement	Vertical placement	Effect of the rim light

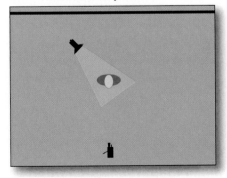  

the rim light is just barely bright enough to visually separate the subject from the background. In some instances, it is omitted entirely.

The brightness of key and fill lights is adjusted by moving the lights toward or away from the subject. Rim light, however, may be controlled by a dimmer, since the warming effect of dimming a light is usually acceptable in this application.

Background Light

Like the key light, the background light is usually "motivated"—that is, it mimics light that would naturally fall on the walls or other background, like a wall lamp, a window light, or spill from a room light (**Figure 16-5**). When working with just a few lights, you can usually achieve background lighting by directing spill from the key and/or fill lights.

Background light intensity should be adjusted so that subject and background seem lit by the same environment, but the subject is slightly brighter. Two or more background lights may be needed to do the job.

Background lights often produce less intense effects because the lighting *instruments* must be placed well away from the background to keep them out of the frame.

With the four lights in place, we can build a complete lighting setup, **Figure 16-6**. Though developed for classic pictorial realism, this basic scheme can be used with any of the four major lighting styles, **Figure 16-7**.

The basic lighting setup demonstrated here uses four lights and only covers a space about the size of a single action area. At large shooting locations, the lighting can involve many more instruments, but they tend to be deployed in multiples of these basic layouts.

An "action area" is a spot within a location that is fully lit because important activity takes place there.

"Natural" Lighting

Because three-point lighting can look somewhat theatrical, many situations call for

---

Figure 16-5 Background light placement.

Horizontal placement

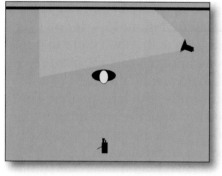

Vertical placement

Effect of the background light

Figure 16-6 Complete lighting setup.

Horizontal placement

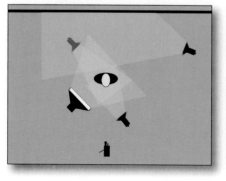

Vertical placement

Combined effect of the four lights

Figure 16-7 Using a basic four-instrument lighting setup.

Naturalism	Realism	Pictorial realism	Expressionism

a more natural, "unlit" appearance. The key to this approach is soft light. Spots and broads can be used if heavily diffused, but large sources, such as umbrellas or softboxes, are often easier to work with.

One-Light Design

For close shots, a single soft source can deliver satisfactory lighting—especially when paired with a reflector for additional fill light. The subject should be close enough to the

"Rugged" vs. "Glamorous" Lighting

How you position and diffuse your lighting instruments often depends on whether you wish to emphasize facial modeling for a so-called "rugged" look, or whether you prefer to de-emphasize it for "glamour."

Rugged lighting exaggerates the planes and angles of the face and emphasizes skin texture. To do this:

- Keep the key light high for more pronounced shadows.
- Reduce or omit key light diffusion. The harder the beam, the more it emphasizes skin and other textures.
- Avoid over-filling to retain enough shadows for pronounced facial sculpting.

Glamorous lighting uses exactly the opposite approach:

- Place the key light lower for moderate shadows.
- Use considerable diffusion (or a softlight) to minimize skin texture.
- Add fill light until the shadows are relatively faint, but avoid over-filling the neck area to downplay aging skin.
- Use a generous rim light to accent hair.

Glamorous lighting.

Rugged lighting.

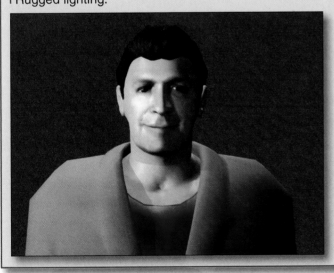

The cinematographer contributes to the effect by using wide angle lens settings for rugged lighting, and telephoto lens settings for glamour.

background so that the light spill can model it. Keep the lighting unit low to simulate window light, and place it at about 8:30 on the clock so that it "wraps around" the subject's face (**Figure 16-8**).

Two-Light Design

A second softlight provides a more versatile fill source. With this design, it often helps to place key and fill lights at the heights you would use for three-point lighting. With two lights, you can move the fill around as far as three o'clock, **Figure 16-9**.

Alternately, you may wish to continue to use the reflector for fill, and bring up the background with the second light.

Three-Light Design

A third light gives you better control over both fill and background, **Figure 16-10**. Studying the light plan in **Figure 16-10**, note that:
- The subject's distance from the background permits the two to be lit separately.

- The fill light is about three times as far as the key light from the subject.
- The background light is far enough to the side so that the hot spot created by the near edge of its beam and the overlapping key light spill (indicated in red) is outside the frame.

The natural style is very popular for lighting interviews because the lighting can match the location, and because the backgrounds are frequently close behind the subjects. Also, soft lighting is fast and easy to work with.

Lighting Backgrounds

If you have the room and the lighting resources, it is often best to move subjects away from backgrounds so that you can light them separately. Spots or floods are effective for background lighting because their longer throw allows them to be placed far enough to the sides to remain out of the frame.

As noted elsewhere, a light's *throw* is the distance between the instrument and the subject or background that it is lighting.

Figure 16-8 One-light design.

Horizontal placement	Vertical placement	Effect of the design

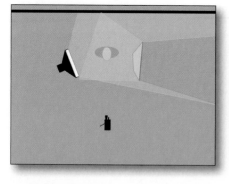 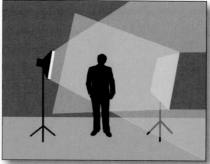 

Figure 16-9 Two-light design.

Horizontal placement	Vertical placement	Effect of the design

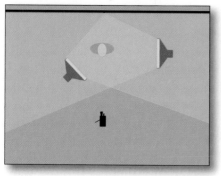 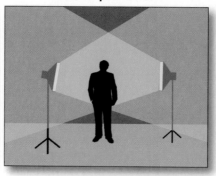 

Figure 16-10 Three-light design.

Horizontal placement

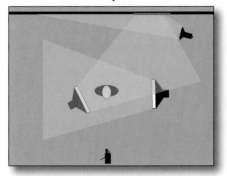

Vertical placement

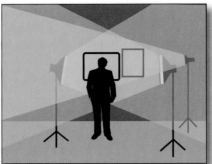

Effect of the design

Lighting Backgrounds for Exposure

When the background is too dark, the image loses apparent depth. So, you may want to wash some light on it to make it more visible. Be careful not to place too many highlights or to make the background too bright, to avoid distracting attention from the subject(s) in the foreground.

Lighting Backgrounds for Texture

You can often make dimensional surfaces (such as plaster or fabric) more interesting by bringing out their texture. For good cross lighting, place spots or broads as close to the backing as possible and rake the light across the surface.

Lighting Backgrounds for Depth

Sometimes, you can enhance depth by highlighting surfaces (like furniture) in front of the background, as well. Generally speaking, lights hung above the frame work effectively.

Adjusting Intensity

When lights are placed close to a background, the beam is much "hotter" near the light. To even out the light pattern, use a half or double-half screen positioned in the spotlight's filter holder so as to reduce light output on the side near the wall.

Outdoors, background lighting is usually created using hard-surface aluminum reflectors, **Figure 16-11**.

Lighting Locations

Few video productions are shot on sound stages (except those made for cable or broadcast), so your interiors are likely to be locations. Every location presents challenges, and meeting them offers the satisfaction that comes from successful problem solving.

Small Interiors

Small interiors are usually difficult to work in, for several reasons (**Figure 16-12**).

Cramped Quarters

Lights require room, not only for the instruments and stands, but for the throws of the lights. Remember: you reduce light intensity by moving the unit away from the subject. Spots and broads are favored for cramped interiors because of their small size. Spots are especially useful in tight quarters because they can be focused to vary the light output without moving the unit. Their intensity can be further reduced by using screens.

Figure 16-11 A hard aluminum reflector.

(Bogen Manfrotto)

Figure 16-12 This interior has problems with space, window light, and irregular ceilings.

Power Supplies

Small interiors are often located in homes or other private buildings where electrical circuits are typically only 15 amps, and an entire room may be served by just one circuit. When working with inadequate power, high-efficiency fluorescent lamps draw less power per unit output than halogens. Consider using the compact units with screw-base lamps, **Figure 16-13**.

Background Spill

Small interiors make it difficult to keep subject light from spilling onto the background. Here again, using the more controllable spots

Figure 16-13 Although it delivers the light output of 375 watts of halogen light, this three-lamp fluorescent is only 78 watts.

(Equipment Emporium, Inc.)

Calculating Power Draw

The formula says that the ***amperage*** (size) of a power load (in this case, a video light) is equal to the ***wattage*** of the load divided by the ***voltage*** of the circuit, or

$$\text{Amps} = \frac{\text{Watts}}{\text{Volts}}$$

It can be difficult to mentally calculate the amps used by a video light (and hence, the ability of an electrical circuit to take the load). Although North American current is nominally 110 volts, the actual voltage in a particular circuit may range from 105 to 130, and typically runs around 115–125. Without troubling to test each circuit, you cannot tell what its true voltage may be. Also, mentally dividing by a number like 117.5 volts is not easy.

To solve both problems, divide by an arbitrary 100 volts simply by moving the decimal. For example, a 750 watt light would draw 7.5 amps (750 watts divided by 100).

This not only simplifies the head math; it also builds in an automatic safety factor, since the nominal amperage will always be lower than the actual. For example, at a true 110 volts, a 750 watt light really draws 6.8 amps, not 7.5.

and broads can help minimize the problem. On the other hand, the gentle spill from umbrellas or softboxes often makes a very agreeable background light.

Ceiling Bounce

The low ceilings of many interiors can actually be a plus, because they make it easy to bounce fill light down onto subjects and background. Too much ceiling bounce, however, puts shadows under subjects' eyes and looks like institutional grid lighting.

Hiding Lights

Small interiors often make it difficult to keep the lights out of the frame. Light stands work well in front and to the sides of the action area, where they are safely off screen.

Study your monitor very carefully for cables, which have a way of creeping into the shot.

To hide lights placed deeper in the set, deploy units that can be clipped or taped to moldings, curtain rods, door frames, or the tops of open doors themselves. Fluorescent ceiling grids are also prime locations for clipping small lamps, **Figure 16-14**.

Some lighting instruments have built-in clips, others mount on posts fitted with clips or flat surfaces for recording.

Large Interiors

Large interiors are more comfortable to work in, but they present problems of their own, **Figure 16-15**. An area, say 50 feet square, cannot be fully lit with the instruments in a typical small production kit. To solve the problem, you need to employ a two-part strategy:

- Light the subjects, not the space (as discussed in Chapter 14).
- Break the action into parts, then light and shoot each part separately.

Managing Barn Doors

The edges of light beams look more realistic when they conform to natural features of the background. In this example, the spotlight is set to light a subject who will appear in the open doorway. The barn doors are set so that the beam edges are hidden by the sides and top of the door.

Without the key light, the doorway is dark.

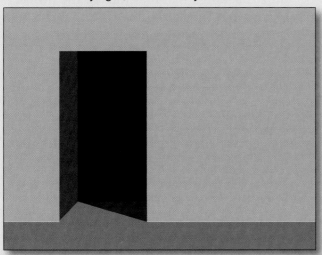

The key light spills onto the walls beside the door.

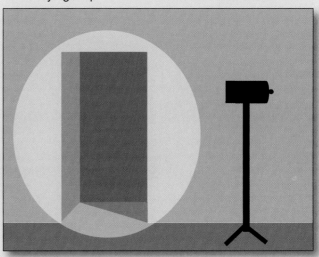

The left barn door masks the light beam to match the left edge of the doorway.

The remaining barn doors cut the beam edges at the bottom, top, and right edge of the doorway.

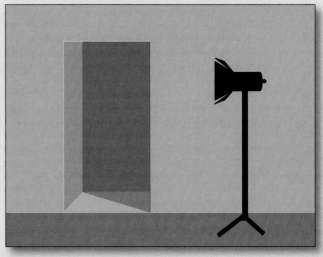

Figure 16-14 Small lights can be gaffer-taped to walls, or supported almost anywhere. Larger lights can be clamped to walls and doors.

Small light taped to a wall

Large light clamped to a partial wall

Small light clamped to a chair

(Lowel Light Inc.)

Figure 16-15 Too large to be lit with a small lighting kit, this room is lit only by its ceiling fluorescent fixtures—with unsatisfactory results.

Here is how to implement this strategy. In this example, we will use four spotlights and three broads, **Figure 16-16**.

- A single broad lights the background—the door area will be covered by a key light.
- A key light is positioned for each action area. The key farthest from the camera is hung overhead to keep it out of frame. A rim light for the closest action area is also added overhead.
- Two broads are placed to provide overall fill light.

Start with the Wide Shots

When the camera will see most of the area, use your lights as follows:

- *Light the background fully.* Since the walls or other backings will fill much of the frame, ensure that they are fully lit. Add lights to bring up furnishings or other contents of the area.
- *Light the action areas to be included in wide shots.* Choose important places within the area for additional key lights. The closest action area also gets a rim light.
- *Add general fill light.* This will bring up the overall light level. Ceiling bounce light works well, although direct fill from broads (as in **Figure 16-16**) is easier to control. Check floor areas carefully to make sure they get enough light.

The idea is to plan camera setups so that each major action area will also be covered in closer angles.

Light the Close Shots

After the wide shots have been recorded, you will re-light for each action area in turn:

- *Light the subject(s).* Typically, this means fine-tuning the key and fill lights and adding some rim light for separation.
- *Light the background.* You can take some instruments away from the background lighting by lighting only the parts that will appear in the close shots.

Take care that the key, fill, and background lights match the appearance of the wide shot lighting.

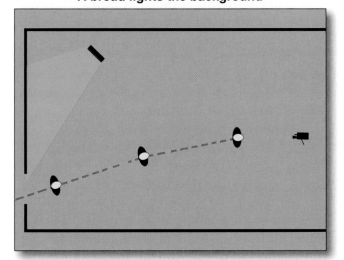

Figure 16-16 Lighting a large area with the instruments in a typical small production kit.

A broad lights the background

Key lights positioned for each action area

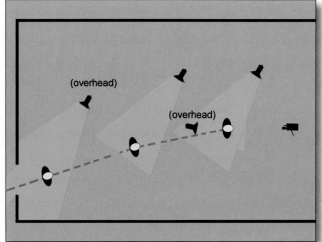

Two broads provide overall fill light

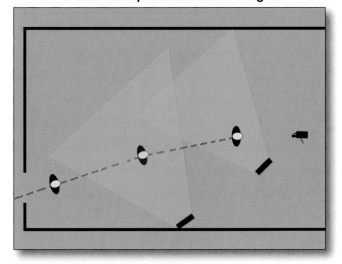

Completed lighting plan

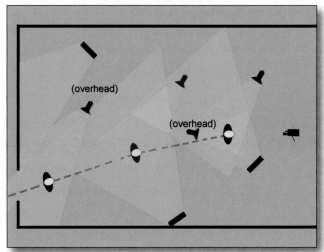

Lighting Moving Subjects

If there is subject movement from one action area to another, you will need to light all of them. Typically, the setup used for the wide shots will work well (**Figure 16-17**).

Exteriors

Chapters 13 and 14 cover, in passing, many outdoor lighting procedures. Following are some additional suggestions for professional looking results.

If the sky is overcast or you are shooting entirely in the shade, you cannot do much in the way of lighting, **Figure 16-18**. In sunny weather, however, you can use your outdoor resources.

Figure 16-17 To fully light each acting area, one fill light (1) is moved closer, an extra fill light (2) provides all-over fill, and two rim lights (3 and 4) accent the first two acting areas.

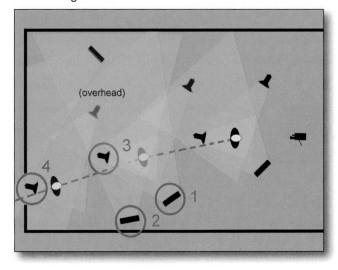

Figure 16-18 Reflectors, screens, and silks are ineffective on sunless days.

(Sue Stinson)

Choosing Reflectors

Except for rim lighting, a rule of thumb for reflectors is "the softer the better," for evenness of coverage and subject comfort.

Choose aluminum reflectors when a long throw is needed, for backgrounds or wide shots. Avoid using them for key lights (except when placed well back) to keep them out of subject's eyes.

White reflectors are excellent for fill. In closeups, subjects can even hold them themselves below the frame line.

Matching Wide and Close Shots

Lighting wide shots is easier outdoors because subjects do not need as much modeling, and because aluminum reflectors can throw effective fill light up to 50′ or more, **Figure 16-19**.

Changing Light and Weather

Reflectors must be tended constantly, especially aluminum units that throw narrow beams. Between the time when a setup is begun and the moment when the shot is recorded, the sun can shift enough to misdirect the reflector light.

Reflectors vs. Screening

Sometimes, you may prefer a screen or even a silk to a reflector. On the one hand, framed screens or silks cannot be used in wide shots, so matching the close shot lighting is more difficult. On the other hand, screens preserve the natural light patterns better than reflectors, and silks can also replace white reflectors when used vertically (**Figure 16-20**).

Figure 16-20 A silk on the left side and a reflector on the right.

(Photoflex)

Figure 16-19 Matching lighting for wide and close shots. A—The wide shot is filled with a reflector at the eight o'clock position, placed 25′ away. B—The closeup lighting matches the wide shot lighting. C—The reflector is 25′ away.

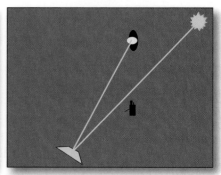

A   **B**   **C**

Lighting Moving Subjects Outdoors

The technique for lighting outdoor movement is the same as for interiors, except that your lighting instruments are reflectors. Hard aluminum surfaces work well because they throw light a long distance.

Also, after traveling 50' or so, the light beam pattern is broad enough to cover a larger area and is reduced to a more manageable intensity.

Multiple reflectors fill a moving shot.

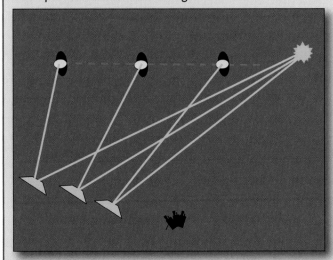

The longer reflector throw provides a wider, less intense beam.

Lighting Problems

All lighting situations have problems, but some are especially common. One of these is subjects who are hard to light pleasingly; another involves specialized light sources.

Subject Problems

Unless you are lighting characters in story videos, you generally want to make your subjects look as good as possible. The following are five of the most common subject problems, with suggestions for solving them.

Heavy Faces

You can use lighting to make heavy faces look slimmer. The trick is to highlight the center of the face and leave the sides in semi-shadow, **Figure 16-21**. To do this, key the lighting with a spotlight and use vertical barn doors (or flags, if necessary) to restrict the light to the center of the face. The light should be placed at or near the six o'clock position. To complete the setup, place soft fill lights at both the three o'clock and nine o'clock positions, moving them away from the subject until they deliver just enough light to reveal details in the shadows.

Thin Faces

The technique for lighting thin faces is just the opposite. Use soft key lights at or near the six o'clock and nine o'clock positions. Often, you may be able to omit the key light entirely, as in **Figure 16-22**.

For easier comparison, **Figure 16-21** and **Figure 16-22** use wide and narrow renderings of the same face.

Figure 16-21 A heavy face appears narrower when its sides are darker than its center.

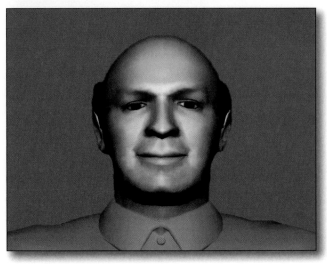

Figure 16-22 A thin face appears broader when its sides are brighter than its center.

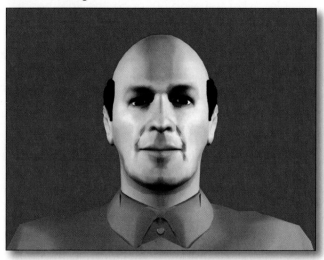

Darker Complexions

Darker facial tones are beautiful when well lit, but they can offer contrast problems—especially in wedding cinematography, when the bride's skin tones are contrasted with a brilliant white wedding dress. The trick here is to get more light on the face than the dress, **Figure 16-23**.

Many on-camera lights can be fitted with barn doors that partially block light from the white wedding dress. With stand-mounted lights, half screens can be added to the barn doors to further moderate the lower part of the light beam.

If your camera light will accept filters, you can buy or make a half screen to reduce the bottom part of the light beam.

Bald Subjects

Balding heads are best handled by the makeup department—a little neutral powder will kill reflections and be quite invisible to the camera. If the subject refuses powder (as men sometimes do) try moving the lights up and then, if necessary, farther to one side to reduce the reflections.

Subjects Wearing Glasses

For reflections on eyeglasses, the solution is similar. In general, small-source spotlights are easier to move out of the incidence/reflection path (as explained in the sidebar). However, reflections from the softlights often used for interviews can be tolerable, because viewers know that the subject has been lit for video.

For brief shots, it is often enough for the subject to lift the earpieces of the eyeglasses slightly off the ears, tilting the lenses downward, and deflecting the reflection. When not overdone, this adjustment is generally invisible to the camera.

Specialized Light Sources

Practicals (such as table lamps) and environmental light sources (streetlights, signs,

Figure 16-23 White clothing against darker complexions creates contrast problems. Reducing the light to the lower part of the frame improves contrast.

Contrast problem

Lighting adjusted

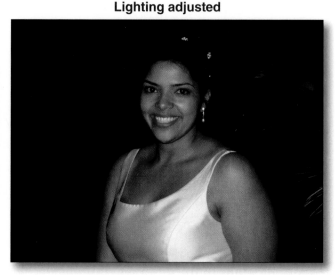

Incidence Equals Reflection

In dealing with bald heads, spectacles, and other reflection problems, remember that light bounces off surfaces at the same angle, but in the opposite direction. So, if a reflection is hitting the camcorder lens, the light is probably too close to the camera position, whether horizontally, vertically, or both. That is why raising a light or moving it sideways will often remove or at least lessen a reflection.

To minimize reflections, place lights at angles of 45° or greater.

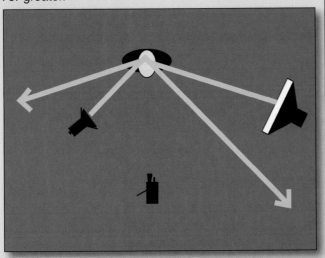

and shop windows) create problems because they are rarely in the right place and/or operating at the right intensity. To solve these problems, it is usually necessary to establish the light source by showing it on camera, and then replace it with a more controllable light.

Practicals

How you handle the lights that appear in the video frame depends on the lighting style you have chosen.

Naturalistic or Expressionistic

Oddly, the opposite extremes in lighting styles can use the same technique: replace bulbs in practicals with screw-base halogen lamps and use them for actual video lighting. The resulting light will be contrasty, but excess contrast is acceptable in these styles (**Figure 16-24**).

Figure 16-24 You can light with practicals if the result fits the lighting style you are using.

Realism or Pictorial Realism

With these more common styles, you may want to establish the practicals and then use camera setups that exclude them, while simulating their light with video lights.

Adjusting Intensity

To balance visible light sources with the rest of your lighting, try fitting larger or smaller lamp bulbs, as needed. Halogen replacement lamps can be dimmed somewhat to reduce intensity, but ordinary lights are already too orange to permit much further color shift through dimming.

On the other hand, if you light a scene entirely with household bulbs, you can simply set the camera's white balance manually to match their 2700K–2800K color temperature.

Moving Light Sources

Subjects often carry light sources, such as flashlights or lanterns. When the practical light is not on-screen, you can simulate its light for better control.

- **Flashlight.** A small spotlight with a handle makes a good simulated flashlight. Focus the beam in the spot position to create a hard edge, **Figure 16-25**.
- **Lantern.** To make a convincing "lantern," clip sheets of diffusion and orange filter material (for "candlelight") to a broad.

Figure 16-25 The practical flashlight is established by showing it on screen. But, the light from the flashlight is provided off-screen by a hand-held spot.

Flashlight beam

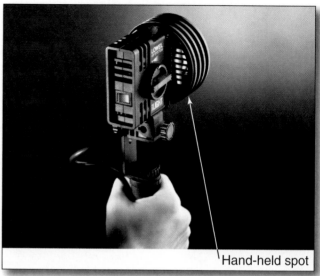

Hand-held spot

(Lowel Light Inc.)

- **Candle.** For a candle effect, omit the diffusion for a harder light. Wearing a leather glove, wave your hand and move your fingers slowly in front of the light to create a flickering effect.

Firelight

Simulated firelight is created in somewhat the same way as moving light sources.
1. Place a broad very low, where a fireplace or campfire would be.
2. Gel the broad with an orange filter sheet.
3. Staple a square of heavy cloth to a stick (denim works well), and slit the cloth at 1" intervals to create a "grass skirt" effect, **Figure 16-26**.
4. Wave this device slowly in front of the "fire" light source to add a convincing flicker effect.

As with most such effects, you can enhance the realism with sound effects—in this case, a crackling wood fire.

Signs

In some night interiors, colored signs and other neon sources tint parts of the subject. If the sign is steady, simply gel a light with an appropriate color. If the sign turns on and off, have an assistant move a flag rhythmically in and out of the light path.

Electronics

Radar screens, computer monitors, and scientific instruments often bathe the faces of the subjects looking at them with light. With some units, the actual screen light may be bright enough, especially if the rest of the lighting is low key. If you cannot show the screens, however (say, because they are supposedly futuristic displays in a spaceship control cabin), place small lights low (at "screen" height) and gel them pale blue or green.

Small LED arrays are often used to simulate electronic screens because their built-in dimmers simplify adjusting brightness.

Figure 16-26 A "grass skirt" cookie can simulate the flickering of a fire.

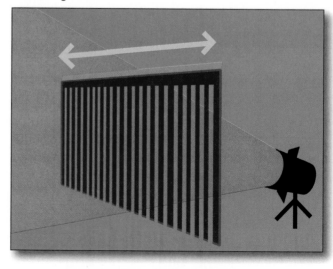

Lighting Night Scenes

Lighting scenes shot at night is challenging because there is little available light to help out. This section suggests some ways to create nighttime designs with relatively few instruments.

Interior Scenes

You can light indoor night scenes by using a few standard techniques.

Use low-key mode. Create a low-key look in which dark and medium values dominate in the background, with brighter accents and a well-lit subject.

Establish practicals. Since room lights are lit at night, establish practicals in the frame, and then mimic their light with video lights.

Control window light. If you can hang heavy screen or neutral density material inside or outside a window, you may be able to reduce its light to a "nighttime" level (the bluish color temperature will look like moonlight). If you do not have the resources to do this, exclude windows and their light from the frame.

Light for the highlights. In low-key lighting, you naturally use less fill light, so that shadows are deeper and show less detail.

Fake the Darkness

It is common to show a subject in bed, turning off the bedside light and going to sleep.

Shot in actual light, the scene would be very contrasty, and then would go black when the light went out. Here is a procedure for lighting this scene more effectively.

1. **Establish the light level.** Fit a halogen lamp in the bedside light and use it to key the scene. Use soft fill from the other side, with a crew member at the light. At this point, set and *lock* the camcorder exposure setting.
2. **Establish "night".** Next, turn off the key and fill lights and set up a very general overall fill, possibly with a pair of fluorescent light banks or large LED arrays. *Without changing the camera aperture,* adjust this fill light until the subject and bed are visible, though dark.

As an alternative, gel a spot or broad pale blue and place it at room-window height for a moonlight look. A window-frame *cookie* in front of the light will enhance the effect.

3. **Synchronize the scene lights.** Rehearse the shot until the crew member at the fill light switch can turn the light off exactly when the subject turns off the bed light, so that the two light sources look like a single light, **Figure 16-27**.

Although the remaining light (from the fluorescent or LED fill) will be somewhat too bright for perfect realism, viewers generally accept it as "darkness."

Figure 16-27 Faking darkness. A—The scene, as fully lit with practical, fill, and pan fill lights. B—After the subject and crew member have simultaneously switched off the practical and fill lights, the pan fill lights provide a very low level of light the viewer will accept as "darkness."

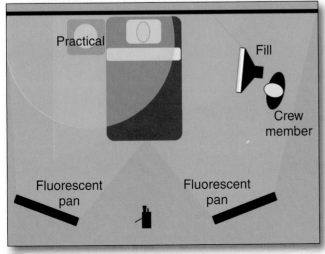

A

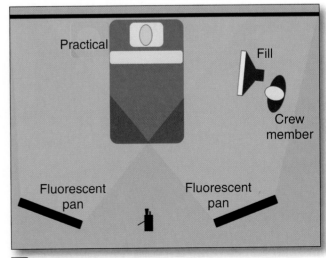

B

Exterior Scenes

It is impossible to light the whole outdoors for night scenes, but a few tricks will yield very satisfying results. First, light only the important action areas, as described previously. Then, try these suggestions.

Use Back-Cross Lighting

Except where the action must be seen clearly, place lights to the side and behind subjects (in the nine o'clock to three o'clock arc) to edge them with rims of light that will separate them from the background, **Figure 16-28**.

Rake the Background

Night scenes are supposed to be dark. So, use just a few lights to pick out features of the background. As usual with background lighting, place the lights to hit the background at oblique angles. See **Figure 16-29**.

Look for Motivation

Outside of urban centers, there is often little actual light. To motivate video lighting, simulate car headlights and house or shop windows with lights placed low and shooting horizontally. For streetlights, move rim lighting directly over subjects to create eye socket shadows.

Do not worry if some of your lights do not have enough motivation. This lighting problem is so common, even in big-budget productions,

Figure 16-28 Lights placed high and close to the building provide back lighting.

Back-cross lighting

Figure 16-29 This large castle background is lit by just three lights.

Rake the background

that viewers have come to accept night exteriors full of unexplained light sources.

Atmosphere

Outside night scenes in movies are often wet because rain or fog (real or fake) picks up and scatters light rays, **Figure 16-30**.

Day-for-Night Lighting

Although you can use a camcorder in very low light levels, there are good reasons for shooting exterior night scenes during the day using *day-for-night lighting* (**Figure 16-31**):
- Daytime shooting is more convenient for everyone.

Figure 16-30 Rain provides effective atmosphere.

(Sue Stinson)

Lighting at "Magic Hour"

Magic hour is the brief period before sundown. On a sunny day, magic hour provides light qualities that look especially attractive on screen. Shadows from the low sun are long, which models objects and enhances the impression of depth. The moisture in the air is often low, so everything appears exceptionally sharp and clear. The color temperature is warmer, lending a golden tone until near sundown, and then a distinctive sunset-orange tint.

McKinley at magic hour.

Lighting for magic hour is simple because the low sun makes reflector placement easy. The problem lies in capturing all the footage required in the relatively brief time before the sun actually sets. For this reason, you may wish to preset and rehearse several different camera setups, so that you can move quickly from one to the next as you shoot.

Another problem with magic hour is white balance, because the color temperature drops continuously as the sun goes down. One way to solve the problem is by compensating for the color shift while shooting. To do this, manually reset the white balance frequently. That way, you will capture the long shadows and clear light of magic hour, but all your original camera footage will have the same neutral color balance.

After you have edited a magic hour sequence, you can apply sunset tint to taste or even warm the images up progressively as the sequence unfolds to simulate an actual sunset. Be aware, however, that a digitally applied sunset color can have a mechanical, too-uniform quality. If you have the skill to capture it, there is no substitute for real "magic hour" light.

The characteristic long shadows and warm light of "magic hour."

- Daytime light levels are high enough for optimal imaging.
- Fewer lighting instruments and accessories are required.
- Though electrical power is helpful, in some cases you can shoot without it.

In video, making daytime shots look like night is easy if you follow a few simple guidelines (the lighting setup is diagramed in **Figure 16-32**).

Set white balance for incandescent. Using an indoor white balance setting outdoors (**Figure 16-33**) will lend an overall "moonlight" bluish cast to the footage, while rendering any

incandescent lights as true white. You may also want to set exposure so that the natural shadows fill in as deep black.

If you will edit digitally (and no incandescent lighting is used) you can create the color, exposure, and contrast of "night" in postproduction. In this situation, it is often safer to get conventionally balanced and exposed footage and then alter it later.

Use back-cross lighting. Instead of directing the main light source at subjects from the front or part-way to one side, position your subjects and reflectors as needed to splash the brightest

Figure 16-31 Day-for-night lighting.

Exterior shot recorded normally.

Completed day-for-night version.

Figure 16-32 The setup for the shot.

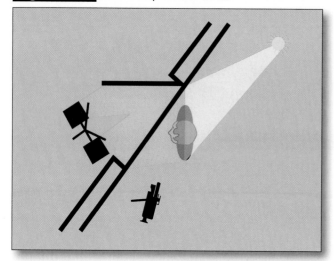

light from the rear (onto hair and shoulders), while the faces remain somewhat darker, **Figure 16-34**.

Frame off the sky. Unless you can darken the sky with a polarizing filter, use neutral or high angles to aim the camcorder away from the sky, which will appear much too light for a convincing night effect (**Figure 16-35**).

In some lighting conditions, a polarizing filter on the camcorder can turn a blue sky dark enough to pass for night, if it is not allowed to remain on screen too long.

Include incandescent light. Try to include some incandescent or halogen lighting in the shot, such as headlights, a street light, or light

Figure 16-33 White balance set for incandescent (indoor) light.

Figure 16-34 Back-cross lighting from the sun creates a rim light of "moonlight."

Figure 16-35 Including the daytime sky reveals the trick.

(oops!)

streaming out of a window or open door. Since the indoor white balance setting will render this light as "white" it will contrast convincingly with the "moonlight" cast of the overall scene, **Figure 16-36**.

Lighting Assignments

In addition to the general lighting situations covered so far, there are a few specific assignments that come up frequently enough to deserve special attention. These include interviews, "stand-up" reports, compositing, very small areas, and graphic materials.

Figure 16-36 Incandescent light completes the illusion.

Interviews

Interviews are among the most common lighting assignments. Typically, they involve two subjects and a moderate amount of background.

One-Person Interviews

The current style of video interview has the subject on-screen all the time. The interviewer's questions are posed as topics to be responded to, so that they can be omitted in editing. In the finished interview, the subject appears to be discussing the subject spontaneously.

Softlights are frequently used because they look natural, they generally light the background as well as the subject, and they are quick and easy to use.

Since interview subjects are rarely media professionals, lighting should be moderate and be kept out of their eyes. Because interviews rarely use angles wider than medium (waist) shots, a reflector can be used opposite the soft key for fill. Even with "natural" lighting, a small spot rigged as a rim light can add modeling and separate the subject from the background, **Figure 16-37**.

Two-Person Interviews

If your interviewer will appear on screen, you must light her or him as well, **Figure 16-38**. In establishing shots, the back view of the narrator in the foreground can often be lit by the lights on the interview subject. The same lights provide rim light to separate the subject and interviewer from the background.

For single shots of the interviewer asking questions or listening to answers, you have two different options: light for real-time recording or re-light for a re-shoot. To light the interviewer separately for real-time recording with a second camera, position the reporter's camera to shoot over the interviewee's shoulder, so that both people are included in wider shots. Place the lights for both people so that they are outside the frame in both setups. If you are using only one camera or if some of the reporter's questions and reactions need to be re-shot, you can re-light and reset the camera after the main interview to pick up this material for later editing into the sequence.

Figure 16-37 Lighting a one-person interview.

Two-light setup

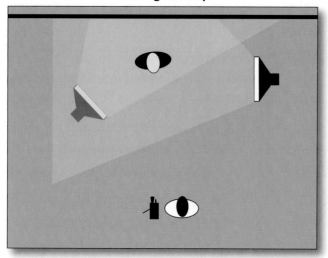

Closeup of the subject

Figure 16-38 Lighting a two-person interview.

Three-light setup

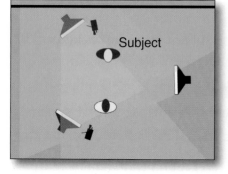

Subject

Shot over the interviewer's shoulder

Reverse setup features the interviewer

If you shoot the interviewer separately, you can place him or her in front of whatever background looks best—even in another location, if necessary. The shift will be invisible to the viewer, **Figure 16-39**.

Stand-Up Reports

Segments presented by reporters standing at the scene of news, sports, or entertainment events are called "stand-ups." Even at the broadcast TV level, reporters often go into the field with a single technician who must function as cinematographer, lighting director, and sound recordist (**Figure 16-40**).

Daytime Reports

To simplify lighting, a one-person crew will often work with just two tools: an on-camera light and a stand-mounted reflector, **Figure 16-41**. The camera-mounted light is typically variable in light output and powered by its own substantial battery. Even in daylight, a front fill light can highlight the reporter's face just enough to emphasize it against the background.

News camcorders are often fitted with batteries big enough to power a light as well.

In **Figure 16-41**, the supplementary light is a stand-mounted reflector. Because the camera person must leave the reflector unattended while shooting, it needs to be heavy and relatively small to resist the wind.

When Cheating Is Legal

Cheating is the common practice of moving a subject between camera setups, usually to increase working room, separate subject and background, or find a better-looking background altogether.

Careful cheating is invisible to viewers, because the lack of true depth in video makes subject-to-background distances hard to judge, and the background "behind" a subject may not appear in earlier shots.

A master two-shot of subject and interviewer, avoiding the windows.

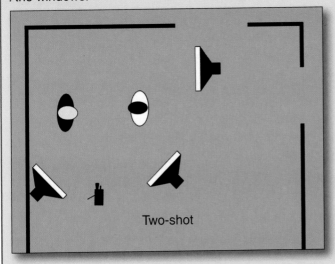

Two-shot

In this example, a subject and an interviewer are placed in the upper-left corner of a room, to keep back and side windows out of the frame for the master shot.

Because there is no room behind the subject to add a rim light for his closeup, the entire setup is "cheated" three feet out from the left wall. From the camera's new position, the move is undetectable.

Lights, camera, and performers are all cheated to the right.

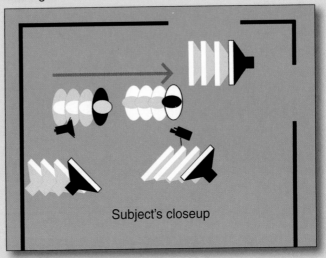

Subject's closeup

Cheating a subject to light the background separately.

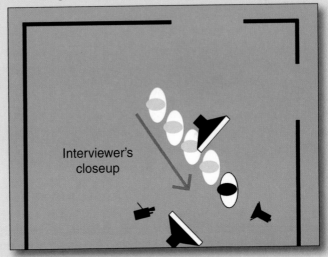

Interviewer's closeup

When it is time for the interviewer's closeup, she is cheated back and to the right (to avoid the right-wall window), and completely re-lit.

Because viewers do not see the right wall in the other shots, this cheat, too, is undetectable.

Nighttime Reports

Night shooting requires a different deployment of lighting tools, **Figure 16-42**. The on-camera light continues to provide much of the illumination. A second, battery-powered light on a stand can fulfill the same function as a reflector, providing more modeling on the reporter's face. Backgrounds at night are generally dark, but look for lighted walls or windows to include in the shot, so that the image behind the reporter has some design to it.

LED lighting units are especially useful for field work because of their very low power consumption.

Figure 16-39 Moving the interviewer farther away from the interviewee allows a better camera position for her closeup.

Subjects are close together in the two-shot

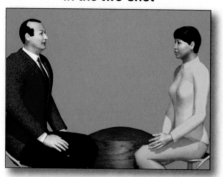

Interviewer is moved farther away

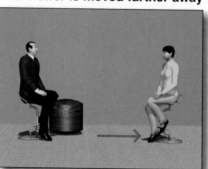

Closeup of interviewer with different background

Figure 16-40 Two-person news teams are typical.

Lighting for Compositing

With the power of today's postproduction software, compositing is used increasingly.

Compositing is the process of video recording subjects against flat, single-color backgrounds, and then digitally replacing the background with other visuals during postproduction.

The more even the background color, the more perfectly it can be replaced. So, the challenge is to keep the background lighting absolutely uniform. To do this, you try to light background and subjects separately, **Figure 16-43**.

Lighting the Background

You can achieve very even background lighting with a pair of softlights (or spots with

Figure 16-41 In this setup for a daytime report, the sun provides the rim light and a reflector fills from the left. An on-camera key light can be used to better highlight the reporter's face.

Sunlight and reflector

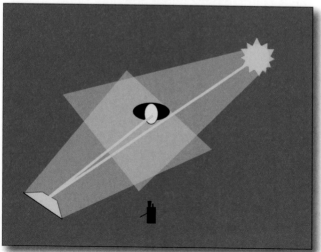

On-camera key light

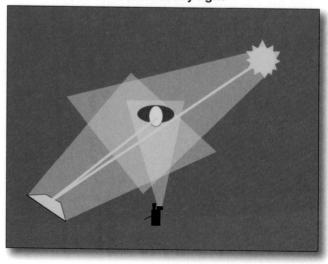

Figure 16-42 Lighting for nighttime reports.

One-light setup

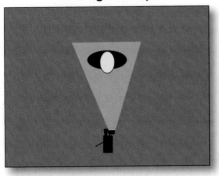

Two-light setup

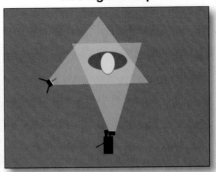

Offset the camera light for better modeling

Figure 16-43 Lighting for compositing.

Spotlights with diffusion evenly wash the background

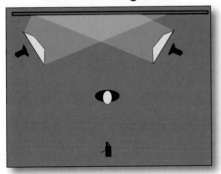

Subject is as far in front of the background as practical

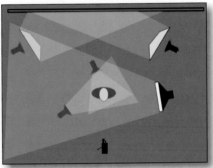

Softlighting on both subject and background eliminates shadows

(Bogen Manfrotto)

heavy diffusion), placed one at each side. Experiment with light positions and throws to create a near-perfect wash of light.

Lighting the Foreground

Place the subjects as far forward of the background as possible. Classic spotlight lighting is easiest to use because the light paths are controllable. To keep foreground light off the background screen, position the lights relatively high and to the side (as far as 8:00 or 8:30) to keep hot spots and shadows below and to the sides of the frame.

Be sure to look at the background footage to determine the quality and direction of its own lighting. If your setup allows, key this footage into the image in place of the composite screen. Adjust subject lighting until it matches the composited background.

Small Objects and Areas

Lighting small objects can be difficult because most professional lights are scaled to illuminate larger subjects. To overcome this problem, several techniques are available.

Tabletop

Tabletop cinematography involves shooting small objects and/or activities on a table, counter, or special photo stand (**Figure 16-44**). It is a common procedure for science experiments, product demonstrations, and how-to training sequences.

In shooting small subjects, camera and lighting problems generally arise from two causes:

- The short camera-to-subject distances (and/or telephoto lens settings) create very shallow depth of field. This makes small objects difficult to keep in sharp focus.

Figure 16-44 A tabletop shot.

(Photoflex)

- The hands, arms, and head of the demonstrator tend to get into the picture.

To help solve these questions, you need lights that are both very bright (to force smaller lens openings and thereby increase depth of field) and very soft (to eliminate or at least minimize shadows in the picture).

Every situation has unique lighting requirements, of course, but **Figure 16-45** illustrates a solution for a typical tabletop setup. The subject and tabletop are bracketed by very large, bright fluorescent softlights. These instruments are excellent for this application because:

- Their 4' square shape provides an extremely large source to reduce shadows.
- Their multiple 40-watt tubes provide a bright light, permitting smaller f-stops to create greater depth of field.
- Their power requirements are low enough for use in most locations.
- Their output is cool enough for subject comfort and for delicate applications, such as food demonstrations and biology experiments.

LED arrays are even cooler and consume less power. Because their light is not as soft, however, they may need added diffusion for truly shadowless lighting.

Tenting

Where you want completely shadowless lighting (and do not require great depth of field), you can use *tent lighting*. By hanging a

Permanent Compositing Studios

Green screen backgrounds that include floors permit the use of "virtual" sets—complete digital environments into which subjects can be composited.

A permanent compositing setup.

(Litepanels, Inc.)

In this studio setup, more than 20 one-foot LED arrays on overhead pipes are used to light the upper part of the background. Large arrays, like the 16-panel floor unit pictured on the right of the studio setup figure, cover the lower walls and floor. Smaller units, like the single and four-panel arrays stored on the left of the studio setup figure, are moved into position to light subjects in the foreground.

LED arrays are very useful in compositing applications where large numbers of lights are required for uniform coverage because they emit almost no heat, they can be balanced by adjustable, remote controlled dimmers, and they consume so little power. When this studio uses 50 panels, all of the panels together draw 2,000 watts.

To achieve a comparable light output with halogen lights would demand over ten times as much power—not counting the wattage of the air conditioning needed to counteract their heat.

Figure 16-45 Tabletop setups use very large, soft lights.

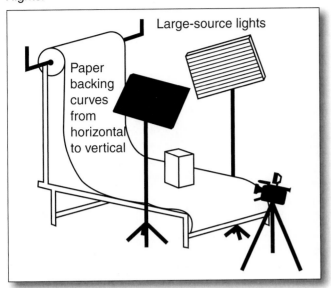

Figure 16-47 A professional tent for small object cinematography.

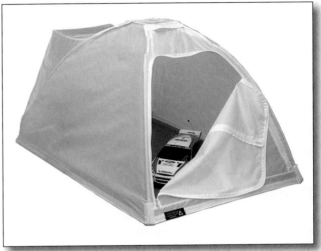

(Photoflex)

white sheet over and around your subject and aiming lights through it, you can create the softest possible lighting effects. Tenting works very well outdoors, with a sheet hung over a pair of clothesline ropes, **Figure 16-46**. Tenting techniques are also useful for lighting products in commercials, **Figure 16-47**.

Ring Lights

Where other lighting techniques are not practical, ring lights can often achieve comparable effects (**Figure 16-48**). Because they surround the lens completely, these lights throw only one soft shadow, which is typically masked by the subject in front of it.

Due to their unique shape and mounting position, ring lights require diffusion and filters that are specifically designed for them.

Graphic Materials

Many programs include two-dimensional subjects—photos, paintings, graphics, letters, book pages—in place of moving subjects. With today's quality equipment, it is often easiest to record subjects with flatbed scanners and import them into video programs during postproduction. If the flat material is larger than about 8″ × 10″, however, scanning is often impractical. You will need to video record these larger subjects directly using a special lighting setup.

Figure 16-46 A simple tent setup outdoors, using a bed sheet and clotheslines.

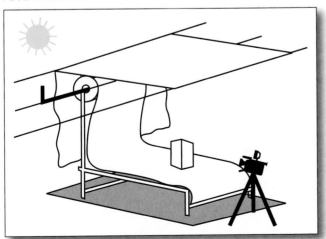

Figure 16-48 Ring lights are available in LED arrays.

(Litepanels, Inc.)

Organizing a Setup

Most often, you will work with the graphic material on a flat surface and the tripod-mounted camcorder is aimed down at it. A sheet of glass will help hold the material flat, but may create reflection problems. The lighting is simple: one unit on each side of the artwork (**Figure 16-49**). Clamp work lights with halogen lamps are easy to position and adjust.

For materials up to about 9" × 12", scanning is now more common than copying.

Dealing with Reflections

With or without a glass cover plate, light reflections are often a problem. To solve them, make sure that the lights are aimed at a 45° angle.

Working Vertically

In many cases, posters, paintings, charts, and other large subjects are best handled vertically. Make sure that the camcorder is centered horizontally and vertically, at a true 90° angle to the artwork. Position the lights far enough back to wash the subject evenly, and keep them at a 45° or less angle from the wall (**Figure 16-50**).

Placing the camcorder far back with a telephoto lens setting will improve the quality of the image recorded.

Figure 16-49 A professional copy stand with light diffusion for larger graphics.

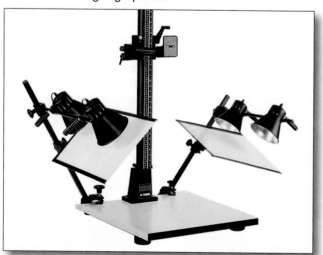

(Bogen Manfrotto)

Figure 16-50 A vertical lighting setup for copying. To avoid reflections, place lights at an angle of 45° or less to the wall.

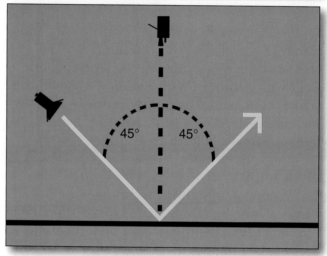

Video Snapshots

When you capture passing events spontaneously (especially with a mobile phone, a personal media device, or a still camera), you have to use whatever light is available. Though you cannot control the lighting, you can make the most of whatever you find. To do this,

- Make sure you have enough light. If the image on your screen looks too dark or lacks detail, see if there is any more light you can turn on. If your recorder's sensitivity can be adjusted, increase it to handle the low-light situation—many still cameras with video capability will do this automatically. A grainy image is better than almost none at all.

- Avoid bright backgrounds and excessive backlighting. If you can change your position, move around until windows and other light sources are out of your frame.
- Look for good lighting. If you have the chance, use the most expressive lights available and, if possible, move your subject into them (**Figure 16-51**).

Figure 16-51 Use the most expressive lights available. A—In this candid video, the down lights over the sink were turned on and the subject was moved to take advantage of them. B—In very dark, contrasty environments, you can sometimes position yourself between your subjects and the available light sources.

A

B

- As always with video intended for e-mail or the Internet, avoid excessive panning, which can cause visible light sources to smear as they move across the screen.

Since the charm of video snapshots lies in their spontaneity, elaborate lighting defeats their purpose. Nevertheless, you can often improve the lighting without making the scene appear "lit."

Internet Lighting

In addition to video snapshots posted to websites, many carefully produced videos are published on the Internet. These programs should be as fully lit as programs destined for other delivery systems. Although many sites now accept high-definition programs, streaming videos can have limitations because they are often intended for less than the full screen area. Here are some tips for lighting Internet videos:

- Use classic four-source lighting (key, fill, rim, and background lights) to model subjects and enhance their apparent depth.
- Pay special attention to rim (back) lighting to separate subjects sharply from their backgrounds.
- Aim for high (or at least "medium") key lighting to deliver bright images. Streaming video, especially when lower quality, can turn dark areas muddy and lacking in detail.
- Keep background lighting simple to avoid shadow patterns and excessive modeling that can compete with foreground subjects. Backgrounds should be just dark enough so that well-lit foreground subjects stand out, **Figure 16-52**.

Figure 16-52 The background should not compete with foreground subjects.

Though this music video was exposed for the subject...

...the very light background may cause too much contrast when the program is streamed.

Other Lighting Applications

The representative lighting solutions in this chapter cannot cover all the situations that you may encounter, but they do demonstrate how to use the basic ideas behind all lighting designs to analyze each situation and create video lighting with style.

Summary

- Classic studio lighting uses three lights on a subject and usually one or more on the background.
- The key to "natural" lighting is soft light.
- Background light intensity should be adjusted so that subject and background seem lit by the same environment, but the subject is slightly brighter.
- Lighting locations presents unique challenges, such as lighting small interiors, having sufficient power supply, adequately lighting a large interior, changing light and weather conditions outdoors.
- In general, lighting should make your subjects look as good as possible.
- Specialized light sources should be established by showing it on camera, and then replaced with a more controllable light.
- Techniques for lighting indoor night scenes include: use low-key mode, establish practicals, control window light, and light for the highlights.
- Tried-and-true techniques are regularly used for common lighting assignments.

Technical Terms

Amp (amperage): In lighting, the amount of electrical current drawn by a lighting instrument.

Day-for-night lighting: A method of shooting daylight footage so that it appears to have been taken at night.

Glamorous lighting: Lighting that emphasizes a subject's attractive aspects and de-emphasizes defects.

Magic hour: The period of time, up to two hours before sunset, characterized by long shadows, clear air, and warm light.

Rugged lighting: Lighting technique that emphasizes three-dimensional qualities and surface characteristics of a subject.

Tabletop: Cinematography of small subjects and activities on a table or counter.

Tent lighting: A lighting arrangement in which white fabric is draped all around a subject to diffuse lighting completely for a completely shadowless effect.

Three-point lighting: Classic subject lighting technique that consists of key, fill, rim, and background lights.

Voltage: The electrical potential or "pressure" in a system—typically 110 or 220 volts in North America.

Wattage: In lighting, the power rating of a lighting instrument. 500, 750, and 1,000 watt lamps are common.

Review Questions

Answer the following questions on a separate piece of paper. Do not write in this book.

1. Classic three-point lighting uses key, _____, and rim lights on the subject, plus a background light.

2. *True or False?* Glamorous lighting exaggerates the planes and angles of the face and emphasizes skin texture.

3. Why are spots and floods effective instruments for lighting backgrounds?

4. A light's _____ is the distance between the instrument and the subject or background that it is lighting.

5. Explain why spots are especially useful in tight quarters?

6. What is the simplified formula for calculating power draw? What are the benefits of using this simple calculation?

7. *True or False?* Reflectors that produce softer light provide even coverage.

8. When a long throw is needed, choose a(n) _____ reflector.

9. How can a heavy face be lit to look slimmer?

10. What are the standard techniques that can be used to light an indoor night scene?

11. What is *magic hour*? What are the qualities of available light during magic hour?

12. Using an indoor white balance setting outdoors will create a(n) _____ cast to the footage.

13. *True or False?* The main goal when lighting for compositing is to keep the background lighting absolutely uniform.

14. What is *tent lighting*?

15. Explain the setup for recording a vertically-positioned, two-dimensional graphic.

STEM and Academic Activities

STEM

1. **Science.** What is Ohm's Law? How does Ohm's Law apply to electrical circuits and power supply?

2. **Technology.** Identify some lighting tools or pieces of equipment that have emerged in the last 20 years. What impact have these items had on the task of lighting a scene or location?

3. **Engineering.** Make a sketch of a single subject on an interior set. On the sketch, draw the position of each lighting instrument for a three-point lighting arrangement. Label each instrument on the sketch.

CHAPTER 17 — Recording Audio

Objectives

After studying this chapter, you will be able to:

- Summarize the characteristics of all the major types of audio tracks.
- Identify the different types and purposes of recording.
- Explain how to solve the problems associated with common recording situations.
- Recall the appropriate use of audio equipment and select equipment appropriately.
- Understand how to record quality audio.

365

About Recording Audio

Audio, as we use the word here, is the recording and reproduction of sound in support of video. Chapter 8 explains how the sound track affects viewers and why it is such an important part of every video program. With this information in mind, we are ready to consider the techniques used to create that sound track—the techniques of professional audio recording.

Professional camcorders allow manual control of audio recording. Consumer camcorders, however, record sound automatically. Automatic recording is simpler, but if you have heard the often mediocre results, you can see why obtaining quality audio requires effort and skill.

Audio recording must always deal with two problems. First, recording captures everything, whether it is wanted or not. Second, on playback, recorded sounds may seem distorted or even quite unlike the originals. Before learning about specific audio equipment and techniques, you need to understand how these fundamental problems affect audio recording.

Problem: Audio Records Everything

Like your own ears, microphones pick up every sound within their range. Sound from an outdoor scene might be a mix of spoken dialogue, background traffic noise, wind, shuffling feet, and even sounds made by the camcorder itself (**Figure 17-1**).

Figure 17-1 A noisy background can degrade audio quality.

The problem arises because of the way in which people process the sounds they hear. You may hear sounds with your ears, but you *listen* with your brain. Without noticing it, you suppress everything in the incoming sound mix except for the parts that interest you. For example, you may focus on the dialogue while blocking out the background noises. Then, if you should notice trees waving in the wind, you might shift your mental focus to the wind sound, to the point where you no longer hear what is being said.

This topic is covered more fully in the sidebar *The Perception of Sound* in Chapter 8, *Video Sound*.

For reasons that are not perfectly understood, this mental sound filtration is more difficult to do with recordings than with live sounds. As a result, the dialogue in a video scene may be ruined by wind or other noises that you would not even notice in the real world.

Problem: Audio Distorts Sounds

Recording a sound does not guarantee that it will be convincing or even recognizable on playback. Many factors affect the quality and character of recorded sound. It takes skill to listen to what the microphone is picking up and then adjust recording conditions to make the resulting signal sound natural.

Even when the sound is not actually distorted, an audio recording can still sound unnatural, especially if it does not match the visual environment. For instance, sound recorded in postproduction and then synchronized with a video scene in a metal aircraft hangar may sound muffled and "dead", when it should have a metallic echoing quality.

The most common problem with sound quality is inconsistency from one shot to the next. The sound character of a wide shot may not match that of close-ups made of the same sequence because the camera's broad field of view has forced the microphone farther away from the performers in order to keep it out of the shot (**Figure 17-2**). In an edited video sequence that alternates between wide and close angles, the constantly changing sound quality will be obvious and annoying.

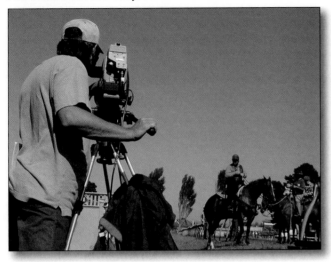

Figure 17-2 A wide shot may force the microphone far back from the subjects.

In the real world, of course, we do not abruptly change position every few seconds, as the camera does from one setup to the next.

Overcoming Audio Problems

To address these problems, professional audio recordists employ a well-proven strategy:
- Identify the separate types of sound that are wanted for a sequence (such as dialogue, sound effects, and background noises).
- Isolate each sound component and record it separately.
- Select, position, and adjust microphones to obtain consistent, high-quality, natural-sounding audio, regardless of the visual perspective.

With clean, separate audio tracks, a sound editor can place and balance audio components to duplicate the function of the listener's brain, calling attention to certain sounds and minimizing others, as required (**Figure 17-3**). A well-mixed sound track directs the ear, eliminates audio clutter, and enhances the realism of the program.

Sound track mixing is covered in the chapters on editing.

Providing the editor with clean, separate components is the sound recordist's fundamental job. To do this job, you need to know the basics of audio recording, the equipment used to do it, and the techniques required to obtain high-quality audio.

Figure 17-3 Digital editing can handle dozens of audio tracks.

(JVC Corporation)

Types of Audio Recording

In recording raw sound for later editing, you must deal with several different types of tracks, each requiring its own set of recording techniques. At different times, you may need to record production sound, background sound, sound effects, voice tracks, and music. The most common type of recording is production sound.

Production Sound

Production sound is, of course, the sound that is recorded with the video. While most amateur sound is recorded by the camcorder's built-in microphone, professional productions generally use separate mikes. Because the camera position is rarely the optimal location for miking the shot, separate video and audio recording devices allow each to be placed in the best spot, **Figure 17-4**.

Production sound calls for special care because mistakes may require another take of the whole shot. As an extreme example, if your first attempt to record the actual sound of a train derailing is unsatisfactory, you cannot re-record the effect without derailing another train.

Because actual recordings are often impractical, train derailments and similar sounds are usually simulated and recorded separately as sound effects and added to the video in postproduction.

Video Digital Communication & Production

"Mit Out Sound!"

Unlike video, film production usually requires two separate machines for recording picture and sound. For this reason, shots in which the audio would be difficult to record adequately or microphones would interfere with the visuals are often made with the film camera only, omitting the sound recorder. Shots filmed without sound are slated M.O.S.

Tradition holds that the acronym "M.O.S." stands for "Mit Out Sound," in the imperfect pronunciation of a famous European director who came to Hollywood at the beginning of the sound era and hated the new technology so much that he did everything he could to shoot "mit out" it.

In video, by contrast, picture and sound are usually captured together, so sound of one sort or another is almost always recorded. Even if you think you will be unable to use the production audio in the finished video, you should still record it, at least with the camcorder's built-in mike. This will create a reference track, also called a "scratch" or "cue" track. In postproduction, you use a scratch track for several purposes: to synchronize replacement ("looped") dialogue with the original, to indicate the exact positions of sounds to be added from separately recorded sound effects, and to provide a record of minor dialogue changes that occurred during shooting. For these reasons, it is always important to record the best quality production audio that is practical under the circumstances.

A waveform display simplifies the task of identifying and editing effects.

Background Sounds

In the real world, individual sounds are heard against a background of more-or-less continuous noise. This noise may come from wind, waves, birds, traffic, heating and air conditioning, or people talking or doing things, like coughing in a theater or rattling dishes in a restaurant (**Figure 17-5**).

Figure 17-4 Fishpole mike supports are sometimes used in studio production.

(Sennheiser)

Even without any obvious background noise, no location is truly silent. If the editor cuts unwanted pieces out of the production track, the "emptiness" of the resulting soundless passages is clearly audible. To prevent this, you record *room tone* at the shoot location. Room tone sounds like nothing by itself, but is used to fill the holes left by deletions from the production track (**Figure 17-6**).

Though in real life, we filter out most background noise, a movie sound track does this job for us. To minimize background noise:

- Turn off controllable sounds, like heating ducts or background conversations, during takes.

Figure 17-5 Heating and air conditioning systems are common sources of background noise.

Figure 17-6 Room tone is used to fill gaps in the production track.

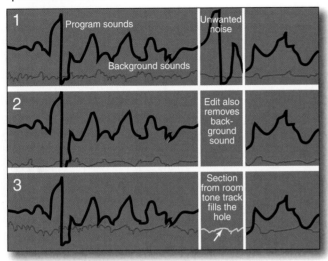

- Minimize uncontrollable noises, like traffic or wind, as much as possible during dialogue or effects recording.
- Make lengthy recordings (up to five minutes long) of ambient (background) sound alone.

In editing, the continuous ambient noise is then laid in under the dialogue and effects tracks, where it helps conceal any background noise that makes its way onto the dialogue tracks, and it minimizes the differences in quality between production tracks made from different mike positions. Professional sound libraries include prerecorded background tracks of a wide variety of locales. These tracks are often substituted for background sound recordings made during a shoot.

An *ambient* track is used in long pieces to blend other tracks. A room tone recording (also called a *presence track*) is used in very short pieces, to fill gaps where the editor has removed parts of the production track.

Sound Effects

Sound effects are often recorded "wild", meaning separately from the production track. This usually happens when using the production track is impractical, often because it is impossible to place a microphone for good quality sound or because the actual sound would be disruptive (like gunshots on a busy street).

Audio was originally recorded directly on film in a machine separate from the camera. To synchronize the two machines, their motors were electrically "slaved" to a "master" motor. Thus, a machine recording sound effects independently of a camera was running "wild", rather than "slaved."

Sound effects that would be hard to get under any circumstances are available in sound effects libraries, some of which contain thousands of different noises (**Figure 17-7**).

In every recording situation, deciding whether to use production track or wild sound effects depends on how many there are and how closely they are synchronized to the picture. For example, the sound of a car starting and driving away could easily be recorded as a wild effect and post-synchronized with the picture. On the other hand, the on-screen feet of a man and two women crossing a marble floor would be extremely tedious to record and individually lay one person, foot, and footstep at a time.

For situations like this, professionals use what is called a *Foley studio*—an environment in which technicians can make sounds while watching a shot on the screen and synchronizing the effects as they are recorded.

Voice Recording

Like sound effects, voices can be recorded separately from the production track. The most

Figure 17-7 Racing car engine sounds are typical effects found in sound libraries.

(Corel)

common application for this is narration recording. In a sound booth or small radio studio, the performer records all the narration in a program, leaving gaps between sections that allow the editor to select the best takes and lay each one in its proper spot in the program.

Dialogue replacement, called *looping*, is also a common procedure. It works like Foley recording, only with voices rather than sound effects. One or more actors record their lines in a sound studio while watching a dialogue scene on screen, and listening to a reference production track through headphones.

Dialogue replacement is often still called "*looping*" because short lengths of movie film were once spliced end-to-end in loops for continuous projection while the watching performers rehearsed synchronizing the new dialogue.

Music

Music is seldom recorded with production sound. Instead, it is captured separately in studios designed to optimize musical sound, and then added to the sound track in editing. Music recording is a complex process that requires sophisticated microphones, mike placement, and electronic mixing and equalizing. As a result, almost all professional music is recorded by studio technicians, rather than by production sound recordists.

However, production people do get involved in playback. *Playback* is the process of reproducing previously recorded vocal music while singers synchronize their simulated on-camera performance with the high-quality recording that they (or others) have made earlier, **Figure 17-8**. In playback, the camcorder will typically lay down a rough recording of the vocal lines sung by the performers on the set for later use as a reference track. In postproduction, the editor will match the prerecorded track to the playback version of it, then remove the rough production mixture from the finished track, leaving only the high-quality original.

Figure 17-8 For vocals especially, performers often lip-sync previously recorded audio. The process is called "playback."

Recorded audio   Lip-sync performance

Microphones

All sound recording must begin with a *microphone*—a device that translates the original noise, created by a pattern of moving air, into an electrical signal that closely imitates that pattern. In order to select and use microphones effectively, you need to know the design basics of the principal types. The design characteristics of microphones include physical form, transmission method, and pickup pattern.

Physical Form

Physically, mikes are designed either to stand independently or to be worn by performers.

Independent Mikes

These microphones are engineered to be held by the camera, by a boom or stand, or by the performer. Sometimes, independent mikes are suspended above the sound source, especially for recordings made in auditoriums.

User-Worn Mikes

Certain very small microphones are designed to be clipped onto the performer's clothing, **Figure 17-9**. Their advantage is that they do not have to be aimed or moved by a technician. Their disadvantage is that they are visible to the camera (unless specially concealed), and their cables or radio signals sometimes cause problems.

Microphone Transducers

The *transducer* in a microphone is the electromechanical system that converts sound energy into electrical energy. Of the several transducer designs developed over the years, the most common are *dynamic* and *condenser*.

Dynamic Transducers

A dynamic microphone transducer is a taut, flat diaphragm, like a drum skin, that is vibrated by incoming sound waves. Dynamic microphones are useful because they are tough and uncomplicated, and the signal they generate is strong enough that it needs no amplification. The handheld mikes used in ENG (electronic news gathering) are often dynamic designs.

Dynamic mikes cannot always maintain a wide and even frequency response, and their sound is often described as "tight" or "less open" than that of the condenser type transducer.

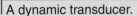 A dynamic transducer.

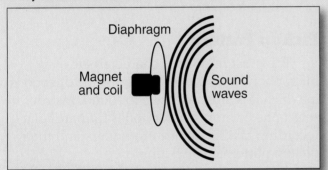

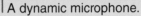 A dynamic microphone.

(Sennheiser)

Condenser Transducers

A condenser transducer is a pair of electrically charged plates with air acting as an insulator between them. The distance between the two plates affects the voltage produced. One plate is rigidly fixed and the other is very sensitive to sound vibrations, which cause it to move toward and away from the fixed plate. The voltage fluctuations that result precisely imitate the sound waves that cause the flexible plate to vibrate.

Condenser microphones have wide, even frequency response and preserve high frequencies better than dynamic transducers. On the other hand, they are more delicate and the battery they sometimes require can fail in the middle of a recording task. Most condenser microphones need batteries in order to apply an electrical charge to the condenser plates and to drive a built-in preamplifier. The preamplifier is necessary because the signal strength of the output from the plates is quite low.

Instead of batteries, many condenser mikes use "phantom power"—electrical current sent through the cable to the microphone from a mixer or preamplifier.

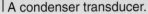 A condenser transducer.

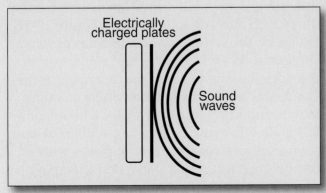

A stereo condenser microphone.

(Sennheiser)

Figure 17-9 A user-worn mike.

(Sennheiser)

Figure 17-10 A wireless microphone and transmitter.

(Azden)

Transmission Method

Cable and radio transmission are the two methods by which microphones send their signals to the camcorder or stand-alone audio recorder.

Cabled Mikes

Traditional microphones communicate via special cables. Because they are shielded against outside electrical interference, these cables provide the cleanest, most reliable signal. On the negative side, cables are difficult to conceal and to keep out of the way of technicians and wheeled equipment, and they require that the recorder be relatively close to the microphone.

Wireless Mikes

Wireless microphones, **Figure 17-10**, send their signals on specialized radio frequencies to receivers cabled to the audio recording system. Inexpensive wireless microphones are prone to interference from unwanted outside signals. High-quality professional types, however, capture sounds extremely accurately, transmit on frequencies free from most interference, and can be set to different frequencies. Using different frequencies, several performers at once can transmit wireless signals to separate receivers (or channels on one receiver). The separate signals can be individually mixed and balanced for greater control.

Many wireless microphone/transmitter/receiver systems run on 120 volt power for use in auditoriums and studios. Battery-powered units are available for use on location.

Pickup Pattern

The most important characteristic of a microphone is its *pickup pattern*—the directions in which it is most sensitive to sound. In order to select one or more microphones for particular recording tasks, you need to understand their pickup patterns. For example:

- A sound effect like footsteps needs a mike with a narrow pickup pattern to exclude unwanted background sound.
- A musical performance needs a broad but directional pattern to capture all the music, but exclude coughs and shuffles in the audience.
- Newscasters' lapel mikes need omnidirectional pickup patterns to capture their voices, no matter which way they turn their heads.

These are the most popular of the many available microphone pickup designs.

Narrow Pickup

"Shotgun" mikes (named for their long, thin barrels) are highly directional, with most of the pickup areas directly in front, **Figure 17-11**. However, the secondary pickup areas of some shotgun mikes will also capture sounds off to the sides; be sure to keep crew members at the camcorder quiet during takes. Shotgun mikes may be mounted on a camcorder, a boom, or a "fishpole." They are especially useful for distant sound sources, for recording in places with competing background noise, and for taping sound effects as cleanly as possible.

Level and Impedance

Level and *impedance* are electrical characteristics that affect the quality of recordings.

Level

Audio products, such as cameras, stereo receivers, and computers, have input and output jacks labeled "line" or "aux", "speaker", "headphones", "mike" or "mic", and sometimes other labels. It is important to understand that each type of jack handles electrical signals of a particular strength, or "level."

In particular, line jacks carry high-level signals that average around one volt in strength. Microphone jacks, by contrast, carry low-level signals measured in millivolts. Microphones plugged into line level jacks will produce only feeble signals, so it is important to use only dedicated mike inputs.

Impedance

Impedance is resistance to the flow of electricity. Most inexpensive microphones are high-impedance types, while most professional units are low impedance. Low impedance translates into better signal transmission and better fidelity, especially in the high frequencies.

For best results, the microphones, cables, and equipment jacks should all be the same impedance. Cables fitted with XLR plugs almost always indicate a low-impedance system.

The impedance on these XLR camcorder jacks can be switched from high to low.

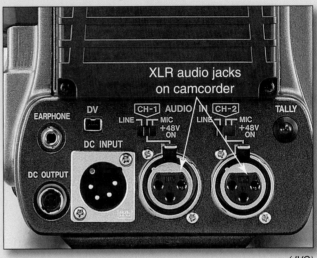

(JVC)

To compensate for impedance differences in various system components, you can use impedance-matching transformers.

Broad, Directional Pickup

Most suspended, stand-mounted, and hand-held performer mikes have a fairly wide, but still directional, pickup pattern. This compromise pattern is less narrowly directional than a shotgun mike pickup, but also less inclusive than an omnidirectional pattern. Because their pickup pattern is shaped more or less like a valentine heart, these are called *cardioid* microphones, **Figure 17-12**.

Figure 17-11 The pickup area of a shotgun microphone lies mainly in front of the mike.

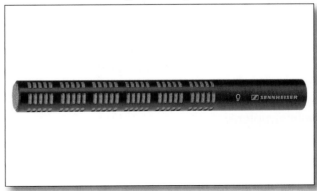

(Sennheiser)

Figure 17-12 The pickup area of a cardioid microphone resembles a heart shape.

Pickup pattern

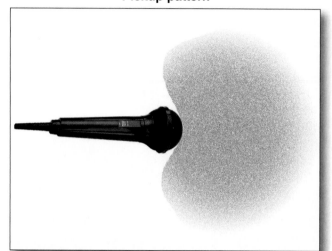

Cardioid microphone (cutaway view)

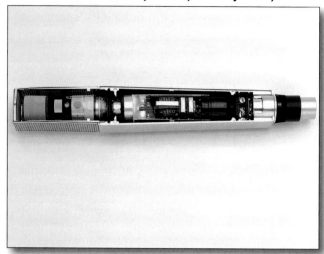

(Sennheiser)

Microphone pickup patterns are actually much larger and more complex than the areas shown in **Figures 17-11**, **17-12**, and **17-13**.

Typically, a camcorder's built-in microphone is a compromise between a narrow shotgun and a broad cardioid pattern. Some camcorder mikes can be switched between wider and narrower pickup patterns.

Omnidirectional Pickup

These microphones are sensitive in a spherical pattern stretching in all directions. The most common models, often called *lavalieres* or just *lavs*, are worn directly on the body. In placing a lav on a shirt front, the goal is to position it high enough to get close to the performer's mouth, but low enough to clear the "sound shadow" created by the chin. Because lavaliere mikes are balanced to compensate for their closeness to the wearer's chest, their relative insensitivity to lower sound frequencies makes them less suitable in some other applications.

In live television broadcasts, you will often see two lavaliere mikes worn side-by-side. One is a backup mike in case the other fails in a situation where the performer must continue speaking uninterruptedly.

Stereo Pickup

Stereo microphones have two separate pickup areas, each feeding its own audio channel. Stereo mikes can add presence and realism to almost any recording, and they are especially effective in studio and auditorium application, where the sound originates from a variety of directions. In studio situations, two mikes are often used, as shown in **Figure 17-13**. Camera-mounted stereo mikes use two pickups in a single housing.

While production sound is often recorded in stereo, full surround sound audio processing is usually created in postproduction.

Figure 17-13 Stereo recording using two monophonic mikes.

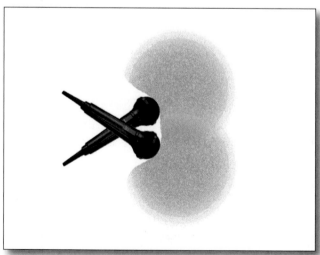

Using Wireless Microphones

Like all radio transmissions, the signals from wireless microphones are at the mercy of their environment. Electrical noise, other radio signals, and signals bounced around by the journey from transmitter to receiver—these and other problems can prevent you from obtaining quality audio recordings. Here are some tips for eliminating these problems:

- Always make the distance from transmitter to receiver as short as possible. If your equipment permits, place the receiver just outside the camera frame and then cable it back to the camera.
- Experiment with receiver position and receiver antenna. Radio mike signals are so sensitive to distance, direction, and reflection that even small adjustments at the receiver end can dramatically improve quality.
- Have backup cabled mikes. Even the finest wireless mikes will not always work. So, be ready to fall back on cabled lavalieres or a fishpole setup, if necessary.
- Before recording audio in a location, scout it for problems, just as you would for video. Experiment with wireless miking to anticipate problems or, if all else fails, settle on an alternative solution using cabled mikes.

Remember that production people may be tolerant of camera, lighting, and performance problems, but they (unfairly) grow impatient if the sound department holds up the shoot.

When the wireless receiver is placed near the camera, the distance to the transmitter may degrade quality. To improve quality, conceal the receiver as close to the transmitter as possible, and cable it to the sound recorder.

Receiver near the camera

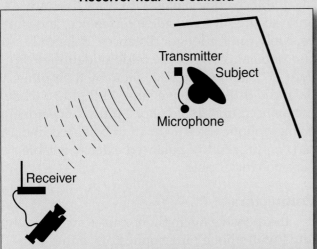

Receiver close to the transmitter

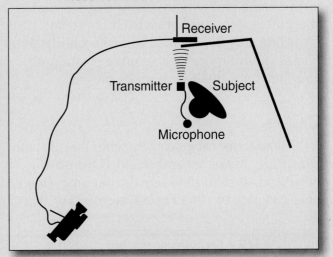

Other Recording Equipment

To complete our survey of audio recording equipment, we can follow the path of the audio signal downstream from the mike on its microphone *support*, along a *cable* (or via *transmitter/receiver*), through a production *mixer*, to its destination at a *recorder*.

Microphone Supports

Microphones can be mounted anywhere and supported by anything that will hold a clip or gaffer tape.

The first task of any support system is to isolate the microphone from bumps and vibrations, which are transmitted as unwanted noises. For this reason, boom, fishpole, and some stand mikes are typically suspended within frameworks by arrangements of shock cords. Microphones mounted directly on the performer cannot be shockproofed, and must be positioned so that they do not pick up the rustle of clothing.

Booms

Mike *booms* are typically used in studios, where their weight and bulk are not inconvenient. Booms are useful because the operators do not have to hold the weight of the microphone and its support. From one place, they can follow a moving performer by pivoting the boom arm, telescoping it in and out, and rotating the mike to point in the right direction.

Fishpoles

Fishpole booms must be held, moved, and aimed manually, but they are far lighter and more compact than booms. For this reason, fishpoles are preferred on many location shoots (**Figure 17-14**).

Stands

Microphone stands are indispensable for recording narration, a musical performance, and any other application in which the subject is off camera and remains in one place. Although short table stands are often used for voice recording, low floor stands with short horizontal booms permit more flexible mike positioning, and they typically accept shock mounts.

Cables

Microphone cables contain wires that are both electrically insulated and shielded from stray noise and even audible signals that unshielded lines can pick up, like radio antennas. For this

reason, microphone connections should always be made with shielded cables manufactured expressly for that purpose. Cables should also be as short as possible in order to minimize signal loss.

Though the plugs may match, never use a headphone extension cable for a microphone, as these lines are unshielded.

Line Balance

Interference and signal loss are affected not only by the length of a microphone cable, but by its "line balance" as well. Cables with two electrical lines are unbalanced, and those with three lines are balanced (**Figure 17-15**).

Unbalanced line cables are typically supplied with inexpensive microphones and fitted with phone- or mini-plug jacks. Though shielded, unbalanced cables still suffer from outside interference. These cables should be used for short runs only—usually less than 20'.

Balanced line cables with three lines are quite secure from outside interference, and can be safely run for longer distances. Balanced and unbalanced lines are seldom identified, but you can usually recognize them by their plugs. Balanced lines will have large XLR plugs, while unbalanced lines have phone plug connectors. A microphone fitted with an unbalanced line can be plugged to a balanced extension cable via a converter plug.

Connectors

Inexpensive microphone cables are typically fitted with phone plug connectors, either full-size or mini-plugs. Mono plugs have a single black band insulating the two connectors, while stereo plugs use a pair of insulators for their three connectors (**Figure 17-16**).

Figure 17-14 A fishpole boom in use.

(Gitzo/Bogen)

Figure 17-15 Unbalanced and balanced audio cables.

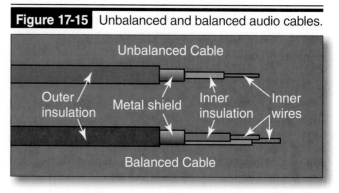

Unbalanced Cable

Outer insulation   Metal shield   Inner insulation   Inner wires

Balanced Cable

Figure 17-16 Mono and stereo mini-plugs.

Mono mini-plug

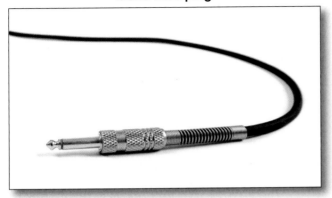

Stereo mini-plug

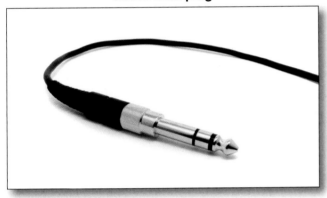

Phone plugs, especially the mini-plug style, are unreliable because they easily lose their electrical connections and stop transmitting the audio signal. This is doubly dangerous if the recordist is not monitoring the input through headphones. When a plug from an external mike is inserted into the camcorder's "audio in" jack, it automatically disables the unit's built-in mike. If the external plug becomes electrically disconnected but still remains in the camcorder jack, neither microphone will transmit an audio signal and the camcorder will capture no sound whatever.

Professional audio setups use XLR connectors that latch together for a reliable connection. You can obtain small, camera-mounted conversion units for some camcorders that allow you to use XLR plugs (with their balanced line cables) on camcorders fitted with mini-plug audio jacks.

The jacks for XLR plugs are labeled in **Figure 17-17**, and in the photo in the sidebar *Level and Impedance.*

Receivers

Wireless microphones incorporate radio transmitters that send the signals to matching receivers, **Figure 17-17**. The receivers, in turn, are cabled to the camcorder or separate audio recorder. Inexpensive transmitter/receiver systems use radio frequencies that can be subject to interference, while professional-grade models use much higher frequencies. Typically, high-quality systems can transmit over longer distances with better signal quality and less interference.

Some receivers can accept signals from more than one microphone, each on its own separate frequency. Some receivers feature switchable frequencies. If one channel is degraded by interference, a different channel can be selected.

Mixers and Equalizers

Whether your microphones are cabled or wireless, their signals must be blended for recording as a combined audio track or balanced as separate stereo channels.

Mixer

To combine the mike signals, you use a *production mixer*, or simply *mixer*, (**Figure 17-18**), which is basically a volume control for each incoming signal and circuitry for combining them. A production mixer is useful for just one microphone too, adjusting the signal to the optimal strength for transmission to the recorder. Almost all professional productions use a production mixer.

Figure 17-17 A receiver for a wireless microphone setup.

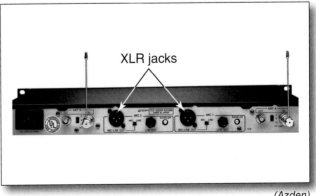

XLR jacks

(Azden)

Figure 17-18 A small production mixer.

(Azden)

Equalizer

An *equalizer* improves overall sound quality by adjusting the relative strength of selected sound frequencies. In production sound recording, it is most commonly used to minimize undesirable sounds like, the rumble of distant traffic or the wheeze of an air duct. An equalizer should not be overused. For example, reducing *all* the sound frequencies that contribute to traffic noise can distort actors' voices so much that they sound like cartoon characters.

Recorders

Production sound is usually recorded by the audio system built into the camera. Backgrounds, sound effects, voice, and music recordings can also be made by other sound recorders, such as digital audio recorders and computers.

Camcorder Audio Recording

In most cases, a camcorder is the best choice for recording audio, even when the video part of the recording will not be used in the program. Camcorder audio recording offers many advantages:

- Because it is in the same format as the program material, the audio recording can be edited or transferred to computer without connecting special-purpose source decks.
- It offers extremely high quality sound. Mini-DV audio can actually be better than CD music audio.

- Even if you will not use the picture recorded with the sound, it serves as a handy visual locator of audio clips, especially if you have identified them with written slates. See **Figure 17-19**.
- Because it is compact and battery-powered, a camcorder is easy to use in the field.

For many field recording applications, then, a stereo-quality camcorder will do the job as well as a special-purpose recorder.

Computer Sound Programs

For studio audio recording, miking directly to the editing computer eliminates the separate step of copying the recording from an outside source (a camcorder or field recorder) into the editing system. Audio editing software typically includes recording controls like level control and equalization. In addition, the wave form displays included with audio software allow you to verify that recording levels are optimal for the system.

Digital Audio Recorder

A good digital audio recorder is useful if you do not want to use a camera. Good quality units are not expensive, but be aware that the tiny models used mainly to record memos and conversations may not offer high enough quality for video work.

Figure 17-19 During field recording, letter and record simple slates.

Automatic Level Control

All consumer camcorders are fitted with automatic level controls—circuits that adjust volume automatically to compensate for different levels of incoming sound. In general, automatic level control works well. Problems arise, however, when a distinct sound is preceded by a long silence. If there is a near-silent period before a distinct noise, the automatic level control may turn the record volume up, in search of a signal. When a normal-level sound begins, the circuits cannot turn down the record level fast enough to prevent the start of that sound from distorting. Automatic level controls can also fail when the sound level rises or falls suddenly and dramatically. Here again, the circuits cannot anticipate the change and react fast enough to accommodate it.

The solution is manual level control, a feature built into virtually all professional-level camcorders. With manual control, the sound recordist can maintain a consistent level through quiet parts and anticipate sudden volume shifts by adjusting gain just before they happen.

Defeating Automatic Level Control

When you cannot disable the automatic level control, you can sometimes work around it. If a loud noise occurs in the middle of a shot, use an external mike (handheld or on a fishpole). That

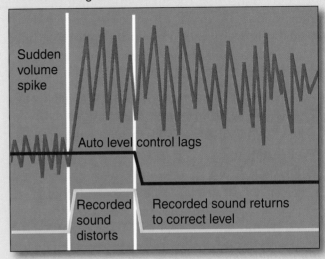

Automatic level control can fail to keep up with sudden volume changes.

Sudden volume spike

Auto level control lags

Recorded sound distorts

Recorded sound returns to correct level

way, you can turn the mike away from the source of the loud sound just before it happens. If you can, use a production mixer. By optimizing sound levels manually before they reach the auto circuits, you can reduce unwanted variations. When recording voice for narration, follow each long pause with a voice slate. The slate, which will be removed in editing, gives the automatic system a sound on which to set a level, so that the narration is all recorded consistently.

Accessories

A professional sound recordist will carry a broad range of specialized accessories. Of these, the following are perhaps the most widely used: headphones, windscreens, furniture pads, plug and cable adapters, and a slate.

Headphones

Headphones are essential for quality sound recording. Like the viewfinder for video, headphones let you preview what you are recording to check the quality (or simply to verify that sound is being properly recorded). In choosing headphones, select the full-ear type that block out extraneous sounds (**Figure 17-20**). Ideally, you want to hear nothing but the sound picked up by the mike.

Figure 17-20   Full-ear headphones.

(Sennheiser)

Windscreens

Wind noise is a common problem with exterior recording. A foam windscreen (sometimes called a "sock" or a "doughnut") reduces wind noise by deflecting the air rushing past the microphone, **Figure 17-21**. A windscreen is also useful indoors for narration or dialogue recording, where performers' breath can create wind noise across an unprotected mike.

The built-in windscreens on (or in) the small mikes of miniature camcorders may not be visible. Some camcorders filter wind noise electronically.

Furniture Pads

To prevent unwanted echoes and boomy-sounding audio, drape blankets or furniture pads (used by moving companies) at strategic points. You can obtain furniture pads inexpensively at do-it-yourself moving rental companies.

Plug and Cable Adapters

If you put together your own audio equipment, everything will be cabled and plugged compatibly. But if you must connect to other equipment, such as a public address system or an auditorium sound setup, you will need a variety of plug adapters.

Slate

When recording audio with a camcorder, use a simple tablet and marker to create identifying *slates*. Visible slates eliminate two major annoyances in audio editing: finding a sound recording and postproduction time. Unlike pictures, sounds cannot usually be scanned in fast-forward or reverse, so finding material can be time-consuming. Even at normal speed, some quite accurate, realistic sounds can be difficult to identify without visual clues. By visually identifying sound effects, slates save a great deal of postproduction time and effort.

For simplicity, you can slate the production and date once at the start of the session, then slate just the particular effect, recording the other information verbally. A typical slate for a field recording might include the type of sound ("wild SFX") and a description of the sound ("Diesel Truck, startup, idle), as shown in **Figure 17-19**.

Audio Recording Techniques

Production, background, effects, voice, and music tracks all require their own recording techniques.

Production Tracks

The audio recorded along with the video is the most important sound track. A good production track is distinguished by its quality, consistency, and freedom from unwanted noise.

Figure 17-21 Windscreens are useful for both exterior and indoor audio recording.

Shotgun mike in a foam windscreen

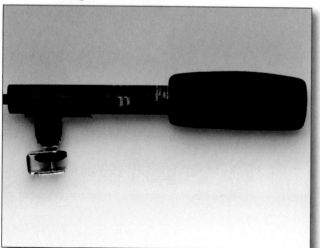

(Azden)

Foam windscreens can be added to most handheld mikes

(Sennheiser)

Quality

The keys to good audio quality are distance and direction. Place the mike as close to the subject as you can, without getting it into the picture. Then, aim the mike directly at the sound source. If two people are talking, point the mike at each one when he or she is speaking. Mikes worn by the subject(s) take care of distance and direction automatically.

If you are using a fishpole boom, begin by deciding whether to place the microphone above or below the frame line. Then, orient the mike on the fishpole, as shown in **Figure 17-22**. That way, you can swing it back and forth between speakers by simply twisting the pole.

If you need to use the camera's built-in mike, place the camera as close to the subjects as practical.

When moving the camera close to the subject, remember that the wide angle lens settings required at very short distances can unpleasantly distort subjects' features.

Consistency

The next task is to keep all the audio sounding alike. To do this, you need to confine the volume level within a relatively narrow range, so that soft passages are well-recorded and sudden loud parts do not sound distorted. If you have manual volume control, maintain sound levels by rehearsing the action and noting where softer or louder sounds occur. During the shooting, anticipate the volume changes by adjusting the sound level just a fraction of a second before the changes happen. If you have only automatic level control, monitor the sound during rehearsals and restage the action (and/or move the microphone), if possible, to minimize excessive volume changes.

In addition to consistent volume, you want to maintain the same audio character, so that the shots recorded from different setups sound alike. If you are using the camcorder's built-in mike, try to keep the camera roughly the same distance from the subjects in every shot. Change image size by zooming instead of moving the camera. With a fishpole mike, experiment with different angles and positions while listening to the effect.

For efficiency, it is important for the boom operator as well as the recordist to listen to the sound through headphones.

Freedom from Noise

Outdoor production recording is a constant struggle to keep wind noise and background sounds out of the dialog track. Wind noise can often be tamed by using foam windscreens on the microphones. Background noise is more difficult. If possible, move the subjects so that noisy traffic or pedestrians or ocean waves are not

Figure 17-22 A fishpole boom is placed above the subjects and the frame line. A—The mike boom is positioned beside the camera (boom operator omitted for clarity). B—By twisting the boom, the operator can aim the mike at each subject in turn.

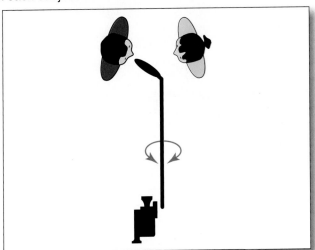

A

B

Recording Stage Performances

Recording plays, musicals, and concerts in auditoriums can be complicated by several factors:

- Audio quality is degraded by the distance between performers and your microphone.
- The width of the stage offers a sound field too broad for the pickup pattern of your mike.
- Audience noise is difficult to keep off the production track.

Unlike musicals, opera performances are often not miked by the theater.

(Corel)

Tapping the Source

If the performance is already miked by the auditorium's own sound system, the best solution may be to plug directly into it. Not only does this take advantage of microphones that are well placed for sound recording, it also eliminates sound from the auditorium loudspeakers.

If you do plug into the auditorium system, it will probably be to a line-level circuit—far too strong for the mike input on your camcorder. To match levels, you need to place a line-to-mike level converter in the signal path.

Following the Action

For a play or other performance where the sound sources move, a camcorder-mounted shotgun mike may offer the best reproduction. Assuming that its camcorder is constantly reaimed to capture the most important action, the mike will be automatically reoriented with it.

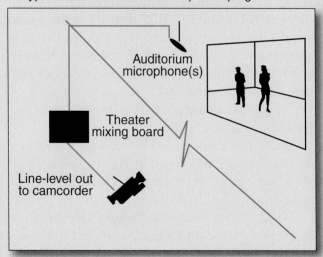

A typical auditorium sound setup for taping.

Auditorium microphone(s)

Theater mixing board

Line-level out to camcorder

Riding Gain

Automatic gain control can be difficult to handle with the widely varying sound levels encountered in plays. During quiet moments, the volume circuits will increase the recording level to the point where audience shuffles and incidental stage noises may grow unnaturally loud. To prevent this, use an external mike (even if it is mounted on the camera) with an audio mixer to control sound level before the signal is recorded.

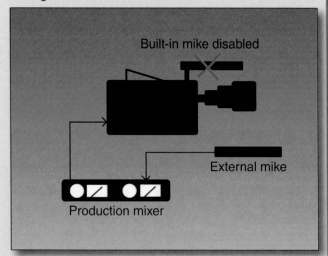

Using an external mike and mixer to control level.

Built-in mike disabled

External mike

Production mixer

behind them. This avoids aiming the microphone directly at the background noise. Watch out for reflective surfaces, like masonry walls, which can bounce noises originating behind the camera back into the microphone, **Figure 17-23**. When practical, place the mike above the subjects, pointed downward. This will reduce its sensitivity to unwanted background sounds.

Background Tracks

The key to good background tracks is consistency—a uniform sound unmarred by distinctive noises that call attention to themselves. For example, background traffic sounds should not include horns, brakes screeching, or the footsteps of occasional pedestrians because audiences will pick them out of the background and think they are part of the program. Background tracks fall into two types: ambient noise and room tone.

Ambient Noise

The sound of traffic or restaurant chatter or ocean surf is fairly easy to record: simply position the mike where the noises sound realistic and record about five minutes worth of material. You want a long recording so that it can be laid under an entire sequence without obvious repetitions or edits.

Figure 17-23 Avoid placing the microphone where it will pick up reflected background sounds.

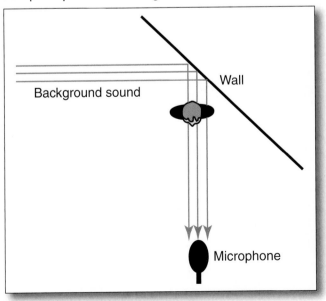

It is possible to edit out unwanted parts if the background is fairly uniform (**Figure 17-24**). Digital sound editing allows undetectable cross-fades between different pieces of background track.

Remember to record at a good strong volume. It is easy to decrease sound levels in editing, but difficult to increase them without obvious quality loss.

Room Tone

As explained previously, room tone is the characteristic sound quality of an environment that is audible (though usually unnoticed) even when there are no overt noises. It is only when an editor cuts pieces out of a production track that the resulting dead silence is obvious. By replacing the dead spots with pieces of room tone, the editor preserves the ambient sound of the environment. Here is the procedure for recording room tone:

- Have the actors, crew members, and equipment remain in place, with the people keeping still and quiet. This is necessary because room tone is so subtle that it is affected by the objects in the environment.
- Place the microphone in a position similar to those used in making the production recordings.
- Record three to five minutes of material. Remember: it does not matter if the near silence is interrupted by an occasional noise. Since room tone is typically used in short pieces, these noises are easily discarded.

Figure 17-24 A factory will often provide fairly uniform background noise.

(Western Recreational Vehicles)

Keep in mind that audio tracks can be reused as required. For this reason, it is not necessary to record every minute of background sound that you might possibly need.

Sound Effects

The keys to good sound effects are isolation and realism. To isolate effects from background noises, position the mike as close as possible to the sound source. Use a highly directional shotgun mike if possible, **Figure 17-25**. If you record wild effects with a camcorder, voice-slate each take so that the effect is identified even when the camera is too close to the sound source to record a recognizable visual.

To obtain realistic sound effects, there is no substitute for listening carefully through good headphones and changing the mike angle and distance until you are satisfied with the sound. As always, record at a good, strong volume. If the script calls for, say, a car driving off outside in the distance, the editor can reduce the volume and adjust the sound quality in postproduction, as needed.

Foley Effects

The same rules that apply to recording good sound effects also apply to the recording of Foley effects. Recorded Foley effects are synchronized to video playback. Prepare for recording by making a duplicate copy of the footage that needs sound effects so that you can record the sound effects on the copy. Later, it will be synchronized with the master video.

Voice Recording

As with all recording, the goal in recording narration or other voiceover material is a clean, quality track. The keys to success are studio acoustics, microphone selection, and mike placement.

Studio Acoustics

Voice tracks should be completely free of reverberation and background noise. If you are not recording in a sound studio, select an area with carpet, draperies, upholstered furniture, and other sound-absorbing materials. Using good quality headphones, listen to the "room tone" of the selected recording area for background noises and sounds, like ventilation system hum. If necessary, change the mike orientation, sound absorption material, or even the location itself to eliminate unwanted background sound.

Microphone Selection

Use a cardioid mike, if possible, because its pickup pattern allows for some movement on the part of the performer.

Shotgun mikes with narrow pickup patterns are usually unsuitable for the close distances typical of studio mike placement.

Figure 17-25 Because of the distance required, recording animal sound effects is especially difficult. Highly sensitive directional mikes are used for capturing nature sound effects.

(Corel/Sennheiser)

Even inexpensive units will perform adequately, but a better grade model adds a professional-sounding crispness that makes voiceover speech stand out.

Microphone Placement

Poorly-placed studio mikes create two problems. First, they cause *plosive* sounds, like "puh" (as in "*p*oorly *p*laced"), to overload the mike and "pop" on the sound track. Second, their stands can interfere with the scripts that are usually used in voice recording. To correct both problems at once, place the microphone at a slight angle to the performer. Positioning a table stand to one side makes room for a script. Placing a boom mike overhead is even better.

Disk-shaped screens placed between performer and mike can function much like foam windscreens.

If you are recording two or more performers at the same time, it is preferable to provide two mikes, separated by as much space as the location allows (**Figure 17-26**). That way, neither mike

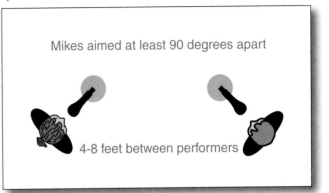

Figure 17-26 Each microphone picks up only one performer.

Mikes aimed at least 90 degrees apart

4-8 feet between performers

Managing Studio Voice Recording

In addition to placing the microphone and controlling record levels, you need to minimize script noises and identify parts of the recording for later editing.

Script Rustle

To eliminate the noise of paper shuffling during narration recording, cover the table or music stand supporting the script with a scrap of carpeting. Place a blanket or rug on the floor beside the performer. Remove any staples or paper clips so that the script pages are loose. To use this setup, the narrator feathers the script pages as they rest on the carpet scrap. As each page is finished, it is picked up and dropped onto the carpet beside the reader.

Scripts can also be read from computer screens, but recording is difficult because even laptop fans make enough noise to affect the background.

Slating Voice Recordings

Normally, a narrator reads through a script in sequence, until making a mistake. When this happens, have the performer repeat the line immediately and then resume reading the next

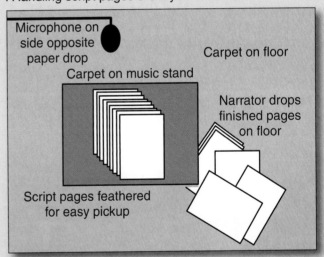

Handling script pages silently.

Microphone on side opposite paper drop

Carpet on floor

Carpet on music stand

Narrator drops finished pages on floor

Script pages feathered for easy pickup

line. To document this repeat for the editor, have the narrator voice slate the retake (for example, "Line 21, take 2"). Place a vertical line (I) in the margin of your own script to stand for the first take. Add a line for each additional take (II, III, etc.). If line 21 has three lines in the margin beside it, for example, then the editor knows that it took three takes to get the line right, and the editor captures only the last one. Or, if you should decide that the second of the three takes was better, circling the second line will tell this to the editor.

will pick up the sounds of the other mike's performer. Use a production mixer to balance the levels between the two mikes.

Dialogue Replacement

Dialogue replacement (**Figure 17-27**) is just a special form of voice recording, and all the suggestions presented earlier apply to this task as well. The major difference is the video monitor, which is placed so that the performers can synchronize their dialogue with the lip movements on screen. Position the monitor so the actors can easily look back and forth between their scripts and the screen.

Here is a simple procedure for replacing dialogue:

1. Edit the original footage, including the poor quality dialogue.
2. Use a copy of the assembled footage to guide dialogue replacement, recording the clean dialogue on an adjacent, unused track.
3. When the dialogue has been recorded satisfactorily, adjust the new dialogue to fine-tune synchronization with the video.
4. Finally, delete the poor-quality production sound.

Music Recording

Professional music recording typically involves many microphones, each balanced and equalized by a large studio mixer and feeding a separate recording track (**Figure 17-28**).

Production music recording is a simpler process. Taping a rock group, string quartet, or other small ensemble, you can obtain excellent results with a single, well-placed microphone. As always, the keys to success are microphone selection and placement. For music recording, use a good quality stereo microphone of condenser design. Place the mike as close as possible to the group, while still keeping the performers inside the primary pickup pattern, **Figure 17-29**.

Camcorders with built-in stereo mikes are engineered to achieve separation automatically.

Without multiple microphones and recording channels, you can still equalize volume levels by moving the performers around, so that the quieter instruments are nearest the mike and the louder ones are farthest away. Listen to the signal during rehearsals and experiment until you achieve a good balance.

Figure 17-27 Scenes in noisy environments often require dialogue replacement.

Figure 17-28 In studio music recording, instruments are often miked individually.

(Sennheiser)

Figure 17-29 Place the mike far enough away to pick up the whole group.

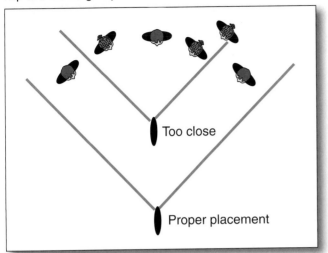

Too close

Proper placement

The Rewards of Audio Recording

If your primary interest is in the video side of production, you may begin with the idea that audio recording is simply a technical process that has to be learned in order to provide the professional sound track that you need. But, as you get into the creative aspects of sound, you may well find that audio is a complex craft that can be rewarding in its own right.

Summary

- To provide the editor with clean, separate components, the sound recordist needs to know the basics of audio recording, the equipment used to do it, and the techniques required to obtain high-quality audio.
- Different types of audio recording include production sound, background sound, sound effects, voice tracks, and music.
- All sound recording begins with a microphone, which can stand independently or be worn by performers.
- The signal from a microphone is sent to the camcorder or stand-alone audio recorder through audio cables or wirelessly, using radio frequencies.
- The most important characteristic of a microphone is its pickup pattern.
- The line balance of microphone cables affects the amount of interference and signal loss.
- Microphone signals must be blended for recording as a combined audio track or balanced as separate stereo channels, using mixers and equalizers.
- The keys to good audio quality are distance and direction.
- The two different types of background tracks are ambient noise and room tone.
- The keys to good sound effects are isolation and realism.

Technical Terms

Audio: Sound as electronically recorded and reproduced.

Balanced line: A three-wire microphone cable designed to minimize electrical interference.

Boom: A studio microphone support consisting of a rolling pedestal and a horizontal arm.

Cardioid: A spatial pattern of microphone sensitivity, named for its resemblance to a Valentine heart.

Equalizer: A device for adjusting the relative strengths of different audio frequencies.

Fishpole: A location microphone support consisting of a handheld telescoping arm.

Foley studio: An area set up for recording real-time sound effects synchronously with video playback.

Lavaliere: A very small omnidirectional microphone clipped to the subject's clothing, close to the mouth.

Looping: Replacing dialogue in real-time by recording it synchronously with video playback.

Microphone (mike): A device that converts sound waves into electrical modulations for recording.

Production mixer (mixer): A device that balances the input strengths of signals from two or more sources, especially microphones.

Pickup pattern: The directions (in three dimensions) in which a microphone is most sensitive to sounds.

Playback: (1) Previously recorded video and/or audio reproduced so that actors or technicians can add to or replace parts of it synchronously, in real-time. (2) Studio-quality music recording reproduced so that performers can synchronize lip movements with it while video recording.

Room tone: A recording of the ambient background sounds in a studio or at a location that is used to fill gaps in the production sound track. Also called a *presence track*.

Slate: A written video and/or spoken audio identification of a recorded program component, such as a shot, a line of narration, or a sound effect.

Transducer: The component of a microphone that converts changing air pressure ("sound") into an electrical signal ("audio").

Unbalanced line: A two-wire microphone cable that is subject to electrical interference, but is less bulky and expensive than balanced line; for use in amateur applications.

Review Questions

Answer the following questions on a separate piece of paper. Do not write in this book.

1. Audio is the recording and _____ of sound in support of video.
2. *True or False?* Microphones pick up every sound within their range.
3. List the steps in the strategy used by sound recordists to overcome audio problems.
4. What is *production sound*?
5. How is an ambient noise track used in editing a program? How is the use of an ambient track different from the use of a room tone recording?
6. What is a *Foley studio*?
7. Describe the process of *playback*, from recording through postproduction.
8. What is the function of a *transducer* in a microphone?
9. The two methods employed by microphones to send their signals to the camcorder or a stand-alone audio recorder are _____ and radio transmission.
10. *True or False?* High impedance microphones have better signal transmission and better fidelity.
11. What is one drawback of using lavaliere mikes?
12. What is the difference between a *mike boom* and a *fishpole boom*? Which is more practical for a location shoot?
13. Microphone cables with two electrical lines are _____.
14. Why are phone plugs, particularly mini-plug styles, unreliable?
15. What is the function of an *equalizer*?
16. *True or False?* A foam windscreen reduces wind noise by deflecting the air rushing past the microphone.
17. What purpose do visual slates serve when recording audio?
18. In voice recording, what problems are created by poorly-placed studio mikes? How are the problems corrected?

STEM and Academic Activities

STEM

1. **Science.** As the diaphragm in a dynamic microphone functions to reproduce sound, the diaphragm in the human body also plays a role in producing sound. How does your diaphragm function to help you speak?

2. **Language Arts.** Choose a single piece of recording equipment that you are familiar with using. Research the features and functions of the model, and exactly how the device operates. Write a review of the device which includes your research information, as well as your opinion of the device based on your experience with it.

18 Directing for Content

Objectives

After studying this chapter, you will be able to:

- Identify the director's areas of authority.
- Understand how to develop a directorial approach for individual projects.
- Explain how to control emphasis and feeling when directing.
- Understand how a director guides on-screen talent in shaping their performances.
- Recall how to solve production problems arising from performance.
- Identify the tasks involved in typical directing assignments.

About Directing for Content

When you direct a video, your job is to determine exactly what is shot and how the result looks and sounds. That makes *you* the final decision-maker—the person in charge. Other people, such as the writer, the videographer, and the editor, make decisions in their own areas of responsibility. In addition, they often make recommendations that a wise director will consider with seriousness and respect. Even though production is a collaboration among many people, the final authority generally rests with the director.

Video directing requires the ability to juggle several different tasks at once, including:

- **Communication:** Presenting the program content effectively to the audience.
- **Performance:** Ensuring that the people in the video are effective on screen.
- **Editing:** Providing the raw materials from which the editor can construct a smoothly assembled program.
- **Camera:** Choreographing the way in which the action is recorded.

This chapter is devoted to *communication* and *performance*. Chapter 19, *Directing for Form* covers *editing* and *camera*.

Taking Charge

Even in professional productions, the powers and responsibilities of the director are not always automatically clear, especially when the director must share authority with the producer, the client, the financial backers, and sometimes others. In order to be in charge, you must actively *take* charge by defining responsibilities, assuming the authority to discharge them, and ensuring that you have the power to enforce your decisions.

Where the Director Rules

Though directors often share overall responsibilities with the producer and others, three major areas should belong to the director alone:

- *Shaping the performances* by blocking the action, guiding the actors, and pacing the scenes.
- *Choosing the setups* by determining where to place the camcorder, where and when to move it, and how to select and change lens focal lengths.
- *Designing the coverage* by choosing how much action to include in each shot, how much overlap to allow between shots, and how many ways to re-cover the action with alternate setups.

You cannot accept the responsibility for directing a production if you lack the final authority over these areas.

Advice and Consent

In three other areas, the director should have "advice and consent" rights (like the U.S. Senate approving judges). These areas are:

- **Videography and lighting:** The aesthetic and technical quality of the video images.
- **Art direction:** The overall design of sets, locations, costumes, etc.
- **Script:** The structure and language of the script that provides the blueprint for the production.

Advice and consent means having the right to give input when these elements are being designed, and veto power over the results. In other words, you can explain what you would like to achieve in these areas and you can express dissatisfaction when the results do not satisfy you, but you should not dictate the ways in which those results are to be achieved. That responsibility and authority should belong to the videographer, the art director, and the writer.

Real-World Compromise

In practice, every real-world professional relationship is different. In many cases, the director trusts the videographer to spot the camera setups, as well as light the shots. In general, the more you can trust your co-creators, the more exclusively you can focus on the core tasks of directing.

Since the realities of production demand constant adjustments and improvisations while shooting, the director must evaluate each new idea to see how it might affect these four tasks. In addition, a good director also imparts an overall style to the material, giving it a distinctive "look and feel." It has been said that a good director combines the talents of a painter, a dramatist, a diplomat, a psychiatrist, an accountant, and an air traffic controller. It is no wonder that it takes time and practice to master all the skills required for successful video directing.

Approaching the Script

Before you can address communication, performance, editing, and camera, you need to master the script. If you are also the writer and/or producer, you have already done this. But, in many professional productions, the director is brought in only after the script is essentially completed, so we will assume that this is when you first encounter it.

The script writing process is covered in Chapter 9, *Project Development* and Chapter 10, *Program Creation*. Parts of the following discussion summarize content from those chapters.

Getting to Know the Script

How you approach a new script depends, of course, on your personal working methods, but here is a system that functions well for many directors:
- **First reading:** Read the script through at a good pace, but without skipping or skimming. The goal is to form an overall impression of it.
- **Second reading:** Review the script thoroughly, making notes about questions, reactions, potential problems, and anything else that catches your attention.
- **Third reading:** Focus on the script's opportunities, problems, challenges—the things that will engage much of your time and effort.

You will continue to study the script, but these first three readings should give you a good understanding of the project you are undertaking.

Understanding the Nature of the Project

If the writer and producer have done their work well, the script should provide answers to five basic questions:
- What is its subject; what is it about?
- What are its objectives; what is it supposed to accomplish?
- Who are its target audience (job trainees, web surfers, older TV viewers, etc.)?
- What is its delivery system (the web, theaters and DVDs, cable TV, a training room)?
- What is its basic concept, its approach to its subject?

If you cannot find the answers to these questions, or if you have issues arising from the answers, you have already taken the first steps in evaluating the script.

Evaluating the Script

There is no such thing as a perfect script. Even if its creators have done an excellent job, they may have taken approaches, employed methods, or included details that are not comfortable for you. (Like any other art, script creation involves many choices that are purely subjective.) To take just one example, you may feel that they have allotted too much space to some parts of the script content and too little to others. Before you begin planning the shoot, you need to resolve matters such as this one.

How you do this depends on the circumstances. If you are accepted as a creative collaborator, you can explain your problems, make diplomatic suggestions for changes, and obtain revisions that solve, or at least reduce, the difficulties you've identified.

On the other hand, if you can do little to fix the problems you see, you will need to pre-plan strategies for addressing each problem as you direct the script. Depending on the individual problem, that can mean you should:
- *Compensate for it.* For instance, if the subject matter is uninteresting, work doubly hard on staging, shot planning, and camera movement to liven it up.

- *Minimize it.* For example, if a proposed shooting location is visually unattractive, plan to frame off much of it by working close to the subjects.
- *Ignore it.* If the top executive appearing in your corporate video cannot read text expressively, you will have to live with the situation. (You cannot say no to the boss.)
- *Work around it.* To continue the previous example, you might save the project from the executive's poor performance by selling the idea of starting the boss on camera, and then continuing the speech as voice-over narration. If the performance turns out to be unusable, the producers might then agree to let a professional re-record the off-camera portions of the text.

It is unfair, but true, that the director gets most of the blame for a bad program (and most of the praise for a good one). Your first obligation is to the project, but you also need to protect yourself.

Choosing a Directorial Approach

Your directorial approach to a script is the set of decisions you make that will determine your directorial style. To demonstrate by example, the chapter on project development mentions one program promoting high-speed power boats and another explaining retirement living options, **Figure 18-1**. The boating subject suggests a dynamic, exciting approach. The retirement video seems to need a quiet, measured approach.

Every subject can be handled in more than one way, and your final job in evaluating the script is to decide how you will treat it. To continue with the powerboat, you might well decide to shoot the material in many brief shots to allow fast edits. You might plan on using wide angle lenses to exaggerate speed and tilted camera angles to add an off-balance feeling. Your script notes might include reminders to get shots of the huge overhead speakers on the powerboats to motivate sound track music. Even a short program involves a great number of these decisions about directorial approach—decisions intended to determine the video's overall style.

When you have settled on your overall approach, you can start planning the details of directing for content.

Directing for Communication

Communicating with the audience means delivering information—selecting, organizing, shaping, and presenting it in a way that helps viewers absorb and respond to it. It is not enough to simply display information and expect the audience to pay attention to it. In directing, you must determine the way in which information is delivered by controlling *emphasis* and *effect*. Before you can shape information, however,

---

Figure 18-1 High-speed power boating and retirement living videos, and related directorial approaches.

High-speed power boating video—Dynamic, exciting directorial approach	Retirement living options video—Quiet, measured directorial approach

A Demonstration Sequence

In order to contrast different solutions to directing problems, we need to apply all of them to the same situation. Throughout the following discussion, we will draw most of our examples from the same source: a sequence about a burglar and her hunt for a key to open a display cabinet.

Why not just break the glass in the cabinet doors? Let us assume that the noise of the shattering glass would arouse the sleeping household, or perhaps the burglar is too nervous to think of this obvious solution.

A burglar.

A locked cabinet full of valuables.

A key to the locked cabinet.

you must see to it that the information actually gets into the program to begin with. That task is not as simple and obvious as it seems.

Information

Perhaps the most common error that inexperienced directors make is to omit important information. For example, **Figure 18-2** shows how our burglary might be staged by an inexperienced director. Where did the burglar find a key to the cabinet? The inexperienced director has not shown us. Though this omission is not critically important, it will bother the audience.

To satisfy the audience's curiosity, we need to include the information about the key, as you can see in **Figure 18-3**. In the added shot, we see the burglar open the top desk drawer and remove a key from it. Now the missing piece of information has been included in the scene. The director's very first job, then, is to identify the information that the audience needs to know, and ensure that all of it is included.

Figure 18-2

The burglar discovers that the cabinet is locked...

...and opens it with a key.

Because the illustrations in this example are similar, it is important to study them carefully for small differences.

Emphasis

The added shot in **Figure 18-3** is just another camera angle with nothing special about it. Although we do see the discovery of the key, it is so unemphatically presented that we might very well miss it. To make sure that the audience gets the information, the director's second task is to emphasize it appropriately, as you can see in **Figure 18-4**. In this version, the audience sees the drawer being opened by the burglar, and then the key inside as her hand enters the shot and removes it. By giving the key its own closeup, the director adds emphasis to the information so that the audience will not overlook it.

In adding emphasis to information, you have several tools, including *image size, composition, camera angle*, and *shot length*.

Image Size

The bigger the image, the more emphatic the information it contains. **Figure 18-5** shows two versions of the insert. Notice how the larger image presents the key more vividly.

Composition

You can call attention to the information you wish to emphasize by the way you organize the composition of the image. In **Figure 18-6**, notice that in the first shot, the key is hard to separate visually from the other clutter in the drawer. In the better-composed second shot, the key has been placed and lit to draw the viewer's eye to it.

Camera Angle

A powerful tool for emphasizing information is camera angle: the perspective from which the audience sees the information. In **Figure 18-7**, both images are inserts of the key in the drawer. The first is an *objective insert*, meaning that the shot is from no particular point of view (neutral). The second is a *subjective insert*, showing the drawer as if the audience were looking at it through the burglar's eyes. Notice that the subjective camera angle is more emphatic than the objective alternative.

Shot Duration

You can control emphasis simply by how long you display a piece of information on the screen. If the audience sees only the single second during which the burglar's hand grabs

Figure 18-3 The added shot of an open desk drawer explains the key.

Figure 18-4 The insert shot emphasizes the key.

  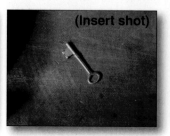 

Figure 18-5   Emphasis through image size. The closer shot adds more emphasis.

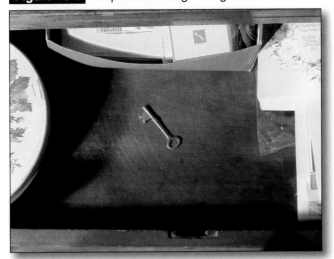

Figure 18-6   Emphasis through image composition.

The key is lost in the clutter.   **The key is emphasized by the composition.**

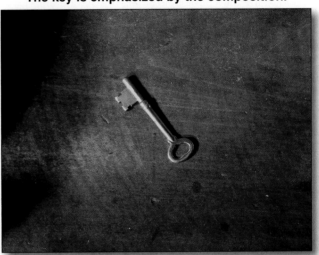

Figure 18-7   Emphasis through camera angle.

Objective insert (neutral point of view)   **Subjective insert (the burglar's point of view)**

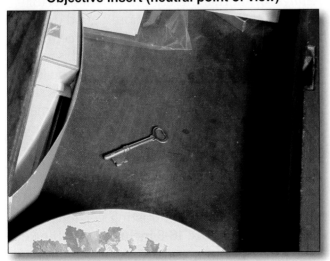
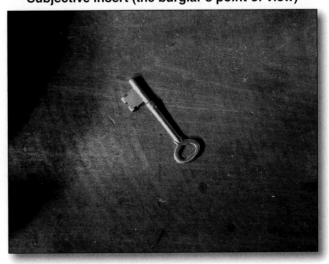

Other Forms of Emphasis

Accomplished directors have used a wide variety of tricks to emphasize elements in their shots.

Brightness and Color

In classic Hollywood lighting of the pictorial realist style, the center of interest was almost always emphasized by brighter lighting—perhaps a hotter key light or an arrow of background light that served as a pointer. In Technicolor musicals, the stars typically wore costume colors that stood out from the overall color scheme.

Today, emphasis using light or color tends to be more subtle. But, color, especially, is still used to distinguish the most important element in a composition.

Though the third subject is small, his orange shirt attracts the viewer's attention.

Movement

Movement is an easy way to draw attention to a subject, especially when the rest of the frame is still. You can also reverse this procedure. In one of his many famous shots, Alfred Hitchcock showed a tennis match grandstand full of spectators whose heads swivel left-right-left-right following a long volley on the court. Only the villain's head is not moving, because he is watching his prey rather than the match. Though the villain is small in the frame, the viewer's eye goes to him immediately.

Audio

Traditionally, audio had only a limited ability to emphasize a certain aspect of a shot. Today, however, most TV sets play stereo audio, and digital postproduction offers complete control over stereo imaging. This ability is especially useful for emphasizing important elements that are off-screen. The opening door, the car driving off in the distance, the incoming mortar shell—these sounds are greatly enhanced by stereo imaging. Surround sound systems take dimensional audio even further.

Although audio emphasis is ultimately created in editing, the director must anticipate it in production. To use the example of the off-screen door, the director may want a performer to react to the sound, even though it is not heard on the set.

the key, they may miss this information. But, if they see the drawer opened to reveal a prominent key for a moment before the hand removes it, they have more time to notice and absorb the information.

Feeling

In the previous examples, the insert shot of the key in the drawer is added purely to convey information. But, a good director presents that information in a way that involves the audience emotionally. By utilizing camera angles, image sizes, shot lengths, lens perspectives, camera movement, and other tools, the director creates

feelings. In short, a merely competent director communicates only information; a good director communicates both information and feelings.

To see how feelings can be created, say that the effect you want for the burglary scene is suspense. Imagine that the burglar knows she has tripped the silent alarm and has only moments to get that cabinet open. But, it is locked! Can the cabinet be opened and how long will it take? Your goal is to make the audience want to say, "Hurry, hurry!"

The two sequences of camera setups presented here show contrasting versions of the sequence. **Figure 18-8** includes only the key

Figure 18-8 Finding the cabinet locked, the burglar opens the top desk drawer. Her hand locates, then snatches the key. She jams the key into the keyhole of the door.

to convey the information that it exists. In **Figure 18-9**, however, the action is broken down into different camera angles and image sizes to emphasize the increasingly desperate search for that key. The resulting sequence adds the crucial effect of suspense.

As you study the second sequence, you will see that the suspense is developed by two principal techniques: *image size* and *sequence length*. These are the same directorial techniques that you use to emphasize information, only now they are employed to create emotional effect.

Figure 18-9 The action presented with different camera angles and image sizes to add suspense.

The burglar sees that the door is locked.

She realizes that she must find the key.

She opens the lower desk drawers.

They are empty.

She is acutely aware that…

…the motion sensor has picked her up.

The last drawer.

Her hand, offscreen, rummages the drawer.

Her hand snatches the key.

She jams the key into the keyhole.

Image Size

Notice that the camera stays close to its subjects: the burglar's face, her hands, the key. Generally, the bigger the image size, the more emotional impact it carries.

Sequence Length

As you can see from the burglar example, the information-only version requires just three shots, while the suspenseful alternative takes ten. By stretching the length of the sequence, the director increases the feeling of urgency.

Composition

Look again at the burglar's closeup, **Figure 18-10**. In the first version, the camera is not level (that is, not parallel to the floor). This compositional effect, often called *dutch*, imparts a sense of imbalance and insecurity. Compare the first shot with a level version of the same angle (in the second shot) to see how tilting the horizon unsettles the composition.

The term "dutch," an informal version of *Deutsch* ("German"), was first applied to off-level angles because they were favored by European directors who brought the expressionistic style of German cinema to Hollywood in the 1920s and 1930s.

Camera Angle

Both image size and composition are created by camera placement. And camera angles, in general, can contribute to the overall emotional effect. In **Figure 18-11**, you can see that high angles often convey a feeling of detachment, while low angles emphasize the power of the image (and the powerlessness of the viewer). Head-on angles feel formal and calm, while diagonal setups convey movement and energy.

In practice, image size, composition, and camera angles are all treated together, as aspects of the same thing.

Lens Perspective

The "feel" of an image is strongly affected by the lens used to record it, **Figure 18-12**. Wide angle lenses exaggerate depth and motion toward or away from the camera. The effect they convey is dynamic and energetic. Telephoto lenses, by contrast, suppress depth and motion, keeping the viewer at a distance and conveying a sense of formality.

Camera Movement

Camera movement can add significantly to the emotional cast of a scene. A stationary camera, panning and tilting from one position, conveys a sense of detached observation. A moving camera, especially when dollying in and out, becomes part of the action. Moreover, a constantly moving camera imparts a feeling of restless energy.

Figure 18-10 Conveying feeling through composition.

Composition with off-level background.

Composition with level background.

Figure 18-11 Conveying feeling through camera angles.

| A high-angle shot feels more detached... | ...than a low-angle setup. | A formal, frontal angle is less dynamic than... | ...a strong diagonal composition. |

Shot Length

To create an effect, a director can control the length of a sequence by shooting it in a greater or smaller number of separate shots. In the same way, the individual shots themselves can be made shorter or longer, depending on the effect desired. In general, short shots with quick cuts convey a sense of dynamic energy, while lengthy shots feel calm and deliberate. That is why fights are typically constructed from many brief shots, while romantic scenes are often assembled from fewer, but lengthier, angles.

The Other Half of Content

Directing to communicate with your audience means managing information, emphasis, and effect. Together, these elements make up the content of your program. However, communicating content is only half the job. Unless you are making a specialized program, like a nature documentary, the content is delivered largely by the people in your video—by the performers.

Directing for Performance

Everyone knows that directors work with actors, but few people understand what directors really *do* with them. This section introduces the craft of drawing effective performances from the talent in your video programs, **Figure 18-13**.

"Talent" is the collective term for the people who appear in films and videos.

What follows is not an extensive treatment of the art of directing professional actors in fiction films, that would require a book in itself (and several are available). Instead, this section focuses on the craft of guiding people through all kinds of videos—from stories to training programs to promotional pieces. Some of your

Figure 18-12 Conveying feeling through lens perspective.

| Wide angle lenses expand apparent depth... | ...while telephoto lenses compress it. |

Fixing Problems in Rehearsal

In working with nonprofessional actors, directors must deal with difficulties that come up again and again. Here are some of these common problems, with some suggestions for solving them in rehearsal.

Context

Videos consist of short individual shots that are often recorded out of order. This choppy, seemingly arbitrary production can leave performers confused about where they are and what they are supposed to be doing. An earlier chapter uses the example of two train station sequences that are shot back-to-back, but are five years apart in story time. In another case, two lovers may be asked to enact their first meeting after playing many scenes in which they already have a romantic relationship.

To avoid problems with this shooting method, begin rehearsing each scene by explaining where it fits into the finished program, and how it relates to scenes that the performers have already completed.

Memorizing

This chapter offers ways to help actors master their *lines*, but movement and activities can be hard to remember too, especially since they must often be repeated precisely from one setup to the next. (A performer's movement within the scene is called *blocking*, and his or her activities are termed **business**.)

To control the amount of memorizing required of the actors, rehearse only a single sequence at a time. If the sequence is a long one, try to break it down further into **beats**—short subsections of, perhaps, two minutes of screen time.

To obtain consistent performances from one setup to the next, consider letting actors refresh their memories by watching playback of previous setups that cover the same material.

Expressing Emotions

As noted elsewhere, amateurs can have problems in expressing emotions convincingly. Some tend to overact, others have trouble making transitions between different feelings, still others cannot handle complex, mixed reactions.

A 12-volt monitor allows playback on location.

Rehearsal is the place to spot these problems and adjust accordingly. Step one is often to help the actor with the problem. If this does not work, you want to plan a strategy for shooting around the difficulty. Develop solutions before you start shooting, whether this involves simplifying the performance requirements, breaking the recording into shorter shots, or thinking up cutaways to help edit out mistakes.

Tuning Performances

In some cases, the quickest way to get the performance you require is by demonstration. Act out the role yourself to show how you want the blocking and business handled, and how you want the lines of dialogue to sound. Professional actors often object to being given these *line readings*, as they are called (though they seldom mind having their movements demonstrated), but nonprofessionals are often grateful for the help.

In modeling movement or speech for an actor, do just enough to get the idea across. If you go through the whole thing, you risk getting performances that are just mechanical imitations.

Charlie Chaplin, however, was famous for acting out every second of every role and then expecting his cast members to imitate him exactly.

Figure 18-13 The "burglar" is, of course, a performer.

performers may be professionals, but others may be amateurs, like corporate spokespersons or workers demonstrating job skills or simply members of your family or community.

Actors and Performance

Directing performers like these means making them appear believable and effective on screen. To be believable, the performers should not appear to be acting, but simply living and behaving normally while the camera just happens to record them. Though a skilled actor is practicing a complex craft, the result looks simple and natural on screen because that craft is invisible to the audience.

To be effective, the performance must be more than simply believable. It must also engage viewers' attention and make an impression on them. Some professional actors achieve this by sheer force of personality, but most performers can do it if they get enough help from the director.

Many people think that performing means creating a character in a fictional program, and that is often the case. But, the people who appear in nonfiction programs are also performing. As a director, you may be coaching a spokesperson for an organization, a person being interviewed on camera, a subject of a documentary program, or a person who demonstrates the skills or techniques being taught in a training video. Even the most seasoned professional actors benefit from good direction, because the director can show them where they fit in the program as a whole and how best to do their part in it.

Some professional directors feel that 90% of directing actors lies in casting exactly the right people for their roles, and then getting out of the way and letting them work.

Outside of Hollywood-style production, most performers are not full-time professionals. Working with less-seasoned talent, directors spend considerable time solving performance problems—problems like self-consciousness, difficulty in speaking lines, ignorance of production techniques, and trouble expressing emotions. The most widespread problem with actors is a general insecurity: a fear that they look ridiculous, that they are inadequate, that they are failing.

Actors and Insecurity

Camera fright is a form of stage fright: a fear of subjecting yourself to public attention, of placing yourself under a spotlight for critical inspection by the audience, **Figure 18-14**. Even professionals can suffer from it, and almost all actors may occasionally feel ill at ease about performing.

Fortunately, camera fright is easier to combat than stage fright, because the performer is *not* actually appearing in public. With reluctant actors (in fact, with *all* actors) it helps to remind them that their mistakes are completely private and will never be seen by anyone except the cast and crew. Shooting multiple takes lets performers practice until they get it right. Assembling the program from separate shots allows the editor

Figure 18-14 Amateur performers are often insecure.

to cut out all the bad parts and show only the best. Finally, reassure the actors that your job is to make them look good on screen, and that is exactly what you are doing.

Helping Actors Overcome Self-Consciousness

Self-consciousness is a painful awareness of face and body that makes the performer behave stiffly and awkwardly. It also makes faces look tense and uncomfortable. There are several directorial tactics for reducing self-consciousness. First, allow plenty of time for rehearsal. The more familiar the performers are with the material, the more comfortable they feel about it. Often amateurs will perform more naturally if they think they are not being recorded. So, it may help to disable the *tally light* on the camcorder (the button that glows red to indicate that the camera is recording). That way, you can record rehearsals without alerting the performers.

If you cannot turn off the tally light, cover it with a small piece of opaque tape.

To reduce physical awkwardness, keep actor movements simple and natural. It helps to check with performers after each rehearsal to see if anything feels uncomfortable to them. If so, correct the problem. In the first shot of **Figure 18-15**, the performer does not know what to do with his hands. The chair provides a natural place to rest them.

Finally, the best way to make performers look natural is by having them do things that they do in real life, so that they do not have to learn new skills. For instance, if you are making a training program on how to operate a fork lift, use an actual fork lift operator as your on-camera demonstrator.

Helping Actors Deliver Lines

All performers can have trouble memorizing speeches, and amateurs have the additional problem of stiffness.

We have all seen local commercials in which the owner of a small business stands woodenly in front of the camera and intones something like, "So-Come-On-Down-To-Billy-Bob's-House-Of-Barbecues-And-Fireplaces..."

If the performer cannot read aloud convincingly, the easiest solution is to allow him or her to improvise. For example, the director might say something like, "Okay, now tell the folks to come visit your store, and be sure to say the whole name." Allowed to say his own words in his own way, the performer/client will probably sound more natural.

If the problem is remembering lines, you can minimize the problem by avoiding long speeches and breaking conversations down into several brief shots. Remind performers that it does not matter if they make a mistake or how many times they make it. In the finished program, only the correct version will appear.

Figure 18-15 Simply standing still can be difficult for amateur performers. You can often use a prop to anchor them.

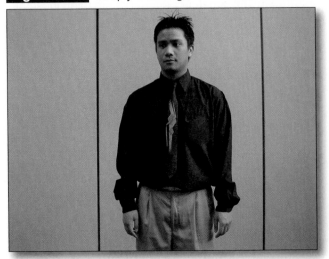

Be prepared to use cue cards so performers can read their lines instead of memorizing them, **Figure 18-16**. Place the cue cards off-camera, but in positions so that the performers' eyes appear to be looking in the right direction. Do not transcribe all the lines onto cue cards before shooting. Instead, have sheets of cardboard and broad markers available at the shoot and make cards only when performers are having real difficulties. Otherwise, your talent will expect cue cards for every scene and refuse to memorize their lines.

With a speech or lengthy remarks by an on-camera spokesperson, you will, of course, need to prepare cue cards in advance.

Helping Actors Master Production Techniques

Many times, amateur performers have problems because they do not understand the technical requirements of the video medium. Here are some of their most common mistakes, along with suggested remedies.

Hitting Marks

Often a performer is asked to move through a shot, and then stop at a predetermined spot—usually because that spot is specially lit for the performer and the camera is set to create a good composition there. Some amateur performers have trouble stopping at the right place, or can do so only by searching obviously for their *marks* on the floor (**Figure 18-17**). The

Figure 18-17   Chalk marks show the performer where to walk and stop.

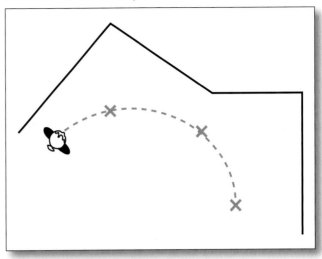

simplest solution is to avoid the problem entirely. For example, instead of having the actor move into the frame and then stop, have the actor stand off-screen while the camera moves to frame him or her.

Looking at the Camera

It is natural for amateurs to look directly into the camera lens, especially after they think they have finished a shot. Their look asks, "Is the shot over?" or "Was that okay?" This can be a problem if you need extra footage at the end of the shot. Explain, diplomatically, that the shot was not usable because the actor looked at the camera. Then, retake the shot. If the actor cannot soon learn to avoid looking at the camera, this is a sure sign that you will encounter other problems, and you had better replace this performer sooner rather than later.

Matching Action

Once you explain the importance of doing (and saying) everything identically for each angle in a scene, most amateur actors will try to do this. To help them out, keep good continuity notes so you can remind them of details before every shot. Use an on-set monitor to play back their previous angles so they can see exactly what they said and did.

Overall, the best solution to technical problems with amateur actors is to remove the problems, rather than ask the performers to

Figure 18-16   A hand-lettered cue card.

Improvised Teleprompters

For reading lengthy texts on camera, professionals use *teleprompters*—special displays that scroll the copy. Commercial models can be expensive, but you can achieve the same results with a personal computer.

The system is more flexible on a laptop model, because the computer screen is easier to position off-camera where the actor can read it comfortably. With a little ingenuity, however, you can also use the monitor of a desktop computer.

A slide program work screen.

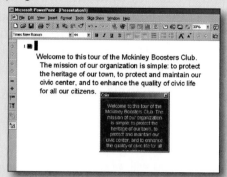

Improvised teleprompter setup.

Preparing the Text

The best applications for teleprompting are slide show programs, such as Corel® Presentations™ and Microsoft Office PowerPoint®, because they can display slides full-screen without surrounding menus. Also, slide programs are specifically designed to display text in large sizes.

If a slide program is not available, you can use a word processor by hiding all unnecessary toolbars and menus, setting the copy in a large type size, and setting the viewing width so that the text fills the computer screen side-to-side.

A word processor display used as a teleprompting device.

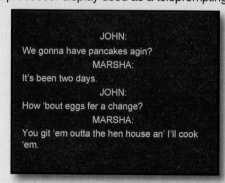

The first step is to key (or import) the text into the program and format it in convenient sections. Set the typeface in a simple font, like Arial or Swiss, at a very large size (try 32, 36, or 48 points).

Unlike true teleprompters, slide programs cannot scroll text continuously. Instead, they display one full page at a time. For this reason, inexperienced readers can be confused by rapidly changing text pages. To solve the problem, repeat the last sentence of each outgoing page at the top of the incoming page. On both pages, set the repeated text in a contrasting color. This makes it easy for the reader to follow continuing copy from one slide to the next.

First slide with text to be repeated in white.

Setting Up the System

Position the computer screen facing the performer and as close to the camera lens as possible. To conceal the fact that the person is looking to one side of the lens, use a telephoto setting and place the camera as far away as practical. (The setup must be close enough for the performer to read the computer display.)

To minimize the angle between the lens and the display screen, set the camera as far from the performer as practical.

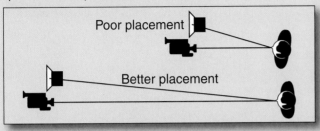

Too close to the camera, the subject is forced to look just slightly to one side.

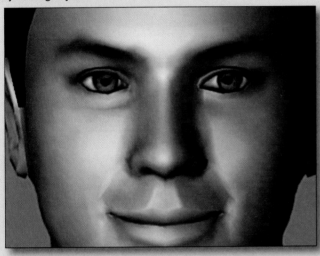

Farther from the camera, the subject appears to be looking at the lens.

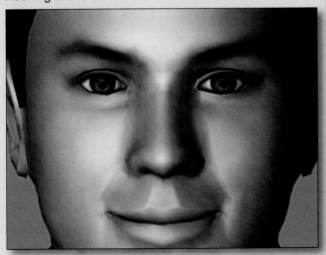

Using the System

To use the program as a set of cue cards, have the continuity supervisor follow the dialogue, advancing the slides to change the virtual "cards."

solve them. It is unlikely that you can turn your talent into professionals during a single shoot.

Helping Actors Express Emotions

Professional actors are skilled at adjusting the intensity of their performances to suit the director. Amateur performers often lack this ability.

This section applies mainly to fiction movies, since nonfiction programs seldom require performers to simulate feelings.

Overacting is often easy to correct. Simply show the offenders some closeups of their work on a monitor, and they will realize that they need to rein themselves in. With persistent overactors, you may want to pull the camera back from them, because the smaller the image, the less intense it appears to the audience. Be aware, however, that this technique has other consequences. If the wider shot of one actor is intercut with closer angles on another, the actor with the larger image may well dominate the scene.

With amateurs, a bigger problem is complex acting involving combinations of emotions. For example, if you say to an amateur performer, "You're delighted that your no-good brother-in-law is leaving, but unhappy that your mother-in-law is staying," the actor may well be unable to show these two opposite emotions at the same time, and the result will be a failure (**Figure 18-18**).

To solve the problem, rework the scene so that the actor can react separately to the good news and to the bad news. When the transition between emotions happens off screen, the actor has to play only one in each shot, **Figure 18-19**. Since you shoot the wife's happy and unhappy closeups as separate shots, the performer can take all the time she needs between setups to change from happiness to anger.

Helping Actors Project Authority

Low angles tend to make a person seem more powerful, more commanding. High angles have the opposite effect. When shooting an executive or a spokesperson for a corporate video, make certain that the camera is at least slightly below the subject's eye level, **Figure 18-20**.

Figure 18-18 Amateur actors may be unable to show opposite emotions at the same time.

HUSBAND: My brother is leaving tomorrow, but Mom is staying six more weeks.

WIFE: (dismayed) What!?

Figure 18-19 Shoot different emotions as separate shots.

HUSBAND: My brother is leaving tomorrow.

HUSBAND: But, Mom is staying six more weeks.

WIFE: (dismayed) What!?

Figure 18-20 A low angle conveys a feeling of authority. Setting the camera just below eye-level conveys the effect more subtly.

Low angle

Just below eye-level

Directing Assignments

So far, we have focused on helping actors perform effectively. However, there are situations where you will be directing the shooting of actual activities, rather than rehearsed performances. Some of the more common types include interviews, documentaries, and "wallpaper."

As noted elsewhere, *wallpaper* is footage designed to fill screen time while the audio delivers the important information.

Directing Interviews

Today, most video interviews show only the subject speaking informally to someone who is positioned beside the camera, **Figure 18-21**. The interviewer is never seen or heard. Though the result looks spontaneous, it is really the result of patient questioning by the director and adroit cutting by the editor.

It is not unusual to shoot two hours of raw footage to obtain 10 minutes of edited screen time.

Setting up the Interview

First, set up the interview in an interior with a suitable background. (Some production teams carry draperies and century stands to create neutral backgrounds when needed.) Test for good acoustics and lack of background noise, because amateurs will be less tolerant of interruptions and extra takes due to sound problems.

Figure 18-21 Most interviews show only the subject.

(Sue Stinson)

Position seats so that the subject is facing the interviewer, who sits close beside the camera. Use a low-backed chair to keep the chair back out of the shot, **Figure 18-22**.

By convention, the subject usually sits slightly screen-left, looking screen-right. However, when presenting opposing viewpoints in separate interviews, have the opposition sit screen-right and facing left. Cut together, the opposing compositions will give the illusion of a discussion between the two subjects, **Figure 18-23**.

Making the Subject Comfortable

To reduce camera anxiety, place the camcorder back from the subject—as much as eight feet away (location permitting). If possible, light with two or, at most, three softlights to simplify the lighting and reduce heat, as shown in **Figure 18-22**. LED video lighting is especially useful for interviews.

Framing Questions and Answers

The key to successful interviewing is good questioning. The objective is to get the subject talking about a topic, rather than answering a question.

In the examples that follow, assume that a video director is interviewing Mr. Winesap for Discovery Channel™. Mr. Winesap, President of the McKinley Boosters Club, tends the flowers in the McKinley Plaza.

Figure 18-22 Plan of a typical interview.

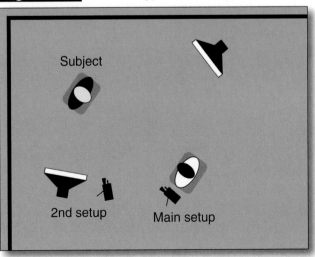

Figure 18-23 Mirror-image compositions give the impression that the subjects are debating.

Positioned slightly screen-left, looking screen-right

Positioned slightly screen-right, looking screen-left

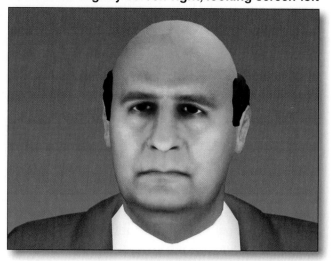

Here is the wrong way to frame a question:

INTERVIEWER: How long have you cared for the plaza flowers, Mr. Winesap?
SUBJECT: I've been running the show for 22 years!

When the question is removed, the answer, "I've been *running the show* for 22 years" will not make sense to viewers.

Instead, ask the subject to talk about a subject. The right way:

INTERVIEWER: Mr. Winesap, talk to us about tending the flowers, including where and how long.
SUBJECT: I've filled the McKinley town square with flowers for 22 years.

Now the editor has an answer that makes sense without including a question.

Warming Up

Most subjects will need to practice this skill, so develop some "throwaway" questions to get the subject comfortable.

INTERVIEWER: Tell us how you feel about McKinley.
SUBJECT: I'm very fond of it.
INTERVIEWER: Great! Now tell us again and this time, include the town's name.
SUBJECT: Oh, I see. (Ahem) I've always been fond of McKinley.
INTERVIEWER: That was perfect. You have a knack for this!

Listening carefully, the interviewer decides that the throat-clearing can be cut from the sound track, so the shot (if used) can begin after the "Oh, I see."

You may also need to gently coach the subject to avoid looking at the lens. Explain in advance that the subject should look only at you. Ask for repeats if he or she forgets and looks into the lens.

Finally, explain that any mistakes or fumbles will be cut out, and only the subject's best answers will be used. As always, remind subjects that your job is to make them look good.

When you feel that the subject is ready, move on to the actual interview topics.

It is common to record during the warm-up (which may produce some usable material), and then simply continue with the actual interview.

Camera Angles

If you have the luxury of two cameras, you can obtain two different angles on the entire interview. These angles can be used to cut back and forth between when editing. With just a single camera, you may want to pause part of the way through shooting and move the camera to another setup to add variety. Make sure that the two setups show a significant difference in both horizontal angle and subject size, **Figure 18-24**.

Figure 18-24 Shooting at different angles and/or image sizes adds variety to the edited interview.

(Sue Stinson)

Directing Documentaries

Documentary programs about events that happen in real time are a special challenge. The director (who is often camcorder operator, as well) must anticipate important shots because the action usually cannot be repeated, **Figure 18-25**.

Talented documentary makers develop a sixth sense that enables them to frame the right subjects at the right moment. Here are some suggestions for obtaining good documentary footage.

See "Around" the Camcorder

If your camcorder has an external LCD viewing screen, use it. If you must use the

Figure 18-25 Recording a famous draft-horse team.

viewfinder, keep your other eye open. Either way, you are using your peripheral vision to keep tabs on the area outside the video frame (**Figure 18-26**). That way, you can spot interesting activities to record.

Since the LCD monitor is on the left side of the camera, it may help to reverse the usual interview positions—place the director/operator behind the left side of the camcorder, and have the subject sit right and look left.

Anticipate the Next Shot

If a softball batter gets a hit, the action will move from home plate to first base. Knowing this, you can re-frame and zoom in on first base, before the runner gets there. The idea is to predict the results of the activity you are shooting, and switch to them in time to capture them as they happen.

Stage Pickup Shots When Possible

Even in real-time shooting, it may be possible to pick up crucial action that you were unable to capture. For example, while recording the team of horses as they pass by, you spot a Dalmatian dog riding on the wagon—but, too late to get a closeup. So, when the team stops for review, frame a closeup of the dog (**Figure 18-27**). The closeup pickup shot can then be cut into the action, **Figure 18-28**.

Figure 18-26 Seeing "around" the camcorder.

Keep your other eye open...

...to see details outside your frame.

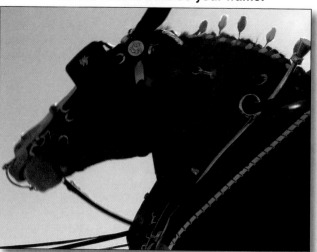

Figure 18-27 Missing a closeup of the dog the first time around, you get a pickup shot later.

Original shot

Pickup shot (closeup)

Figure 18-28 The pickup shot is inserted into the edited sequence.

If you pan across the dog from left to right, you can simulate the movement of a rolling wagon (since there's nothing in the background to reveal the trick).

Get Cutaways

In any documentary situation, there are moments when nothing important is happening. Those are the times to get *cutaways*—shots other than the main action. By using these cutaway shots, the editor can make widely separated actions look continuous. During Jersey heifer judging at the Leibniz County Fair, for example, there are long, uninteresting stretches while the judge walks around looking at the entries. By using cutaways of the spectators, the editor can invisibly cut out these passages (**Figure 18-29**).

Good documentary makers are supreme opportunists—ready to think and move fast to record all the pieces the editor will need.

Cutaway shots are covered more fully in the next chapter.

Directing "Wallpaper"

Abstract ideas are impossible to visualize (see the *Creating Wallpaper* sidebar in Chapter 10, *Program Creation*). Instead, you must fill the screen with images that seem to fit the words on the sound track, even if the relationship is not very close. This filler footage is often called *wallpaper*.

Some wallpaper can be prescripted, but there are times when you will have to improvise with what you find on the scene. Once again, the trick is to look for cutaways to fill the screen during necessary, but nonvisual, narration.

In **Figure 18-30**, the narrator's introduction to the annual McKinley kinetic sculpture race is covered by establishing shots of the opening parade.

Figure 18-29 While the judge (blue shirt) inspects the entries...the editor can shorten the process with a cutaway.

(Sue Stinson)

Figure 18-30 Using cutaways during narration.

NARRATOR: Established two decades ago by the McKinley arts community...

...the great annual kinetic sculpture race...

...has become an internationally famous event...

...with imitators around the world.

When confronted with static material, you can compensate somewhat by panning and/or zooming—anything to get movement into the wallpaper, **Figure 18-31**.

If you shoot wide shots, each with several areas of interest, movement can be added in editing by digital panning and zooming.

Other Assignments

Pickup shots, inserts, cutaways, color shots are collectively known as *cover*, the extra content a good director captures to provide the editor with full *cover*age of the action. These and other types of shots are the subject of Chapter 19, *Directing for Form*.

Figure 18-31 Zooming to create movement in wallpaper.

REPORTER: Meanwhile, at City Hall...

...the mayor refused to comment on the allegations...

...saying only that "appropriate administrative interventions...

...are ongoing at this point in time."

Summary

- Even though production is a collaboration among many people, the final authority generally rests with the director.
- Even a short program involves a great number of these decisions about directorial approach—decisions intended to determine the video's overall style.
- To make sure that the audience gets the information in a program, the director must emphasize it appropriately.
- By utilizing camera angles, image sizes, shot lengths, lens perspectives, camera movement, and other tools, the director creates feelings.
- Directors shape performances by blocking the action, guiding the actors, and pacing the scenes.
- When working with less-experienced talent, directors spend considerable time solving performance problems.
- Common directing assignments include interviews, documentaries, and "wallpaper."

Technical Terms

Beat: A short unit of action in a program; often, though not always, corresponding to a scene.

Business: The activities a subject performs during a shot, such as writing a letter or filling a vase with flowers.

Cover: Additional angles of the main subject or shots, like inserts and cutaways, recorded to provide the material needed for smooth editing.

Cutaway: A shot other than, but related to, the main action.

Dutch: Referring to an off-level camera. The phrase, "Put dutch on a shot" means to purposely tilt the composition.

Line reading: A vocal interpretation of a scripted line that includes its speed, emphasis, and intonations. In simple terms, "You came back!" is one reading of a line, and "You came back?" is a different reading.

Lines: Scripted speech to be spoken by performers.

Marks: Places within the shot where the performer is to pause, stop, turn, etc. These spots are identified by marks on the floor made of tape or chalk lines.

Objective insert: A detail of the action presented from a neutral point of view.

Subjective insert: A detail of the action presented from a character's point of view.

Tally light: A small light on a camera that glows to indicate that the unit is recording.

Teleprompter: A machine that displays text progressively as a performer reads it on-camera.

Review Questions

Answer the following questions on a separate piece of paper. Do not write in this book.

1. The _____ is generally the final authority on decisions for a production.
2. What are the five questions that a well-prepared script should have answers to?
3. Identify strategies for addressing problems in a script as you direct the script.
4. *True or False?* The bigger the image, the more emphatic the information it contains.
5. A(n) _____ insert is a shot from a neutral point of view.
6. How is the *dutch* compositional effect created? What does it add to a shot?
7. What are *beats* in a program?
8. What are the effects of self-consciousness on an actor's performance?
9. *True or False?* A performer's marks are usually indicated using tape or chalk.
10. Explain the effect created by using low angles to shoot a person.
11. Pickup shots, inserts, cutaways, color shots are collectively known as _____.

STEM and Academic Activities

1. **Technology.** Research the impact of technology on mass communication. Identify an early technological advancement that changed mass communication and explain its impact. Identify a recent technological advancement that changed mass communication and explain its impact.
2. Social Science. While watching television programs, commercials, movies, etc., make note of instances where color is used for emphasis. Explain how color is used for emphasis in each example. Is a certain color used for emphasis more often than others?

CHAPTER 19

Directing for Form

Objectives

After studying this chapter, you will be able to:

- Understand the importance of directing to edit.
- Recall the qualities of good coverage.
- Summarize the important areas of continuity in a program.
- Explain the relationship between screen direction and the frame.
- Recall the principles in staging action for the camera.
- Recognize different types of camera movement and the distinctive visual effect each creates.

417

About Directing for Form

Complementing Chapter 18, *Directing for Content* (communication and performance), this chapter covers directing for form—the techniques involved in staging and recording the actual program footage. Mastering the video form requires technical knowledge because of the special nature of the medium. Every program consists of hundreds of very short pieces (shots) that are often recorded out of order. When rearranged by the editor, these pieces must fit together seamlessly, even though each is made quite independently of the others. Also, because space and time on the two-dimensional screen follow their own special rules, action must be carefully staged to appear as if it is happening in the real world.

The world on the screen is introduced in Chapter 3, *Video Communication*.

For these reasons, a video director needs a thorough understanding of the formal aspects of the craft, beginning with the fundamental principle that *absolutely everything recorded, both visually and aurally, is nothing more than raw material for future editing*.

Covering the Action

As you direct a production, remember that you are not making a video program, but only its separate component parts. The actual program will be assembled from your shots at a later time by the video editor. What you are doing is providing the building blocks, the component materials from which the editor will construct the final program.

For clarity, we will treat the editor and director as two different people, though directors often edit their own programs.

To help ensure a successful result, the components you supply must provide the editor with everything that is needed, in a form that the editor can use. The process of shooting the material to anticipate the editor's requirements is called *directing to edit*. To begin with, directing to edit means delivering full coverage of the action so that the editor has enough material to work with.

Coverage

Coverage involves more than just capturing all the action. It also means ensuring that the action is recorded and then rerecorded in multiple shots that repeat all or at least part of the content. Good coverage typically requires *repetition*, *overlap*, *variety*, *protection*, and *cutaways*.

Repetition

Repetition involves recording the same action more than once by shooting it from more than one point of view. For instance, you might record the entire scene from three different camera setups: a two-shot, a closeup of one actor, and an opposing closeup of the second actor (**Figure 19-1**).

A Demonstration Sequence

The discussion of coverage uses a simple sequence of a couple picnicking on a wooded bluff above the ocean. The basic actions in the sequence are summarized in the following illustrations.

The couple picnicking.

He takes a drink of soda.

He points out a storm over the ocean.

She doubts that it will reach them.

In actuality, this short sequence might be covered by just three or four setups. But, for demonstration purposes, we will use many more angles than usual.

Figure 19-1   Repetition.

| A neutral two-shot covers the entire sequence. | Her closeup repeats the entire sequence. | His closeup repeats the entire sequence. |

Multiple angles allow the editor to cut between one shot and another at any time to improve performance, adjust timing, and control style.

Overlap

Even when you do not rerecord the entire action in each of several setups, it is usually important to overlap action from shot to shot. This is illustrated in **Figure 19-2**. Overlapping means beginning a new shot by repeating the last part of the action in the previous shot.

By overlapping action, you allow the editor leeway in setting the edit point between the two shots. Even if you think you already know exactly where you want the edit to be, protect yourself and the editor by providing extra footage.

In addition to overlapping parts of the action, you should begin recording at least five to ten seconds before *any* action begins, and continue shooting for the same length of time afterward as well. These extra seconds allow the editor more leeway in choosing edit points.

Variety and Protection

To offer the editor a broad range of options, you should provide coverage from a variety of angles. Varying angles helps make edits invisible and also tends to hide small mismatches between different takes (recordings) of the same action, **Figure 19-3**.

Sometimes mismatches or other mistakes are too large to hide, so the editor cannot use the shots in which they appear. That is another reason

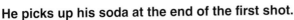

Figure 19-2   Overlap.

He picks up his soda at the end of the first shot.

He picks up his soda at the start of the second shot, drinks, then puts it down again at the end.

Figure 19-3 His action is covered in three different styles of closeup.

for covering the same action from different angles. In fact, if you are not sure about something in a shot, obtain an extra shot from another angle.

Such an alternative setup is sometimes referred to, literally, as a **protection shot**.

Cutaways

Cutaways are shots that show information other than the main action. Typical cutaways (**Figure 19-4**) include:

- *Inserts*: Closeups of details.
- **Reaction shots:** The responses of subjects other than the one who is currently the center of interest.
- **Object shots:** The second half of a glance-object pair—an object shot shows what a subject is looking at from his or her point of view.
- *Color shots*: Shots that show expressive details of the environment in which the scene is unfolding.

Nowadays, using a shot for color alone is considered somewhat old-fashioned. It is better to relate color cutaways to the main action.

In addition to delivering extra information, cutaways are extremely useful for interrupting other shots, so that they can be invisibly shortened or so that other changes can be made. For example, suppose that the two-shot of the couple includes more dialogue than necessary, **Figure 19-5**.

The trouble is that the basket is already packed. Discovering this potential problem, the director adds a cutaway shot of the ocean. By inserting this cutaway into the closeup, **Figure 19-6**, the editor can snip out the line about packing the picnic basket without leaving a hole in the original shot. Of course, a reaction shot of the woman could be used as a cutaway instead.

Maintaining Continuity

Continuity is the complex process of ensuring that all the small pieces of a video (the shots and

Figure 19-4 Cutaways.

Insert of the soda being set down.

Reaction shot of her while he talks offscreen.

Color shot of the ocean view.

(Sue Stinson)

Figure 19-5   HE: I don't like the looks of that squall coming in. I'll start packing the picnic basket. You'd better finish your soda.

the details recorded in them) add up to a smooth, consistent whole. A program with good continuity appears to be a single *continuous* presentation that has no mismatches in information, action, or screen direction.

On larger productions, the script supervisor keeps track of these elements, but the director still needs to be keenly aware of continuity.

The script supervisor's job is not to create and maintain continuity, but only to monitor it.

Achieving good continuity is the director's job. In smaller productions, script supervision may be only a part-time responsibility of a production manager or an assistant. In such a situation, the director is the only person who can ensure consistency.

Continuity of Information

In shots that include duplicate or overlapping action, the details should all be the same. For example, dialogue should be identical. Do not allow an actor to say "Hello!" in the two-shot and "Yo!" in his closeup. Though an editor can cut mistakes and trim pauses (if given enough coverage), it helps if the actor's lines are consistent from shot to shot.

Physical details should match. If the actor is holding a drinking glass in the two-shot, then in the closeup, he or she should continue to hold the glass, it should be in the same hand, and the liquid in the glass should remain at the same level.

This may seem obvious, but remember that shots appearing back-to-back in the program may be recorded days or even weeks apart. For example, an actor may park a car in one location and then enter a building hours later and miles away, **Figure 19-7**. Between the two shooting sessions, the clothing carried on hangers has changed.

Continuity of Action

You also need to ensure that action is repeated accurately from one shot to the next. In our demonstration picnic, for instance, the man puts down his soda bottle, points offscreen, and talks about a squall over the ocean.

Now, suppose the editor wants to use a wider shot of his warning about the squall. The problem is that, in the performer's first shot, he says the line *after* he puts down the bottle. But

Figure 19-6   Inserting a cutaway.

HE: I don't like the looks of that squall coming in.

You'd better finish your soda.

(Cutaway)

Styles of Coverage

Although most directors have their own working methods, general approaches to ensuring full coverage fall into three broad types: classical, contemporary, and personal.

Classical Coverage

In Hollywood before about 1960, it was standard practice to record a scene like this:

Notice several things about this classical approach to coverage. Starting with a wide view, it gradually moves closer and closer to the characters. The complete scene is recorded at least five times, offering the editor a wide choice of shots. Finally, the pairs of over-the-shoulder two-shots and closeups are symmetrical, in that they are almost mirror images of one another.

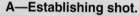
Classical coverage.

A—Establishing shot.

B—Master shot.

C—Over the shoulder CU of her.

D—Over the shoulder CU of him.

E—CU of her.

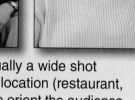

F—CU of him.

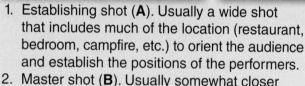

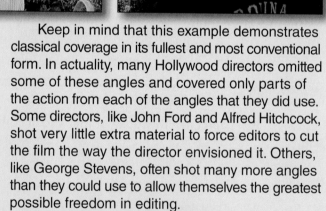

1. Establishing shot (**A**). Usually a wide shot that includes much of the location (restaurant, bedroom, campfire, etc.) to orient the audience and establish the positions of the performers.
2. Master shot (**B**). Usually somewhat closer than the establishing shot, but still includes all the principal performers and all the action and dialogue.
3. Over-the-shoulder two-shot, favoring one character (**C**). This angle moves closer, but still holds more than one person.
4. Over-the-shoulder two-shot, favoring the other character (**D**). This angle is the reverse of the previous over-the-shoulder two-shot.
5. Closeup of one character (**E**). Closer now, and only one person on screen.
6. Closeup of the other character (**F**). The reverse of its predecessor.

Keep in mind that this example demonstrates classical coverage in its fullest and most conventional form. In actuality, many Hollywood directors omitted some of these angles and covered only parts of the action from each of the angles that they did use. Some directors, like John Ford and Alfred Hitchcock, shot very little extra material to force editors to cut the film the way the director envisioned it. Others, like George Stevens, often shot many more angles than they could use to allow themselves the greatest possible freedom in editing.

Contemporary Coverage

Today, the establishing and master shots are sometimes omitted, or at least combined. In these wide shots, the actors may begin the scenes with the opening lines, then jump directly to the closing ones, knowing that the remainder will be captured by other camera setups. The contemporary

Contemporary coverage.

coverage photos present an approximation of the same scene shot in contemporary style. Notice that the angles are not symmetrically opposed and that the coverage tends toward tighter angles for the smaller video screen.

Personal Coverage

In the personal style, typical of commercials and music videos, each part of the action is treated separately and composed without apparent regard for a formal plan. (Notice that screen direction is reversed in the third and fifth shots.)

Even with the most personal style coverage, however, an astute director will give the editor options by shooting at least some material from multiple angles and by overlapping action from shot to shot.

Personal style coverage often does not show a formal pattern.

(Sue Stinson)

Figure 19-7 Since the first shot shows the light shirt on the outside, the same shirt must be outside in the next shot.

when he repeats the action for his next angle, he points offscreen with his soda bottle while he says the line—he never puts the bottle down (**Figure 19-8**).

Figure 19-8 Error in continuity of action.

He sets down the bottle before he says his line.

He gestures with the bottle as he says the line.

The problem arises because the next shot starts with the bottle already on the bench, **Figure 19-9**. If the editor cuts from the closeup to the two-shot, the bottle will jump from the man's hand to the bench.

But, since the alert director has shot an insert of the bottle being set down on the bench (**Figure 19-10**), the editor can use it to cover the mistake.

Now we have seen cutaways used for two different reasons: first to cover the omission of the line about the squall, and then to conceal the action mismatch between two other shots.

Continuity of Direction

The third and final consistency is continuity of direction. It is called *screen direction* because actors' looks and movements are measured in reference to the borders of the screen. Although direction is just one of three principal types of

Figure 19-9 The bottle is already down when the next shot begins.

Figure 19-10 Correcting the error in continuity of action.

HE: I don't like the looks of that squall coming in.

(Insert shot)

You'd better finish up your soda.

continuity (along with information and action), it is such a large, complex, and important subject that it deserves a more extensive discussion.

Managing Screen Direction

Screen direction is the concept that people and objects on *screen* should usually point in a consistent *direction*. This means that they should aim, more or less, at the same edge of the frame (usually left or right) from one shot to the next, until either the sequence (scene or continuous action) ends or the director indicates a purposeful change in screen direction. **Figure 19-11** illustrates a mistake in screen direction. In **Figure 19-12**, the screen direction is consistent.

The Importance of Consistent Direction

It is important to maintain screen direction for three reasons. First, whenever you present a new subject, your viewers respond by establishing their basic point of view: where they "are" in relation to the scene. They do this automatically and without thinking about it. As long as people and objects on screen continue to face approximately the same direction, viewers do not have to make the subconscious effort to reorient themselves. But, if someone suddenly switches screen direction, viewers are forced to mentally relocate themselves to match. In **Figure 19-11**, for example, the viewpoint suddenly jumps to the opposite side of the bench. Though viewers can easily figure out that the second woman has not really moved, the brief effort required to do so diverts their attention from the content of the program and reminds them that it is only a video, rather than reality.

A second reason for maintaining screen direction is to enhance the illusion that an action is unfolding continuously in real time, instead of in numerous separate shots.

Finally, assigning different screen directions to different actions allows you to alternate between them without confusing the audience.

Figure 19-11 A two-shot establishes the screen direction. The woman on the left looks to the right in her closeup, but the second woman also looks screen right, when she should look left.

Figure 19-12 Now, the second woman's closeup matches her direction in the two-shot.

Keep in mind that screen direction is only a convention. It can be successfully broken when you have a specific need to do so.

The Power of the Frame

Screen direction can be confusing because you must establish and maintain it in the real world, where you are shooting your program. But, screen direction is not determined by events in the actual world. *It exists only in relation to the frame around the image.* (To see how this works, study Figure 19-14, in which a subject walks around three sides of a house.)

For example, in the real world, a subject walks in three different directions: south, then east, then north (as shown in the composite image, **Figure 19-13**). In the video world, however, she never changes direction, always moving from screen-left to screen-right, **Figure 19-14**.

Figure 19-13 In the real world, the subject walks in three different directions.

But, if the sequence includes an angle from the point of view of the man on the front stoop, the screen direction must be temporarily reversed (**Figure 19-15**), even though the subject follows exactly the same route in the real world.

In short, screen direction has nothing to do with the action itself and everything to do with the viewpoint of the camera.

There are three types of screen direction: look, movement, and convention.

- *Look*—Actors that remain in one place stay facing the same direction from shot to shot.
- *Movement*—From shot to shot, moving actors and objects (like vehicles) always go across the screen in more or less the same direction.
- *Convention*—Subjects face and/or move either toward the left or toward the right in accordance with the conventions established by maps.

Screen Direction: Look

When two people are speaking to each other, good screen direction calls for one person to face one side of the screen and the other person to face the opposite side. **Figure 19-16** illustrates correct screen direction looks. Note that screen direction does not have to be exactly identical from shot to shot, as long as the subject is oriented, more or less, toward the same screen edge. Note, too, that the subject is positioned toward the screen edge opposite the direction of the person's look.

Screen direction is often preserved even when the subjects are not together. In a telephone sequence, for example, it is customary for one caller to face screen right and the other to face

Figure 19-14 In all camera setups, the screen direction remains left to right.

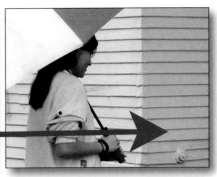

Figure 19-15 Same sequence, different point of view.

Basic left-right screen direction.

Shot from the man's point of view reverses screen direction.

The basic screen direction resumes.

screen left. As you cut back and forth between the two, the opposing screen directions reinforce the idea that the two people are talking to each other, **Figure 19-17**. It is also common to divide the screen with a partial wipe and show both callers at once, **Figure 19-18**.

You can also use screen direction to indicate that subjects are *not* related. For example, suppose

Figure 19-16 As long as the camera stays in front of the action line, the subjects maintain the same screen direction.

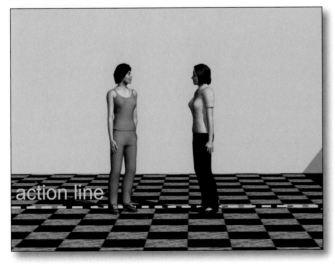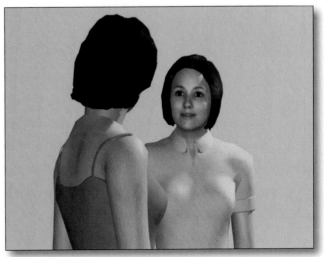

(Continued)

Figure 19-16 *(Continued)*

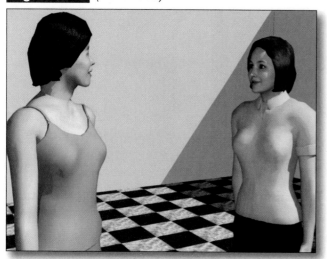

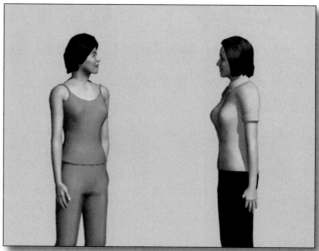

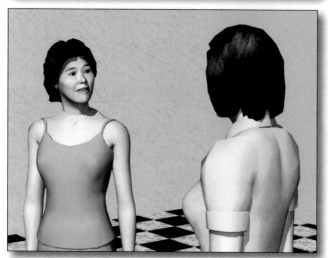

Figure 19-17 Opposing screen directions of subjects speaking to each other.

Figure 19-18 Split-screen conversation between two subjects.

Figure 19-19 A split screen showing two different phone conversations.

you show a radio show host offering prizes for the first person who calls in. The shots in **Figure 19-19** show two different callers responding. Notice that this split screen contains exactly the same shots as the previous example. But, because the callers are facing opposite directions, it appears that they are *not* talking to each other.

Screen Direction: Movement

When movement of a subject continues through several shots, the screen direction should be essentially the same in every shot, **Figure 19-20**.

Frequently, you will need to manage the screen direction of more than one subject at a time. In doing so, you help the audience remember who is who in a sequence. You can also use screen direction to show what type of action you are presenting.

Parallel Movement

When two subjects start at the same point and head for a common destination, you keep the screen direction of both subjects the same, **Figure 19-21**.

Opposing Movement

When subjects are moving simultaneously, but toward different destinations, their screen directions are opposite (**Figure 19-22**).

Figure 19-20 Shot-to-shot screen direction maintained.

| Figure 19-21 | When one subject is following another, their screen direction is the same. |

| Figure 19-22 | When subjects are headed for different destinations, their screen directions are opposite. |

Divergence

You can also use opposing screen directions to indicate that two people are starting from the same point and heading for different destinations.

Random Movement

You can use screen direction to indicate that a succession of shots is not a continuous action. To do this, you purposely vary screen direction so it shows no pattern at all. **Figure 19-23** shows a subject headed in different directions, as if to present an anthology of her activities.

It helps, of course, that the backgrounds and costumes are different, as well.

Screen Direction: Convention

Finally, some screen directions are set by convention. As noted in the sidebar *Scenes in Automobiles*, a car traveling from New York to California is established as moving from right to left because on a map, convention places New York to the right of California (**Figure 19-24**). For the same reason, a person watching a sunset will often face screen left because left is the "west" side of the screen.

Today, it is common to establish the conventional screen direction with one or two shots and then ignore it during the rest of the sequence, even when not shooting inside the vehicle.

Controlling Screen Direction

To create and maintain a screen direction, use the concept of the *action line*—an imaginary line drawn to divide the camera from the subject(s). To see how this works, study the camera ground plan (in **Figure 19-25**) for the walk around the cottage presented earlier. The action line is represented as an arc between the subjects and camera setups A, B, and C. As long as the subjects and camera remain on opposite sides

| Figure 19-23 | A complete lack of consistent screen direction clearly says that the shots are not continuous. |

Scenes in Automobiles

Shots made inside automobiles follow special screen direction conventions because movie makers have no choice. If you shoot the driver from the passenger side, the performer will face screen right. The passenger, shot from the driver's point of view, will face screen left.

If you establish that a car is headed from New York to California, convention dictates that it travel from screen right ("east") to screen left ("west"). By the conventions of normal screen direction, that would make the driver and the passenger travel in opposite screen directions, and the driver would be moving in the opposite screen direction from the car. However, since viewers have ridden in cars, they recognize the passenger's viewpoint of the driver and are able to ignore the apparent violations of screen direction.

of the action line, the screen direction will remain consistent (**Figure 19-26**). But, if the camera crosses the action line, screen direction will be reversed. From camera setups D and E, the subject reverses screen direction, **Figure 19-27**.

Of course, there are situations in which you *want* to reverse screen direction. As noted earlier, the man watching from the front stoop of the cottage would see the woman with the umbrella walking from the other side of the action line. For this reason, a shot from his point of view would correctly reverse screen direction, **Figure 19-28**.

Changing Screen Direction

It is neither possible nor desirable to continue a screen direction indefinitely. When a sequence ends, the next sequence will naturally start with

Figure 19-24 Convention dictates some screen directions, such as a car traveling westward should move from right to left.

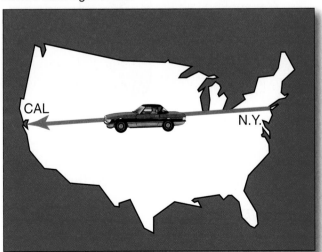

Figure 19-25 Ground plan showing camera setups and the action line.

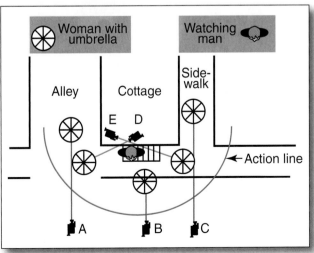

Figure 19-26 Consistent screen direction.

Shot A

Shot B

Shot C

Figure 19-27 Reversed screen directions.

Shot D

Shot E

Figure 19-28 From the point of view of the man on the front stoop.

Setup A sets the screen direction.

Setup B establishes the man looking.

Setup E represents his point of view.

Setup C resumes the original screen direction.

its own new screen direction. But, it may also be necessary to change direction within a single sequence. When you need to do this, you have four options:

- Show the subject changing direction.
- Use neutral screen direction.
- Insert a cutaway.
- Change direction off screen.

On-Screen Change

One way is to show your subject changing direction in the middle of a shot, **Figure 19-29**. This establishes a new screen direction.

Neutral Direction

Another way to soften a change in screen direction is by using a neutral screen direction

Figure 19-29 The audience watches as the subject reverses screen direction.

(either toward or away from the camera), as you can see in **Figure 19-30**.

Cutaway Buffer

You can conceal a direction change by inserting a shot of something else between shots with different screen directions, **Figure 19-31**.

Empty Frame

Finally, you can have the subject change direction while off the screen entirely. In the first shot, the subject leaves the frame. The camera holds on the empty screen for a moment, then we cut to a second shot in which the subject is moving in a different screen direction (**Figure 19-32**). This technique works equally well in reverse: cutting to an empty second shot, which the subject then enters.

For the smoothest possible transition, use both techniques at once: have the subject exit shot one, hold the empty shot, cut to empty shot two, have the subject enter the shot.

In general, strict observance of screen direction is considered less important today; some sequences (or even whole programs) pay little attention to it.

Staging for the Screen

Maintaining screen direction is just one part of a larger task: staging action for the screen. It has been said that half a director's job is to put

Figure 19-30 A neutral screen direction softens the change from one direction to another.

Figure 19-31 Cutaway buffer.

The subject walks screen-left to screen-right.

Cutaway shot.

Subject has reversed screen direction.

Figure 19-32 Empty frame.

The subject leaves the frame.

The empty frame holds a moment.

Then, the subject is seen moving in the opposite screen direction.

the camera in exactly the right place, at the right time, with the right lens pointed in the right direction—in short, to select the ideal camera setup for each shot. What is the most effective setup for each shot, and how does the director know it? Partly by instinct—an artist's gift—and partly by practice.

Some would say that there is no such thing as an "ideal" or "right" camera angle. They argue that the "right" angle is whichever one seems to work the best.

The craft of staging action for the camera is grounded in four basic principles concerning the special world on the screen:

- The actual world is nothing more than raw material for the creation of the screen world.
- The only unchangeable part of the screen world is the frame around it.
- The dimension of depth in the screen world is controlled by the director.
- The window on the screen world is small and the view is somewhat indistinct.

In short, the director does not record the real world, but creates an artificial world on a relatively small, two-dimensional screen that is surrounded by a frame. No matter what you are shooting, your selection of camera angles should be guided by these four principles.

Creating a Screen Geography

The actual world is nothing more than raw material for the creation of the screen world.

Whether you are shooting in a tiny room, on a large soundstage, or on location in an endless landscape, your actual surroundings are nothing more than raw material that you use to create screen images. In the real world, you can see where everything is and how it relates to everything else. But, the audience watching the world on the screen cannot do that without your help. So, your first job is to create a screen world geography that shows viewers where the action is taking place.

Depending on your needs, you may wish to show the whole environment or you may want to reveal only enough of the locale to let your audience know where they are. Since **Figure 19-33A** was shot at an actual farmer's market, it can include a considerable amount of real-world space. But, the stall in **Figure 19-33B** is actually a small set erected on a quiet street. So, we see only enough to tell us what we are looking at.

If needed, you can combine two or more separate real-world places into a single screen environment. For example, the graveyard in **Figure 19-34** is actually many miles away from the church.

As these examples suggest, you create a screen world environment by controlling what you show (and how much of it), and how you combine multiple real-world environments into single screen locales.

Working within the Frame

The only unchangeable part of the screen world is the frame around it.

Figure 19-33 Creating screen geography.

A—On location at an actual farmer's market.

B—Closer view of a vendor's stall.

Figure 19-34 By establishing the church in the first shot, the editor makes the graveyard shot seem to be in the same location.

(Sue Stinson)

The frame is by far the most powerful tool that a video director has to work with. It can hide anything that you do not want viewers to see. For example, **Figure 19-35B** seems to be the gloved hand of the burglar shown in **Figure 19-35A**. But, as **Figure 19-35C** reveals, the sequence's inserts were actually shot with a different performer. The frame around Shot B, with the gloved hand, hides the deception by framing off the second actor.

By hiding the edges of sets or unwanted details of locations, the frame can also suggest that the world displayed on the screen continues indefinitely beyond the borders of the image,

Figure 19-36. This, of course, is what makes the farmer's market stall set seem convincing.

Managing Depth

The dimension of depth in the screen world is controlled by the director.

We have repeatedly emphasized that depth in the world of the screen is only an illusion. Controlling this illusory third dimension is an important part of staging for the screen. Sometimes, you may wish to suppress the feeling of depth. But, more often, you are trying to enhance it.

Figure 19-35 The power of the frame to conceal. A—This shot is from the main burglar sequence shown in Chapter 18. B—Inserts of the burglar's hand...C—were shot later with a different performer.

A   B   C

Figure 19-36 Once the actual farmer's market has been established, the close shots of the set seem to be part of it.

Farmer's market location

Close shot

Enhancing Depth

To strengthen the feeling of three-dimensional space, a director can use several strategies:
- Stage the action to move toward and away from the camera, rather than from side to side. That way, your subjects appear to move through the depth of the scene.
- Set up shots so that the camera looks diagonally at walls, roads, and other surfaces with parallel edges.
- Look for multiple planes parallel to the camera.
- Stage movement against stationary backgrounds, for contrast.
- Use wide angle lenses. In conjunction with action staged to flow toward or away from the camera, wide angle lenses exaggerate apparent depth and enhance the sense of dynamic movement, **Figure 19-37**.

Suppressing Depth

There are times when you want to reduce apparent depth instead of enlarging it; when you want to turn the action into a design painted on the flat surface of the screen (**Figure 19-38**). To do this:
- Reverse all the previous suggestions for enhancing depth.
- Keep action parallel to the camera.
- Avoid diagonals and converging lines.
- Omit foreground objects that would reveal the scale of the action.
- Use telephoto lens settings—the longer the better. They suppress the sense of apparent depth, and emphasize that what the viewer is really looking at is a video painting with the screen as its canvas.

Figure 19-37 To increase the sense of depth, the second image was taken closer to the sign, using a wider angle lens.

Scaling for Video Displays

The window on the screen world is relatively small and the view is somewhat indistinct.

Today, full high-definition video is essentially as sharp as film. And, larger screen sizes offer a viewing experience comparable to theater projection. On the other hand, computer screens and mobile devices are smaller and some have lower resolution. Since many programs are now created for Internet distribution and/or desktop display, you need to know how to compensate for these constraints, as the *Size Matters* sidebar explains.

There are strategies for combating the effects of low resolution and small screen size.

Figure 19-38 A telephoto lens squeezes the image into a design on the picture plane.

To compensate for low resolution, work close to your subjects in order to make details more visible. To offset the loss of dynamic impact on a small screen, use lenses and stage action to emphasize depth, **Figure 19-39**.

Remember that video's smaller scale is not always a drawback. Well-handled by a director who understands it, video has an intimate quality that welcomes viewers and makes them feel comfortable. This intimacy and comfort is part of the reason for the enormous psychological power of TV.

Moving the Camera

In considering staging for the screen, we have assumed that the camera remains essentially in one place for each shot. In actual video directing, that is often not the case, because the camera is almost always in motion to one extent or another. Sometimes the movement is as obvious as a dramatic crane shot that swoops down from high above the action. At other times, the movement may be confined to subtle panning and tilting. But, even when the camera appears to be still, the operator is usually making small adjustments to keep the shot well-composed—this technique is often called *correcting*.

To make these continuous small corrections, a good camera operator never locks the movement controls on a tripod, even if the shot is supposed to be a fixed one.

Size Matters

In recording theater, opera, and dance, TV directors have traditionally mainly used close shots that look like this:

Two shot.

Red's closeup.

Orange's closeup.

But, rarely showing the whole stage, like this:

Design for Molière's *The Bourgeois Gentleman*, by Terry Gates.

This is because in the full-width shot, the performer's expressions would be unreadable, even on large (27″) standard definition TV sets. Framing off much of the stage was especially unfortunate with dance and other spectacles that depend on the overall design for their effect.

Today, however, most consumer TV sets are 36″–72″ or more in size and display 1080p high-definition images. With this extra scale and resolution, directors can (and should) loosen up compositions to convey more of the original theatrical effect.

But, if you intend for your program to display on small hand-held devices, you will need to go in the opposite direction—working even closer than the traditional TV image sizes. For example, if the full-stage design above represents, say, a 60″ screen, then the same image on a 10″ screen would look like this:

If the previous image = 60″, this one = 10″.

And, if your project will be published to a variety of display formats, it may be best to play it safe and shoot it like a traditional, standard-definition program.

Camera movement is one of the director's most powerful and versatile tools. Used well, it can greatly enhance your programs. Used badly, it can consume valuable production time and resources while contributing little or nothing.

Moving shots, especially shots using a dolly, crane, or stabilizer, can be expensive and time-consuming to set up, rehearse, and execute.

Types of Camera Movement

There are several different types of camera movement, and each type creates a distinctive visual effect.

Panning and Tilting

With rotation, the camera support (tripod or dolly) remains in one place while the camera is

Figure 19-39   As soon as you have set the scene, work close to your subjects. Use wide angle lenses to enhance impact.

 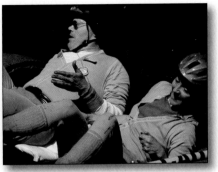 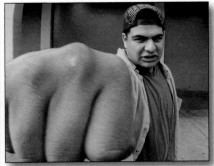

pivoted on it, **Figure 19-40**. Horizontal rotation is called *panning* and vertical rotation is *tilting*.

Dollying, Booming, and Pedestaling

Dollying shifts both the camera and its support from one spot to another during the shot. Dollying in and out moves the camera into the action, while dollying laterally moves the camera beside the action. Moving the camera up or down in an arc is called *booming*, while raising or lowering the camera on a central column is termed *pedestaling*. See **Figure 19-41**.

Except for "pan" and "tilt," there are no standard terms for camera movement. You will see "truck," "track," and "dolly" used for horizontal movement, and "boom," "crane," and "pedestal" applied to vertical shifts.

Zooming

The term *zooming* refers to changing the lens focal length during the shot to show a wider angle with smaller subjects, or a narrower angle with larger ones. Because a zoom out reveals

more of the scene, it somewhat resembles a dolly out, and the opposite is true of a zoom in. But, a zoom is not a true movement. Because zooms do not involve changing the position of the camera, they do not shift the point of view of the lens. Instead, zooms merely reduce or enlarge the same image (**Figure 19-42**).

The distinctions between rotations, moves, and zooms are not merely theoretical, because each type of movement produces a different visual effect. In professional production, dollying is often preferred to zooming, though zoom lenses are, of course, indispensable for real-time coverage of news and sports.

Composite Movement

Though types of movement can be separated for discussion purposes, real-world videography typically mixes several types at once. For example, a crane designed for dollying back and forth and booming up and down is also fitted with

Figure 19-40   Panning and tilting.

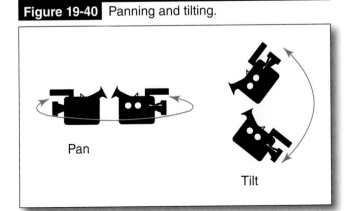

Figure 19-41   Pedestal, boom, and dolly.

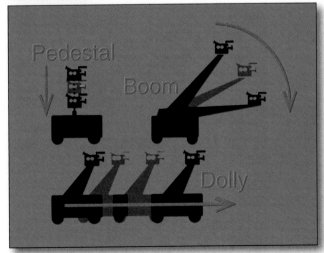

Improvised Dollies

Almost any moving (or movable) object can be pressed into service as a camera dolly. Here are some suggestions.

Chairs

A thrift-shop wheelchair makes an excellent dolly because it is light, collapsible, and fitted with very large wheels. With larger cameras, the system is limited to low angles because the operator shoots sitting down. With an external LCD screen, however, you can hold the camera above your head for a more normal height.

Office chairs also make smooth-rolling dollies, especially on the hard, even floors of commercial buildings. The sound recorded during moving shots may not be usable, however, because the small wheels are noisy.

For this shot, the videographer was pulled along in a child's wagon!

People Movers

When possible, take advantage of systems that do your moving for you. Airport beltways allow the camera to keep pace with subjects moving along side. Escalators are naturals for improvised crane shots. You may wish to follow a subject on the escalator, or else shoot down on a subject walking on a lower floor.

Some hotels, casinos, and even parking structures have glass-sided elevators that allow dramatic booming up or down. It is difficult to conceal the camera's means of transport, however, because the passing floors interrupt the view.

An available-light moving shot in a mall.

Vehicles

Vehicles make good camera platforms. Big-budget productions use camera cars and trucks, as well as special suction-cup camera mounts. But, you can obtain excellent shots out car windows by framing shots in an external viewfinder and using your arms as shock absorbers. To reduce vibrations that shake the image, make sure that the camcorder does not touch any part of the car.

The camera car dollies with the motorcyclist.

(Sue Stinson)

Hand-Held Crane Shots

Camera stabilizers permit hand-held moving shots that closely imitate the movements of large and expensive studio cranes. For example, a moving shot of a truck entering a campground can actually be made using the steps of a forest cabin.

Remember: in the screen world, if it is not in the frame, it does not exist!

This "crane shot" was made with a hand-held camcorder.

Figure 19-42 Zooming in progressively enlarges a portion of the original image to fill the frame.

(Sue Stinson)

a pan head so that the camcorder can pan and tilt, as well as change position. Mount a zoom lens on the camera supported by this system, and you can combine all types of movement at once. Stabilizer rigs and remote-controlled jib arm booms also combine zooming with all forms of camera movement.

Today, directors tend to think less in terms of panning, tilting, and dollying. Instead, they simply want the camera to flow freely with the action. For clarity, we will continue to organize camera moves into separate types (rotate, dolly, and zoom), but keep in mind that real-world videography is moving toward a fully integrated form of composite motion.

Reasons for Moving the Camera

Why move the camera at all? Why not compose the action in a succession of fixed shots and then edit them together? That approach is simpler, faster, and cheaper. There are several good reasons for camera movement. Though some people use it just because "it feels right for this shot," thoughtful directors understand what a moving camera can do and why they want to use it.

Traveling shots take extra time to set up and rehearse. They also tend to require more takes because they are more difficult to execute properly.

A Typical Composite Movement

Here is a single shot that includes several different camera movements. As a truck pulls off a forest road and into a campground, the camera:

1. Dollies to the right to hold the truck in frame.
2. Dollies forward while booming downward.
3. Makes a right angle and continues forward and downward to the side of the truck.
4. Booms down to a low angle shot.
5. Dollies back to hold the subject as she gets out of the truck.
6. Pans with her as she walks away.

Notice how the foreground tree acts as a fixed reference that helps enhance the feeling of forward and downward movement.

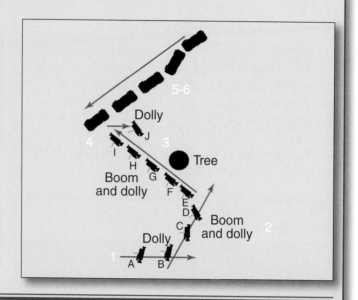

The truck and camera movements. As the truck appears, the camera dollies right (A). At the end of the dolly, the tree appears (B). The camera dollies and booms down (C). The dolly/boom continues (D). The foreground tree enhances the sense of movement (E). The truck stops as the dolly/boom down continues (F), and ends framing the driver (G). The camera dollies right to hold the driver (H), and booms down to a worm's-eye angle (I). The camera dollies back while panning to hold the subject (J).

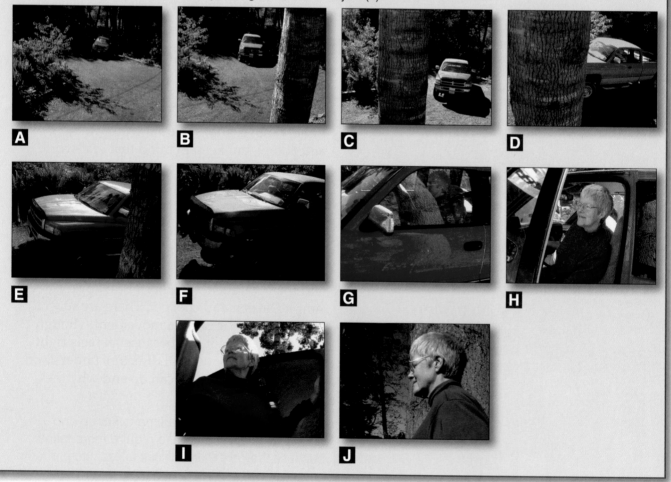

Moving to Follow Action

The most obvious reason for moving the camera is to follow action that is also in motion. To keep a walking subject in a full shot, for example, the camera has to dolly along beside her (**Figure 19-43**).

Moving to Reveal Information

Because the audience cannot see anything that is outside the frame, moving the frame by moving the camera delivers new information. For example, **Figure 19-44** shows two successive shots revealing that what looks like a statue is actually alive.

As an alternative, imagine that the camera frames the lower robes of the "statue," and then slowly tilts up to reveal the living eyes in its white face, **Figure 19-45**.

Although both approaches deliver the same information, the slow revelation by the moving camera is more subtle and perhaps more effective. In this example, the camera tilts up to reveal the human eyes in the "statue's" face. When the movement is a dolly or crane shot, the new information may be about spatial relationships, which the camera reveals by rolling through the acting area.

Moving to Emphasize Depth

By dollying into or out of the action, the camera strengthens the feeling of a third dimension in the scene. As explained in Chapter 6, *Video Composition*, moving sideways instead of in or out can also enhance the sense of depth by changing the relationships of different elements within the image.

Figure 19-43 By dollying along with a moving subject, the camera maintains the same perspective.

Figure 19-44 Shot 1 establishes the "statue." Shot 2 shows her moving.

Shot 1   **Shot 2**

(Sue Stinson)

Figure 19-45 Alternative camera movement.

Beginning of the tilt.

End of the tilt.

Camcorder Stabilizers

Today, perhaps a majority of moving shots are made with a Steadicam®—the industry standard in camera stabilizing systems. (Glidecam® and other designs are also available.)

(The Tiffen Company)

Moving to Involve the Viewer

As the camera functions as the viewers' "eyes," it carries them with it into the center of the action, involving them kinetically in the story.

Moving to Increase Energy

By its very nature, movement is more dynamic than stillness. A good director knows when to keep the camera still and let the action move across and through it, and when to move with the action, so that the background changes behind it.

Moving to Enhance Continuity

Above all, a moving shot strengthens the feeling that events are unfolding right before the audience's eyes. No matter how smoothly separate shots are cut together, they are still unconsciously perceived as assembled pieces. A continuous moving shot, however, feels like "the real thing."

Techniques for Moving the Camera

Like other aspects of directing and camera work, good camera moves are carefully planned and executed. The audience may be unaware of a well-designed move and yet unconsciously pleased and satisfied by it.

In professional productions, responsibility for camera moves may be shared among the director, the videographer, the camera operator, and often the chief of the dolly crew.

In planning camera movements, you need to consider composition, motivation, and speed.

Composition

A moving shot consists of three parts: before, during, and after the actual motion. Of these, the start and finish are especially important. In composing them, remember that both the beginning and ending frames should be pleasing compositions. When practical, the ending composition should be the stronger of the two. By following these suggestions, you give viewers a satisfying feeling of a beginning, middle, and end—of progress to a completion.

Motivation

Earlier, we identified a number of reasons (motivations) for moving the camera. Here, we use the word *motivation* in a special sense: the visible justification for making the move. The most common justification is subject movement. When performers walk across the screen and the camera starts moving to keep them in the frame, the audience takes that motion for granted. When the camera dollies forward to get into the center of the action, the viewer instinctively understands and accepts the move. In the red truck example in the sidebar *A Typical Composite Movement*, the camera move brings the viewer forward to greet the truck and its driver. The audience will often accept a move that lacks clear motivation, but the best moves justify themselves to the viewer.

Speed

A well-executed camera move begins slowly, accelerates smoothly to its top speed, and then decelerates as it comes to a stop again. Keep in mind that the best camera moves are unnoticed by the audience. The audience may be aware that their viewing position has changed, but they do not notice all the video craft required to change it.

Exceptions occur in programs like commercials and music videos, where dynamic camera moves may be used purely for dramatic effect.

A Reminder

This chapter concentrated on the "cinematic" part of the video director's job—the design and execution of images and shots that succeed individually, while providing suitable material for editing. As you think about these visual aspects of directing, remember the other half of the job (covered in Chapter 18): directing to communicate with the audience and drawing effective performances from the cast. Directing at the professional level requires all these skills, all operating together.

Summary

- The process of shooting program material to anticipate the editor's requirements is called directing to edit.
- Good coverage typically requires repetition, overlap, variety, protection, and cutaways.
- A program with good continuity appears to be a single continuous presentation that has no mismatches in information, action, or screen direction.
- It is important to maintain screen direction to establish the subject's basic point of view, to enhance the illusion that an action is unfolding continuously in real time, and to avoid confusing the audience.
- Screen direction exists only in relation to the frame around the image.
- In staging action for the screen, the director must create a screen geography, work within the frame, manage depth, and scale the image for video various displays.
- The camera is almost always in motion to one extent or another. In planning camera movements, consider composition, motivation, and speed.

Technical Terms

Action line: An imaginary line separating camera and subject. Keeping the camera on its side of the line maintains screen direction.
Booming: Moving the entire camera up or down through a vertical arc. Also called *craning*.
Color shot: A view of the scene, or a detail of it, not directly part of the action.
Correcting: Making small continuous framing adjustments to maintain a good composition.
Dollying: Moving the entire camera horizontally. Also called *trucking* and *tracking*.
Insert: A close shot of a detail of the action; often shot after the wider angles, for later insertion by the editor.
Pedestaling: Moving the camera up or down on its central support.
Protection shot: A shot taken to help fix potential problems with other shots.
Screen direction: The subjects' orientation (usually left or right) with respect to the borders of the screen.

Review Questions

Answer the following questions on a separate piece of paper. Do not write in this book.

1. What is involved in directing for form?
2. *True or False?* While directing a production, you are *not* making a video program.
3. The process of shooting material to anticipate the editor's requirements is called _____.
4. What is the role of repetition in covering action? How does it affect editing?
5. What is a *protection shot*?
6. *True or False?* Inserts are closeups of details.
7. What is *continuity* in a program?

8. Why is continuity of direction called *screen direction*?

9. What are the three types of screen direction?

10. By convention, what direction does a plane traveling from Los Angeles, CA to Albany, NY move across the screen?

11. What is an *action line*? What is the purpose of an action line?

12. Identify the four methods used to change screen direction within a single sequence.

13. The only unchangeable part of the screen world is the _____.

14. List some of the strategies used to enhance depth.

15. *True or False?* Horizontal rotation of the camera is called tilting.

16. Moving the camera up or down in an arc is called _____.

17. What are some of the reasons for moving the camera while shooting?

STEM and Academic Activities   STEM

1. **Technology.** Investigate how telephoto lenses work and explain how they change the appearance of an image. What do the millimeter marks on the lens mean? What is the typical millimeter range for telephoto lenses?

2. **Engineering.** Develop plans for a camera dolly system that is constructed of everyday items. Explain how your dolly system works, make a detailed sketch of the dolly system and its components, and give an example of its use.

The lioness and lion seem to be looking at each other. Although they were together in reality, they were facing opposite directions. How did the videographer relate them to each other (without reversing one of the shots)?

20 Editing Operations

(Corel)

Objectives

After studying this chapter, you will be able to:

- Recognize the difference between subtractive editing and additive editing.
- Explain the tasks involved in each of the five major phases of editing.
- Understand the importance of systematically organizing raw materials.
- Identify how elements of a program can be enhanced.
- Recall the steps in archiving a video project.

About Editing Operations

Editing is the task of taking the materials recorded during production and transforming them into a finished audiovisual program. Like videography, simple editing can be performed by a beginner with almost no instruction. At the professional level, editing is a rich and satisfying art that requires mastery of the many tasks, skills, and techniques of video postproduction. These tasks and techniques are the subject of this chapter.

Because this is the first of four chapters on editing, it will be useful to start with an overview of postproduction as a whole to see where editing operations fit into the larger process.

Video Postproduction

Postproduction is the general term for the fascinating process of turning raw video footage into a finished program, ready to play on a movie screen, TV set, or through a website.

The terms *editing* and *postproduction* are more or less interchangeable. We prefer the shorter "editing."

Editing deserves extensive coverage for three reasons. First, it is a complex process made up of many different operations that require explaining. Next, editing is performed almost entirely with computer programs—powerful software that can be daunting to approach and time-consuming to master, **Figure 20-1**. Most importantly, editing is the production phase in which everything

finally comes together to create the very different "video world" that has preoccupied us throughout this book.

Creating the Video World

Chapter 9, *Project Development* shows how the writer designs a blueprint for the video world to be constructed for a program. Chapter 19, *Directing for Form* explains how the director and production staff build the pieces that will be assembled to make this world. Now, Chapters 20–23 on postproduction reveal how the editor assembles the video world from the raw materials designed by the writer and constructed by the director.

This three-stage process is *additive editing*, as explained in the text sections that follow and the *Subtractive and Additive Editing* sidebar.

An edited program consists of many layers all working together, and all transparent to viewers (who notice only the program content). These layered components, **Figure 20-2**, typically include:

- Production video and audio.
- Audio effects and background tracks.
- Music.
- Transitions.
- DGEs (digital graphic effects).
- Titles and other graphics.

Historically, different editing tasks have been performed by different specialists,

Figure 20-1 A video editing work screen.

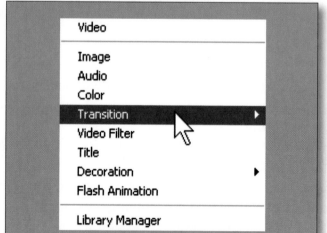

Figure 20-2 A program contains many elements.

particularly music and sound effects editors, audio mixers, special effects companies, title houses, and laboratories. Today, however, every one of these specialties can be performed right on the desktop—and a single editor is expected to master all of them.

The Challenge of Digital Post

If digital editing is challenging because it demands mastery of so many specialized operations, it is made doubly difficult by the peculiarities of editing software.

"Post" is widely used shorthand for the longer "postproduction."

To begin with, software program components simply *must* be organized—distributed logically through a structure of windows, icons, drop-down menus, keyboard commands, and other controls. Without organization, no one could use these scores of tools, or even find them. The problem is that the organization scheme is always somebody else's and never your own. You are forced to follow thinking and adopt work methods that may not be anything like yours.

Secondly, features must be labeled—often with unfamiliar names and/or tiny pictures (icons) that fail to illustrate their functions. Learning new software is often like learning a foreign language.

In addition, many of these functions are buried in menus or sub- and sub-sub menus, where their relationship to the main menu is not intuitive, **Figure 20-3**.

Moreover, editing software is complicated by its own versatility. To fill a variety of needs, the programs offer more features than most individual editors will use. And, to accommodate different work styles, the software may include several different ways to activate each function. In many cases, you can perform an action via a pull-down menu, a screen icon, a keyboard shortcut, or even more than one of each!

Finally, no two editing programs look or work exactly alike, and some are highly individual. Others are not really unique, but their quirky appearance (sometimes called a "skin") conceals their similarities to other programs, **Figure 20-4**.

As a result, learning editing software can take hours of exploration and frustration. Sometimes, high-end professional packages cannot be mastered completely, even after months of full-time effort.

The Convergence of Film and Video

As noted in Chapter 1, the technologies of video and film have been growing ever closer together. During the actual production phase, the two media use different lighting and recording procedures. But in postproduction, most films are converted to very high quality video (often directly from the developed negative) for editing with video software. This video is called a *digital intermediate (DI)*.

From that point on, projects follow essentially the same procedures, regardless of whether they were originally video or film. Operations, like transitions, titles, effects, and color management, are all completed digitally. At the conclusion of postproduction, the high-quality video is used to create different release media—theatrical film and digital prints for movie release, and differently transcoded digital media for disc, TV, and Internet display.

Today, people move comfortably between film and video recording, choosing one or the other according to the needs of each production. Professionals trained originally in video can feel equally comfortable in film production, because so much of the process is now the same in both media.

Today, American Cinematographer, the distinguished journal of the American Society of Cinematographers (ASC), even-handedly covers both film and video.

Figure 20-3 A drop-down menu.

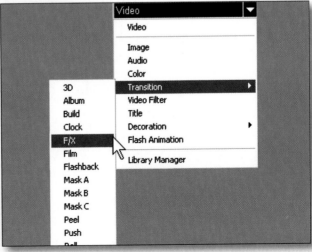

Figure 20-4 Cinelerra is a public domain editing program for Linux systems.

(Cinelerra.org)

These difficulties are common to almost all complex and powerful software applications, whether designed for video, graphics, word processing, or management applications.

The Underlying Concepts

With all these obstacles, how do editors learn a software application? And, how do they transfer their hard-earned skills to a different editing program when needed? They do it by grasping the concepts that lie under the program skins, behind the menus, below the icons.

As we will see in later chapters, all editing software packages share the same basic workflow, from starting a new project to publishing the finished program. All of them use one of two graphic metaphors for assembling footage: an on-screen "slide tray" storyboard, or a timeline. (Programs above the simplest beginner level favor timelines, and some offer slide trays as well.) The tabs and buttons and icons address the same features, and the various working windows display the same kinds of information, no matter how different everything may first appear, **Figure 20-5**.

Once you have learned a program, you should have little trouble in using another one of similar complexity. That will free you to move up to a more powerful and versatile software package.

The Range of Editing Software

It is probably easier to start with entry-level software, and then progress to more sophisticated applications. To accommodate this growth path, editing programs are available for several different kinds of users:

- Casual amateurs are often satisfied with the simple applications bundled with their computer operating systems, such as *Windows*® Movie Maker.
- Hobbyists may want easy-to-use software with features that create more professional looking programs.
- "Prosumers" (advanced amateurs and beginning professionals) can master the more sophisticated features of mid-level software, or move up to top-level programs.

Figure 20-5 A storyboard sequencing approach uses a virtual slide tray metaphor.

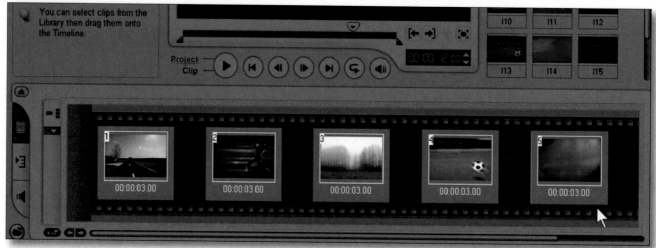

- Professionals usually use the most powerful (and difficult) programs, typically working with software optimized for the computer systems preferred by their production companies.
- Some editors prefer stand-alone systems consisting of software designed for and built into special-purpose hardware that is used only for editing, **Figure 20-6**.

The four chapters on postproduction are designed to move from the general to the specific:
- Chapter 20, *Editing Operations* (this chapter) lays out the basic processes of video editing (and film, too, for that matter).
- Chapter 21, *Editing Principles* shows how to build the special world of video programs from the materials created during preproduction and production.
- Chapter 22, *Digital Editing* explains the concepts behind digital editing activities that are common to all software programs.
- Chapter 23, *Mastering Digital Software* uses Corel® VideoStudio® Pro to demonstrate an approach to learning any editing program.

The Craft of Editing

At its simplest, video editing means placing one shot after another to create an organization that makes sense while it presents information or a story. At the next level of complexity, editing achieves that organization and presentation invisibly. The audience perceives the program as a simple continuous flow, without noticing that it is carefully built up, one piece at a time, of many separate units of picture and sound. At its most sophisticated, editing does more than organize information and present it invisibly. It does those jobs with *style*, with an emotional character that touches viewers' feelings, as well as their minds.

A two-hour program is a mosaic of hundreds or even thousands of individual pieces of video and audio.

Every one of a video's many pieces appears in the program as the result of an editorial decision. The editor has determined to use that particular element at that particular point and for that length of time. Multiply this three-part decision by the hundreds or thousands of elements in a professional program, and you can imagine how much artistic control is in the hands of the editor.

To illustrate the number of choices required for even the simplest edit, look at **Figure 20-7**, which shows the raw materials for a typical sequence. Taking one approach, the editor might open with an establishing shot, cut to a two-shot of the first employee, and then show the other employee (**Figure 20-8**). Or, perhaps it would be better to feature the first employee before revealing the whole scene, and then show the executive's closeup (**Figure 20-9**).

Which choice is right? There is no "right" choice. Using the six different camera setups, there are 150 *different ways* to order the first three shots in the sequence. How do you decide, then? Ultimately, you do it by an instinct for the right thing at the right time. At its highest level, editing is an art that cannot be fully described.

Like all arts, however, it is created through craft. The craft of editing consists of the *operations* of postproduction and the *principles* that guide them. The editing operations discussed in this chapter are the things that you do to create a program. Editing principles (covered in Chapter 21, *Editing Principles*) are your *reasons* for doing those things. If you master these editing operations and principles, you will command the craft. And, once you have mastered the craft, you are ready to practice the art.

Figure 20-6 A stand-alone editing system. (The monitor(s) is not shown.)

(Edirol)

Figure 20-7 Shots for a typical sequence.

A—Establishing shot of executive and two employees.

B—Two-shot favoring the executive.

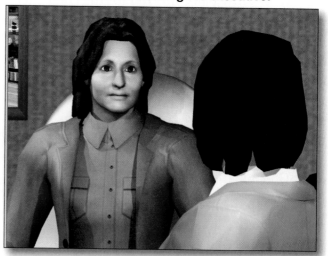

C—Two-shot favoring the first staff member.

D—Closeup of the executive.

E—Closeup of first staff member.

F—Closeup of second staff member.

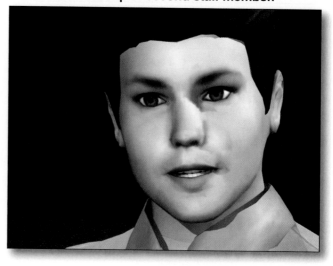

Figure 20-8 Shot sequence option.

A—Establishing shot of executive and two employees.

B—Two-shot favoring the first staff member.

C—Closeup of second staff member.

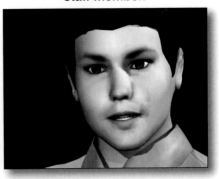

Figure 20-9 Another shot sequence option.

A—Closeup of first staff member.

B—Establishing shot of executive and two employees.

C—Closeup of the executive.

Since editors are dependent on their directors for competent footage, we will occasionally look back at the director's job of recording appropriate raw material for editing.

In editing, as in directing, many people evolve through three stages:

- Beginners often record material without planning the shooting and without modifying the raw footage. The result is called "home movies."
- Experienced amateurs record material from selected viewpoints and cut out the mistakes and repetitions in the footage.
- Professionals create and record material in short, well-planned pieces, and assemble programs by selecting and sequencing those pieces.

To put it another way, amateur video editing tends to be *subtractive*, while professional editing is *additive*. In order to understand the nature of true editing, you need to recognize the fundamental difference between the subtractive and additive approaches.

Subtractive Editing

Subtractive editing means removing elements from a shooting session by copying the original footage, while omitting the unwanted parts. Editing subtractively resembles the way you might organize a folder of still photos for mounting in an album: first you discard the obvious mistakes; then you remove duplicate or near-duplicate shots. Removing the mistakes improves the look of the album page, and cutting duplicate shots makes it more interesting. For the same reasons, videos created by subtractive editing are certainly more interesting than raw footage.

Video snapshots recorded with mobile phones and other appliances can be interesting as raw footage—pure documentations of actual events. However, they are generally more successful when kept short.

Nevertheless, a video *program* (as distinct from a video *snapshot*) is not an album of pictures that happen to move, but is a presentation of actions and/or ideas with a beginning, middle, and end—in short, with a *continuity*. Since video shots are not always recorded in their intended sequence, showing them in their original order can deprive the shots of some of their meaning and much of their effect.

Additive Editing

Even when a video records real events, like a vacation or family party, its component shots can often be taken out of shooting order and rearranged to make better sense or tell a more coherent story. Used in this way, shots become self-contained building blocks that can be selected and placed in any order to construct a program from the ground up. This is *additive editing*.

With additive editing, a video program starts as an absolute blank, like a canvas before an artist puts paint on it or a foundation slab before a house is constructed, **Figure 20-10**. On that empty canvas (a blank computer timeline), the editor places pictures, sounds, graphics,

Figure 20-10 Like this empty work screen, an additive editing project begins with a perfect blank.

(Ulead)

titles, and effects one piece at a time, building up a structure from nothing until it becomes a complete program.

In fabricating your program, you are limited only by the availability of material. You can ignore shots that you think are unnecessary, arrange shots in whatever order you choose, repeat shots or pieces of them, and include material that was not originally shot for the program. In short, you can do anything you like, as long as you remember that you are not improving an existing program that was created in the camcorder. Instead, you are creating a *new* program, completely from scratch.

Without the key concept of additive editing, you can only make photo albums that move. But, if you understand the basic idea that editing means creating something quite new, then you can master the five phases of postproduction.

Editing Phases

Here are the five major editing phases.

- **Organizing.** First, the video shots, sound components, and other raw materials are catalogued and filed so that each separate piece can be located for use.
- **Assembling.** The pieces of the program are chosen and arranged in the proper order.
- **Enhancing.** The selected picture and sound elements are modified, where necessary, to improve their quality or adjust their effect.
- **Synthesizing.** The separate layers of material (for instance, live action and graphic visuals, plus production sound, sound effects, and music) are blended together, and the shots and sequences are linked by transitions to create a single, seamless program.
- **Archiving.** Finally, the finished video is stored on tape or disc, **Figure 20-11**.

We will look at each operation in detail. Though we must examine them one at a time, keep in mind that a video editor typically works on several of these operations at once.

Remember, too, that "video" is the general term for electronic motion pictures. So, the phrase "video postproduction" covers sound, as well as image editing.

Subtractive and Additive Editing

Here are sequences from two videos edited from the footage shot when a famous team of Clydesdale draft horses visited the town of McKinley. Both programs were constructed from essentially the same raw footage.

Subtractive Editing

The first program was edited subtractively. Although two near-duplicate angles have been cut, the remaining footage is presented in the order shot. Notice that the effect is repetitive, and arbitrary changes of screen direction give the program a choppy look.

Editing subtractively means cutting bad or repetitive shots.

Additive Editing

The second video was created by additive editing. This opening sequence was built shot by shot to introduce the team and establish the locale. Notice that screen direction is changed by cutting in a head-on shot of two horses before reversing the orientation of the team in the final shots.

Editing additively means creating a sequence from scratch.

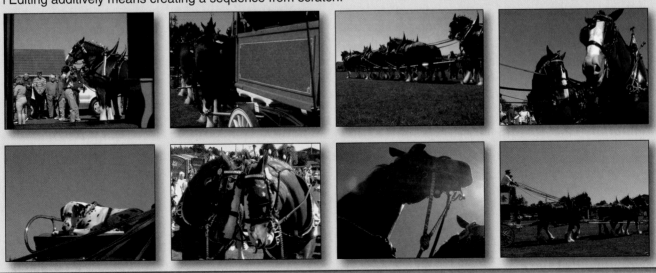

Figure 20-11 Digital storage is preferable for archiving programs.

Figure 20-12 Slate information is often written with a felt marker.

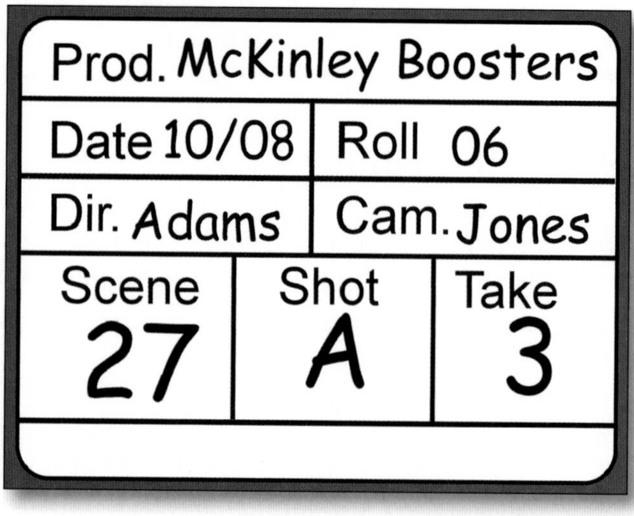

Prod. McKinley Boosters		
Date 10/08	Roll 06	
Dir. Adams	Cam. Jones	
Scene	Shot	Take
27	A	3

Organizing

All video projects produce a great deal of picture and sound material that is complicated to sort out. Shots and whole sequences are often shot out of order. Most sequences include multiple camera setups, and often multiple takes of shots that look quite similar. For these reasons, materials to be edited can be time-consuming to locate and difficult to identify. To reduce confusion and speed up the process, it is useful to organize your raw materials systematically before starting to use them.

Organizing audiovisual materials involves *identifying* and *describing* them.

Identifying

Today, state-of-the-art productions identify audiovisual materials electronically. (See the sidebar, *Digital Asset Management*.) However, even the most up-to-date operations also mark the start of each take of every shot by recording a slate, **Figure 20-12**. There are good reasons for continuing this traditional identification method:

- Unlike digital data, a slate is part of the visible shot, making it easy to identify on screen.
- Because of its distinctive appearance, a slate is easy to spot when viewing footage at high speed. (Tape-based recordings are generally up-loaded in long batches and then divided into individual shots by starting each new clip file with its slate.)

- The shot identification code used on slates is also referenced in the production's *continuity script*—the master document that shows, at a glance, what has been recorded, how it has been covered, and which takes are preferred.

When a shot cannot be slated before it starts, an "end slate" is recorded afterward. In this case, the slate is held upside down to signal that it goes with the preceding shot, instead of the following one.

Professional slates generally show the program title, director, and videographer—information that remains the same from shot to shot. If the recording medium is a tape or disc, it too is identified by a simple number. The individual shots are usually identified by a code in which the scene is numbered, the shot is given a letter, and the take is numbered. (A *shot* is an uninterrupted length of audio/video footage from camera start to camera stop. A *take* is one attempt to record a shot, an attempt that is often repeated until the director is satisfied.) In this coding system, "27 A 3" translates to "scene 27, shot A, take 3." As you can see, a slate labels each program element, no matter how small, so that it can be catalogued and tracked.

In professional news, sports, and documentary programs, as well as in most amateur videos, shot-by-shot slates are not generally recorded during production. Instead, the editor identifies each shot in postproduction. Even when applied by the editor, the production slating system of sequence/shot/take ("27 A 3") is highly effective because it both describes the piece of material and names it.

Describing

Typically, a shot description includes several items.

- **Angle:** The image size and camera position, such as "neutral angle medium shot" (or in brief form, "Neutral MS").
- **Content:** What the shot depicts, such as "Tammy takes sip of coffee."
- **Quality:** "NG" for no good, "OK" for acceptable, or perhaps "B" for best.
- **Notes:** It is often helpful to add notes about the shot, such as "she smiles in this take only."

Spending the time to describe every program component before starting to edit can save far more time later when locating shots.

Admittedly, the identifying and describing of your material is not a very exciting process, and you may be tempted to skimp on it or even omit it. But, after you have spent a tedious hour in a frustrating search for a shot that you know exists but cannot locate, you will appreciate the importance of systematically organizing your editing components.

When you have organized your raw footage, you are ready to work with it. You start by assembling the pieces of your program.

Assembling

Even when it is well organized, unedited video footage is still just a collection of separate shots, takes, and pieces of audio. At its simplest level, editing involves putting together these pieces of raw material to create a finished program. This assembly operation includes *selecting*, *sequencing*, and *timing* each individual shot.

For simplicity, we will treat each step individually—as if you selected all the shots in a sequence at once, then put all of them in order at once, and so forth. In actual practice, you are likely to select, sequence, and time each shot in order before proceeding to the next shot.

Some software products encourage you to select and sequence whole groups of shots at a time.

Selecting

Since all professional projects start with far more footage than will appear in the finished program, the editor begins by picking the shots to be included, and selecting the best take of each. The *Raw Footage* sidebar shows all the unedited shots in a demonstration sequence. This simple action of pouring and offering a drink has been covered by five separate shots:

- *Slate 15 A 1 and Slate 15 A 2.* A two-shot favoring Frank, the pourer (takes 1 and 2).
- *Slate 15 B 1.* A reverse two-shot favoring Jack.
- *Slate 15 C 1.* A medium shot of Jack that zooms to a closeup.
- *Slate 15 D 1.* An extremely high ("birdseye") insert of the drink being poured.
- *Slate 15 E 1.* A somewhat lower alternative insert of the drink being poured.

In reviewing this raw material, the editor notes several things:

- Shots A, B, and C all cover the entire action. So, any one of them alone could deliver all the information in the sequence.
- Frank's two-shot offers two slightly different takes to choose from.
- Jack is covered in two completely different angles.
- The pouring insert has been shot in two different versions.

Accordingly, the first task is to decide which shots to use. In this case, the editor chooses *Shots 15 A 1, 15 B 1,* and *15 E 1.* See **Figures 20-13** through **20-15**.

Sequencing

With the preferred shots and takes chosen, the next task is to put them in order. On the camera recording, the insert of the pitcher and cup follows the other shots (**Figure 20-16**). Now the editor makes the decision to move the insert to its proper place in the action, **Figure 20-17**.

Raw Footage

The discussion that follows refers to the shots in a single sequence that shows the pouring and offering of a drink. The three images in each shot represent the beginning, middle, and end of the shot content.

SLATE: 15 A 1—Two-shot favoring Frank (take 1).

SLATE: 15 A 2—Two-shot favoring Frank (take 2).

SLATE: 15 B 1—Two-shot favoring Jack.

SLATE: 15 C 2—Zoom in to closeup of Jack.

SLATE: 15 D 1—Birdseye insert.

SLATE: 15 E 1—High-angle insert.

Figure 20-13 Shot 15 A take 1 covers the action more completely.

Figure 20-14 Shot 15 B balances the two-shot of 15 A, and is technically better than 15 C.

Figure 20-15 Shot 15 E is a less extreme angle than 15 D, and it covers more of the action.

Figure 20-16 The inserts were originally recorded after the main shots.

Digital Asset Management

All digital editing is performed using computer files, which are known collectively as *digital assets*. (The term covers still, as well as moving images.) Since every file must have a unique *file name*, it is labeled automatically. Unlike a slate, however, a file name label does not encode information about a video recording, such as its program, sequence, shot, and take.

For example, the file name of one actual video shot is "P1120211.MOV," which identifies the recording codec (QuickTime Movie), but nothing else.

To add this information, different approaches have been developed:

- Metadata
- Editing software database
- Asset management database
- "Homemade" database

Metadata

Metadata is supplementary information stored in an image file, but not visible in the actual image. Shot metadata recorded by a video camera may include, for example, the date and time of recording and the video resolution.

Advanced recording systems allow users to add extra, customized metadata to help identify and describe each recording. This data may include slate identification ("27 A 1"), shot quality ("good," "NG," "best take," etc.), setup description ("MS, Marcie"), and any other information desired.

In a well-planned project, the categories of metadata are set up during preproduction.

In some cases, the custom metadata is added when the recording is transferred to a computer from the flash card or drive that is removed from the camera. In other setups, the metadata is added in real time, as the shot is recorded. For example, the high-capacity hard drive shown here attaches to a camera to record video, either as a backup to the camera's internal system, or as the primary recording medium. Using a laptop computer or handheld device, you can enter metadata during recording by communicating wirelessly with the drive.

A wireless metadata recording setup.

(Focus Enhancements)

Editing Software Database

Professional-level software editing programs typically include specialized database functions for identifying shots. In this approach, shots are provided with extra information as they are logged into the actual editing computer. Since some cameras do not assign a filename to each shot, starting *timecode addresses* are generally used in those cases. Of course, once these long recordings are broken into "clips" that usually (but not always) correspond to individual shots, individual shot names can be assigned.

Asset Management Database

Some editors prefer to work with specialized asset management databases, organizing shots in these separate programs and then dragging-and-dropping them into the editing program. ACDSee™ Pro, the program shown here, lets you select information ("LS," "good," etc.) from a custom list that you prepare in advance, and key in custom notations ("elephant family passing") and a slate ("seq 27 SHOT 12").

Note that the thumbnails indicate the content of their shots by showing the first and last frames, plus two in between.

In addition to programs like ACDSee™ Pro, open-source video content management programs are available on the Internet.

"Homemade" Database

Since many software suites already include a database program, you can use yours to create your own shot database. The sample shown here has been set up to log tape-based footage, but you can create the database fields best suited to your system.

The asset management approach you take will depend on your available equipment and software. However you do it, organizing your raw footage is absolutely essential to efficient editing.

A professional asset management program.

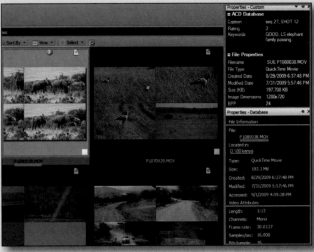

(ACD Systems)

A shot list in an office program database.

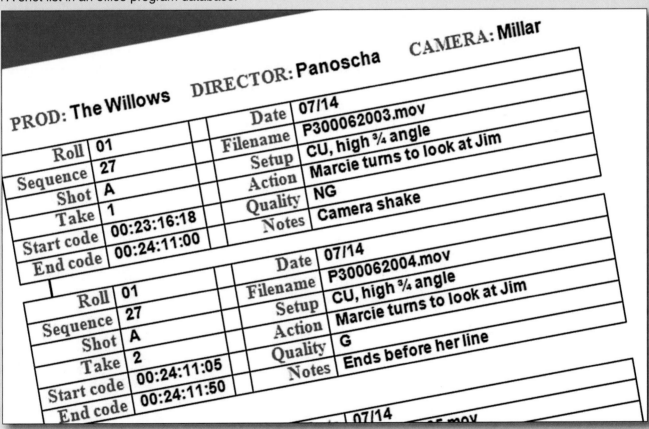

PROD: The Willows	DIRECTOR: Panoscha	CAMERA: Millar	
Roll	01	Date	07/14
Sequence	27	Filename	P300062003.mov
Shot	A	Setup	CU, high ¾ angle
Take	1	Action	Marcie turns to look at Jim
Start code	00:23:16:18	Quality	NG
End code	00:24:11:00	Notes	Camera shake
Roll	01	Date	07/14
Sequence	27	Filename	P300062004.mov
Shot	A	Setup	CU, high ¾ angle
Take	2	Action	Marcie turns to look at Jim
Start code	00:24:11:05	Quality	G
End code	00:24:11:50	Notes	Ends before her line
			07/14

Figure 20-17 The insert is moved to its proper place in the action.

Timing

In our example, the action is repeated three times. So, the next task is to time each shot by assigning it just one part of the action, and discard the overlapping parts. Each shot is trimmed to its usable section by setting its in-point (start point) and out-point (end point).

Enhancing

When you have assembled your images and sounds, you can enhance them to better suit the requirements of the program. Enhancing raw editing materials can mean *improving* them, *conforming* them, and/or *redesigning* them.

Improving

Even in high-level professional production, some footage may not measure up to the quality standards of the program.

Video

A shot may have luminance and chrominance problems that make it too dark or too light or degraded by an unwanted color cast, **Figure 20-18**. Its white balance may be off due to a mistake or to a camera movement from, say, a window-lit subject, to a scene lit by incandescent lighting. Contrast is often a problem, especially in outdoor locations. Framing may need improvement to recompose a shot or to eliminate a visible microphone, **Figure 20-19**. With digital manipulation, you can often fix all of these problems completely.

Speeding or Slowing Action

Digital processing allows you to change the speed of a shot—not only to create obvious fast or slow motion, but also to make subtle adjustments. You can usually alter speed as much as 10% without making the shift noticeable (though you need to try different modifications and judge the results on a case-by-case basis).

The simplest reason for changing speed is to adjust the length of a shot for timing purposes, usually to fit an available slot in the program.

Slowing an action slightly can also lend it added significance by giving it greater visual weight.

Any shot can be stretched to fill the required time.

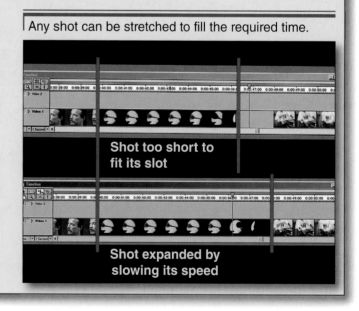

Shot too short to fit its slot

Shot expanded by slowing its speed

Figure 20-18 Correcting exposure.

Original footage too dark.

Exposure corrected digitally.

(Sue Stinson)

Audio

Audio problems may include incorrect volume levels, as well as poor sound equalization, presence, and mike-to-subject distance. With digital sound editing, you can make substantial enhancements.

Even the simple software included with Mac® or Windows® operating systems can make respectable improvements in sound quality.

Conforming

Even when various components are all equally acceptable for quality, they may not always match. Frequently, you need to adjust video and audio to eliminate differences between one piece and the next.

Video

Often, white balance is not consistent from shot to shot, especially when different setups are shot at different times of day, or even on separate days (**Figure 20-20**). When intercutting these shots, you tweak the color balance to conform their look.

Audio

Since every new camera angle demands a different microphone position, equalizing characteristics like distance, presence, and background noise, is a constant chore.

Figure 20-19 Eliminating a distraction. A microphone showing in upper-right part of frame is removed by digital cropping.

Figure 20-20 Late afternoon light can change color temperature between one setup and the next.

Redesigning

Digital processing allows you to change picture and sound character, as well as quality. Sometimes this involves computer-generated sounds and images, but more often it involves processing raw footage for editing.

Video

On the video side, you can adjust overall color to provide a sunset glow, create a nostalgic sepia monotone, or create a black and white program. You can apply slow- or fast-motion effects, or strobe. You can transform the image with computer graphics effects, and apply any of the dozens, or even hundreds, of image filters available for use in computer-based editing.

Audio

In the sound department, you can alter material to create any effect you choose.

Today, with digital editing, image and sound enhancement is so widespread that many editors process most of their shots and audio clips as a matter of routine.

Synthesizing

Synthesizing means fusing separate elements together to create a single, new product—in the editor's case, a video program. To understand synthesizing, it helps to think of programs the way many computer editing applications represent them: as multilayered timelines.

A *timeline* is like a very long, thin matrix, with the program's shots represented in order on a horizontal axis. The program begins on the left and the timeline scrolls leftward as you move along it to the end, which appears on the right. The timeline is several levels high, usually with separate lines for two or more video streams, transitions, titles, and filters. The audio component may appear as a dozen or more separate tracks of production sound, background, effects, and music.

Integrating material horizontally from moment to moment through the length of the program may be called *connecting*; while blending the many levels of picture and sound may be thought of as *layering*.

Connecting

Creating a seamless horizontal flow from start to finish requires connecting shots to form sequences, sequences to form sections, and sections to form whole programs.

In short videos, the entire program may be a single section. Also, in fiction programs, sections are often called "acts," like the acts in theatrical dramas.

Typically, you select shots and choose their in- and out-points to make invisible edits, so that each sequence seems to the audience like a single presentation rather than a collection of individual pieces.

You connect sequences with transitions, like fades, dissolves, wipes, and digital effects. The decisions you make in selecting and timing the transitions between sequences will determine the coherence of the program as a whole.

Layering

Where connecting joins elements end-to-end, layering melds them from top to bottom. Even a simple program may involve working with several visual layers at once, including:

- **Composites.** Pieces of different images may be combined into a single picture, **Figure 20-21**.
- **Superimpositions.** Two or more images may be layered so that all are visible at once, **Figure 20-22**.

In *compositing*, pieces of one image completely replace pieces of another. In *superimposing*, each image is displayed in its entirety, but at partial intensity, so that both pictures are visible at the same time.

- **Multiple images.** The screen may be divided into separate areas to display one or more small images in a larger one ("picture-in-picture"), as in **Figure 20-23**, or to create a mosaic of images, as in **Figure 20-24**.
- **Titles.** Titles are often layered over live action, **Figure 20-25**.

Audio layering is often even more elaborate. Usually, the visual track presents only one image at a time. But, a sophisticated audio track is

Figure 20-21 The flag was composited onto the fireworks shot.

(Sue Stinson)

Figure 20-22 A closeup of the horses was superimposed on the long shot.

Figure 20-23 The reporter in the small picture comments on the race.

Figure 20-24 All the animals in these four images are moving.

Figure 20-25 A title superimposed over live action.

almost always a mix of many elements, including dialogue, background effects, synched sound effects, and music. To permit more precise control, these components are usually divided into subcomponents. Often, the dialogue will include separate tracks for each speaker. Sound effects tracks may number two, four, or even more to permit multiple overlapping effects. Music may be distributed among multiple tracks to allow cross-fading between sections and stereo imaging.

Historically, video has been limited in its ability to synthesize program material out of multiple streams of picture and sound. But, the development of digital, nonlinear postproduction has offered editors almost limitless possibilities in integrating program components.

Archiving

The final task in editing a program is storing it in its finished form, so that it can be presented, duplicated, and preserved. In digital format, archiving is the essential last step in the post-production process. Archiving involves *rendering*, *storing*, and *presenting*.

Rendering

In *rendering* a finished program, a computer takes the low-quality editing version that you have assembled and uses it as a blueprint for automatically creating a high-quality version. To save hard drive space and editing time, some digital editing applications work with low-resolution images and crudely rendered effects and titles. The resulting "work print" contains all the instructions necessary to make a duplicate that is exactly the same, except the images, sounds, and effects are all full quality. Rendering is the process of creating this finished-quality copy, **Figure 20-26**.

Other editing software, designed to run on fast computers with large memory capacity, can render programs in real time, as each subcomponent is completed.

Storing

The rendered version of a digital program is automatically stored on a hard drive, but that storage is only temporary. Digital video requires so much space that programs can quickly fill even very large arrays of hard drives. To overcome

Figure 20-26 In rendering your completed project, you can specify its video and audio parameters.

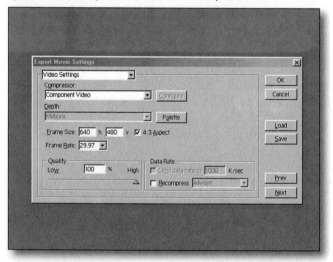

this problem, the rendered program is exported for permanent storage on a high-capacity medium, like a tape or disc. Because of the long potential life of the program signal, the program is truly archived.

No medium is absolutely permanent. Digital tape signals fade like their analog cousins, and discs are eventually degraded by handling. But, as long as a complete signal can be read from a digital tape or disc, a perfect duplicate can be made. For this reason, well-maintained digital video libraries are indeed archives.

Publishing and Presenting

The last step in postproduction is preparing the program for presentation in a particular medium. Digital movies are transferred to film for making release prints, or set up for direct digital projection. Most other programs are prepared for viewing on DVD and/or the Internet via streaming.

Creating ("authoring") a DVD means organizing the program into chapters and providing a menu for accessing the chapters (**Figure 20-27**). These menus often include live action or moving graphic backgrounds and elaborate audio. DVDs prepared for commercial release usually have multiple language tracks and/or subtitles, copy protection, and regional coding.

DVD authoring is covered in Chapter 23, *Mastering Digital Software*.

Figure 20-27   A DVD menu.

Operations and Principles

In this chapter on video editing operations, we have reviewed the fundamental editing procedures that are common to all programs. The next step is to see why we carry out these operations—what we are trying to achieve in editing. That is the subject of Chapter 21, *Editing Principles*.

Summary

- An edited program consists of many layers all working together, and all transparent to viewers.
- Editors learn a software application by grasping the concepts that lie under the program skins, behind the menus, below the icons.
- Subtractive editing means removing elements from a shooting session by copying the original footage and omitting the unwanted parts. With additive editing, a video program starts as an absolute blank and the editor places pictures, sounds, graphics, titles, and effects one piece at a time to build a complete program.
- The five major phases of editing are organizing, assembling, enhancing, synthesizing, and archiving.
- Organizing audiovisual materials involves identifying (slate, filename, metadata, timecode address) and describing (angle, content, quality, notes) them.
- Enhancing raw editing materials can mean improving, conforming, and/or redesigning both video and audio.
- A program timeline is several levels high, usually with separate lines for two or more video streams, transitions, titles, filters, and audio.
- Archiving involves rendering, storing, and publishing.

Technical Terms

Additive editing: Creating a program from raw footage by starting with nothing and adding selected components.

Continuity script: The master document that shows, at a glance, what has been recorded, how it has been covered, and which takes are preferred.

Digital assets: Electronic files of still and moving images.

Editing: Creating a video program from production footage and other raw materials.

File name: The identification of an editing element as it is stored in a computer. File names may or may not be identical to timecode addresses or slate numbers.

Metadata: Supplementary information stored in a still or video image file that is not visible in the actual image.

Rendering: Creating a full-quality version of material that was previously edited in a lower quality form.

Subtractive editing: Creating a program by removing redundant or poor-quality material from the original footage, and leaving the remainder essentially as it was shot.

Timecode address: The unique identifying code number assigned to each frame (image) of video. Timecode is expressed in hours, minutes, seconds, and frames, counted from the point at which timecode recording is started.

Timeline: The graphic representation of a program as a matrix of video, audio, and computer-generated elements.

Review Questions

Answer the following questions on a separate piece of paper. Do not write in this book.

1. What are the layered components typically included in an edited program?
2. *True or False?* All editing software packages use an on-screen "slide tray" storyboard.
3. What is the difference between *subtractive* editing and *additive* editing?
4. Identify the five major editing phases and explain the tasks involved in each phase.
5. When recording both professional programs and amateur videos, shot-by-shot slates are not generally recorded during production. Instead, the _____ identifies each shot in postproduction.
6. How is metadata used for video footage?
7. The video footage assembly operation includes selecting, sequencing, and _____ each individual shot.
8. Timecode is expressed in hours, minutes, seconds, and _____.
9. What is a *timeline*?
10. Describe some common uses of layering in video programs.
11. What is the process of *rendering*?

STEM and Academic Activities

STEM

1. **Technology.** What advancements have been made in video editing software products available on personal computers in the last 10 years? What video editing functions are you familiar with on your own home computer?
2. **Mathematics.** A shot of the slate is recorded for 10 seconds at the beginning of every take of every scene. In a production with 25 scenes and 3 takes of each scene, how much footage (in minutes) is dedicated to recording the slate?
3. **Social Science.** Discuss how the ability to publish videos on the Internet has affected the quantity, quality, substance, and availability of video programs.

A "Red Scarlet" digital cine camera, set up for studio production. This camera can record up to 30 uncompressed images per second on a nine megapixel sensor large enough to use full-size 35mm film lenses. Red cameras, and cameras from the Arri Corporation and other companies, are replacing film cameras in an ever-growing percentage of professional productions.

(Red Digital Cinema Camera Company)

21 Editing Principles

Objectives

After studying this chapter, you will be able to:

- Identify important editing principles and explain how they contribute to the creation of an effective program.
- Summarize methods for creating continuity and identify different types of continuity.
- Understand the editing processes used to shape performance.
- Explain the role of controlling emphasis in program editing.
- Recall the impact of a program's pace in delivering content.

About Editing Principles

Editing principles determine the qualities that you want in your finished programs. In effect, editing *operations* (covered in Chapter 20) are what you do, and editing *principles* are what you want to achieve by doing it. Important editing principles include:

- *Continuity:* The information or story should be presented in a coherent order that the audience can follow.
- *Performance:* The people in the program should appear believable, and they should create the intended effect on the audience.
- *Emphasis:* Information should be presented with an impact proportional to its importance, and the audience's attention should be directed to the most important aspects of the program.
- *Pace:* The program should flow briskly enough to maintain interest, but deliberately enough so that viewers can absorb the content.

While performance, emphasis, and pace are very important, the principle of continuity is absolutely essential—without it, your program cannot communicate adequately.

Continuity

Video is usually a linear form of communication. When telling a story, explaining an idea, teaching a skill, or presenting a person or place, you take the viewer through one thing after another, in a predetermined order from start to finish. The principle that governs the order of presentation is *continuity*—the art of organizing and sequencing program content so that it makes sense to the audience.

We are concerned here with the continuity popularized by Hollywood, which is now more or less standard. There are, however, other forms of visual narrative; and where personal videos on websites are concerned, classical continuity may be abandoned entirely.

Methods of Organization

Continuity can be achieved through different methods of organization, depending on the type of program. These methods include *narrative*, *argument*, *association*, and *subjective*.

Narrative

Stories are usually organized in chronological order. Documentaries and training programs (especially those about events, tasks, or processes) are also chronological. The program illustrated in **Figure 21-1**, for instance, shows the building of a fifth-wheel trailer from chassis to finished RV.

Argument

Programs intended to persuade viewers are typically organized by logical argument. For example, an argument for recycling is expressed in **Figure 21-2**.

Figure 21-1 A narrative (chronological) sequence on building a trailer.

(Western Recreational Vehicles)

Figure 21-2 A sequence organized by argument.

We produce a lot of trash.

Discarded trash degrades
the environment.

A degraded environment
harms wildlife.

Disposing of trash is difficult.

We should recycle trash instead
of disposing of it.

Recycled paper saves
a lot of trees.

Association

Programs (or program segments) intended to convey an impression of people, places, or events may be organized by association (*associative continuity*), with similar things grouped together. At the Leibniz County Kinetic Sculpture Race, for example, one sequence might associate the attempts by several different extravagant vehicles to climb the notorious "slippery slimy slope," **Figure 21-3**.

Subjective

The most elusive form of thematic continuity is purely subjective—*subjective continuity*. Like a poet or a painter, the editor presents content in a structure that viewers can feel, even if they cannot clearly explain it.

Subjective continuity is not illustrated here because it is impossible to suggest on the printed page.

In all its forms, continuity is evident at several levels of organization: the individual shot, the sequence made up of shots, and the program made up of sequences. In addition, longer programs are often divided into *acts*.

Shot-to-Shot Continuity

You establish basic continuity at the shot level by putting shots in order. Returning to an example introduced previously, **Figure 21-4** shows three shots in the order they were recorded by the camera. Since the insert of pouring the drink is out of order, the editor establishes correct continuity by placing it between the pair of two-shots, **Figure 21-5**.

Matching Action

When inserting the closeup between the main shots, the editor has to match the movement in order to preserve the illusion that the audience is watching a single continuous view, rather than three independent shots. In matching action, the editor often has three different options:

- *Start or end off screen.* The easiest method is to make the transition off screen, where the audience cannot tell whether or not the action matches. **Figure 21-6** shows the sequence starting closer on Frank, the pourer. Because the camera frames off the pitcher, viewers cannot see where it is when the insert begins.

Figure 21-3 A sequence organized by association.

Figure 21-4 The insert was originally shot out of order.

Figure 21-5 The editor moves the insert to its proper place.

Scenes and Sequences

The words "sequence" and "scene" can cause confusion because they are sometimes used interchangeably. Strictly speaking, though, sequences and scenes are not the same. A *scene* is a single action or process. A *sequence* is a single unit of fiction or nonfiction content. (The word "scene" itself can cause confusion because in stage plays, the word is used where movies use "sequence." Also, some discussions of films use "scene" as a synonym for "shot.")

A Scene

A scene usually has the following characteristics:
- Consists of a small number of individual shots (typically fewer than ten).
- The shots all cover (or at least relate to) one short piece of a larger action.
- The shots are all in a single area of the set or location, even though they may be made from different camera setups.

A Sequence

A sequence often consists of several separate scenes, all of them related so that they build a continuous action. A sequence will often (but not always) end with a transition (like a dissolve or wipe) to the new sequence that follows.

An Example

To demonstrate the difference between scene and sequence, imagine a short training program titled, *How to Make Your Own Firewood*.

The sequences. This process consists of three major parts: cutting down a tree, cutting the felled tree into firewood, and building a wood pile. In the edited training video, these parts will become separate sequences.

The scenes. The first sequence (cutting down a tree) breaks down into individual scenes: preparing the chain saw, selecting the tree, and felling the tree. Each scene covers a small part of the action that unfolds continuously in a single place and at a single time.

The shots. The first of these scenes (preparing the chain saw) is made up of several individual shots.

Production vs. Editing

Another difference is that scenes are units of organization used during production, while sequences are created during the process of editing.

Note that camera slates carry scene numbers, rather than sequence numbers. (Sequences are not identified by number or letter.) So, "27A3" on a slate identifies the third take of shot A in scene 27.

Shots in the scene "Preparing the chain saw".

Filling the fuel tank.

Adding chain oil.

Setting the choke.

Checking chain tension.

- *Cut during a pause.* In **Figure 21-7**, the wide shot continues until the static moment when the pitcher has descended, but the water has not yet started to flow. Matching is relatively easy because movement is minimal in both shots.

- *Cut during the movement.* Though this is the most difficult option, it is often the most effective, because the seemingly continuous motion helps sell the illusion that there is a single, uninterrupted view, **Figure 21-8**.

Figure 21-6 Because the pitcher is not in the first shot, its movement need not be perfectly matched.

Figure 21-7 At the end of the first shot, the pitcher has already been tipped. So, the insert begins after the movement has stopped.

Figure 21-8 The pitcher is descending as the editor cuts to the insert.

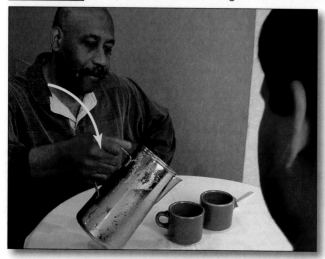

Except in special circumstances, matching action is usually done with every shot in a continuous sequence.

It is not always necessary or even desirable to match action exactly, as we will see.

Maintaining Screen Direction

In addition to seeming uninterrupted, continuing action usually needs to maintain a consistent screen direction. If the director has done a good job, the raw shots in a sequence will have the screen direction already established.

Screen direction is covered in Chapter 19, *Directing for Form.*

If there are inconsistencies in screen direction, you may want to remove them. To do this, you have three available methods:

- *Use only correct shots.* If you can cover all the action adequately without the incorrectly oriented shots, simply omit them from the sequence.

- *Insert cutaways.* If you must use an incorrect screen direction, precede it with a shot that does not show the action. **Figure 21-9** shows two mismatched shots. By cutting away from the movement, you soften the fact that, when you return to the movement, the screen direction has changed. See **Figure 21-10**.

- *Flip the incorrect shot.* If you lack material for a cutaway, you can sometimes reverse the shot. With digital processing, it is simple to correct screen direction by reversing the left/right orientation of an image, **Figure 21-11**. Of course, reversing the shot will not work if details in it reveal that it is backward.

Film editors often use the term "flopping" for reversing an image because in film this was accomplished by turning a strip of film over and copying it through its transparent backing. Today, the process is digital and is usually called "flipping" instead.

Figure 21-9 In the second shot, the screen direction is reversed (and the cup is in the wrong hand).

Figure 21-10 The cutaway to the other subject softens the mismatched screen direction.

Figure 21-11 The closeup of the cup is flopped (reversed).

Intentional Mismatches

In narrative and other linear types of continuity, matching action and direction is usually important. But in associative continuity, when shots are linked only by theme (as in the kinetic sculpture race example), there is no need to match action or screen direction. In fact, it is often better to juxtapose shots with clearly different actions and directions, to emphasize their independence.

Sequence Continuity

Sequence continuity is the organization of all the shots in a sequence. Though there is no precise definition, a sequence may be thought of as an assembly of shots organized to present a single action, process, or idea. The continuity of a sequence may be *linear*, *associative*, or *subjective*.

For a fuller discussion, see the sidebar *Scenes and Sequences*.

Linear Sequences

The majority of sequences employ *linear continuity*, with the shots lined up in temporal order for stories and training programs, and in logical order for editorial presentations.

In creating a linear sequence, the challenge is to decide how much detail to include. That is:
- How many sub-steps in the action do you need to select for showing?
- How much of each step should be shown?

The answers to these questions depend on the purpose of the sequence.

Timing each shot is addressed in the discussion of emphasis, later in this chapter.

Since the editorial decision process is easier to illustrate than to explain, consult the sidebar *One Action, Three Sequences* later in this chapter for a demonstration.

Associative Sequences

Sequences organized by theme can be more difficult to arrange because they lack the built-in constraints of chronological order. For this reason, it is helpful to discover some other organizing principle in the raw footage and use it to govern the sequence. The possibilities depend entirely on the material you have to work with. Here are some examples from a documentary about the weekly McKinley farmer's market:
- *Color.* Sequence the shots to begin with muted reds and gradually work up to bright, dramatic scarlet, **Figure 21-12**.
- *Subject.* Often, you can find several shots related by the fact that they contain similar subject matter—in this case, hats. See **Figure 21-13**.
- *Size.* An alternative approach is to begin with small subjects that grow in size as the sequence progresses, **Figure 21-14**.

As you can see, there is no "logical" reason to group shots by size or color or any other characteristic. The point is to give the audience a feeling, however unconscious, of coherence and order. Without consciously thinking about it, viewers generally want to feel that there is a purpose behind the program they are watching. A well-ordered sequence conveys that sense of purpose.

Figure 21-12   Shots associated by color.

Figure 21-13   Shots associated by subject matter.

Subjective Sequences

In specialized programs, like certain music videos or very personal expressions, sequences may have no organizing principle other than the editor's purely subjective preferences. Though this is a perfectly legitimate approach to creating sequences, it can be difficult to handle because there are no organizational guidelines. The only test is, does the result succeed—that is, does the intended audience find satisfaction in watching and listening to it? If viewers sense an underlying coherence in a sequence, however obscure, they feel reassured that they are not just watching random footage.

Figure 21-14 Shots associated by size.

Program Continuity

We have noted that the basic building block of a video is the individual shot, and a number of shots that are related, ordered, and matched is a sequence. Sequences, in turn, are organized into a complete program.

This three-level organization (shot/sequence/program) is often enough for short works. Longer videos usually play better when the sequences are first combined into larger groups that, collectively, make up the program as a whole.

These larger groups of sequences have different names in different program types. The scripts of feature films and TV shows are typically organized into acts. Long nonfiction programs, like documentaries, infomercials, and training videos, are also constructed in multi-sequence groups that are referred to simply as "parts" or "sections". Like individual shots in a sequence, these larger groups are also organized to create a coherent, effective continuity of presentation.

Parallel Sequences

Often, separate sequences are presented concurrently in alternating pieces. This technique is known as *parallel cutting* or *cross cutting*. Classic parallel cutting is as old as the movies:

- Our hero hangs from the roof railing of a tall building. CUT TO:

- Our heroine races up to the roof to rescue him. CUT TO:
- He looks down at… CUT TO:
- …the ground far below. CUT TO:
- He reacts with dismay. CUT TO:
- Our heroine races onto the roof…

…and so on. In Chapter 5, *Video Time*, parallel cutting is illustrated by the shots repeated as **Figure 21-15**. For simplicity, this example consists of single shots. However, each segment may actually consist of a short scene, say three or four shots of his struggle to climb up over the railing, then several more of her racing up flights of stairs to reach him, and so forth. Also, complex actions may involve intercutting several different sequences at once.

To choose a more complex example, the conclusion of *Star Wars: Episode I* intercuts four separate actions as the princess, the boy, the Jedi knights, and the alien army lead separate attacks at the same time on their common enemy. Each visit to each of these actions consists of a full sequence made of many individual shots.

Acts

As noted previously, the sequences in longer programs are often grouped into larger units called *acts*. An act is a major division that

Figure 21-15 Two sequences presented through parallel cutting.

is traditionally opened with a fade-in effect and closed with a fade-out.

In commercial television, act breaks are also punctuated by commercials.

In programs that are not constrained by TV time requirements, act divisions are flexible. Typically, an act will end for one or more reasons:
- The preceding group of sequences creates a cumulative effect on the audience (like the suspenseful effect of an elaborate robbery or escape).
- The audience needs a moment to relax and/or absorb the information in the previous group of sequences.
- The material that will follow is different in style or tone.

- The material that will follow is different in content, whether a distinctly new part of a story or a major subject change in an informational program.

Transitions

The conventions of traditional continuity separate sequences and acts by transitions. A dissolve signals a change from one sequence to the next. A fade-out/fade-in marks the transition between acts. Today, however, it is common practice to speed up the pace by joining sequences with direct cuts. Act breaks, however, are usually still signaled by fades or dissolves, **Figure 21-16**.

Figure 21-16 A fade-out.

(Sue Stinson)

With the availability of digital effects (DVEs), new transitions are beginning to acquire standardized meanings. A slow wipe, especially a horizontal one, can replace a fade-out/fade-in, creating a major pause without fully closing and reopening the program. (Of course, this alternative will not work if the program must stop for a commercial.)

The *Star Wars* movies are also notable for using soft-edged horizontal wipes for transitions.

Special DVEs, like flips, fly-ins, or hundreds of other choices, can signal a change in the program without indicating whether the audience should consider it minor (like a dissolve) or major (like a pair of fades).

One Action, Three Sequences

Here is a single action edited into three very different sequences. The action covers the process of connecting a travel trailer to its tow vehicle, and driving off with it. The three sequences show how this action might be presented in three very different types of video programs:

- *Instructional.* A training program on how to operate a trailer.
- *Promotional.* A video is designed to sell potential buyers on the delights of trailer travel and recreation. The objective is to show how easy it is to manage a fifth-wheel trailer (the kind that overlaps the pickup truck that tows it).
- *Dramatic.* The story of a woman who sets out to see the world on her own.

Instructional Version

In this version, the objective is to show the steps in hitching up as clearly as possible. The shots detail mechanical and electrical hookup procedures.

Hitching the trailer.

Operating the jacks.

Latching the hitch.

Attaching the safety cable.

Plugging in the electrical cable.

Hand position for backing the rig.

Choosing and timing transitions is a subjective process that depends greatly on the taste of the editor.

Sound and Continuity

In this discussion of continuity, we have repeatedly mentioned audio. Of all the tools for creating continuity at the shot, sequence, and program levels, sound is by far the most powerful. Audio can be flexible, subtle, evocative, easy to work with, and very inexpensive. For a more comprehensive treatment of audio's expressive capabilities, see Chapter 8, *Video Sound*.

Promotional Version

The purpose of this version is to suggest that hitching up is a quick and easy process.

Simply hitch the truck and trailer.

Operate the electric switches…

…to retract the support jacks automatically.

And, drive away.

Dramatic Version

The goal of this version is suspense—Can she succeed in the tricky business of mating truck and trailer?

The bystanders add suspense to the sequence.

Performance

After continuity, the next major editing principle involves performance, since almost all video programs involve people. Though you cannot transform an awkward amateur into an award winner, you can often rescue a poor performance, improve an adequate one, and fine-tune even an accomplished job of screen acting. Editing performance involves four major tasks:

- *Selecting* the best parts of the best shots.
- *Adjusting* the timing of dialogue and action to create sequences that move effectively.
- *Enhancing* the content by underlining, shading, or even changing the actors' original meanings.
- *Directing* viewers to the most important aspects of the performances.

To understand how these four processes shape performance, imagine a sequence in which Bob is desperate to buy a car and Bill is reluctant to lend him the down payment, **Figure 21-17**.

Selecting the Best Performance

Typically, each moment of an actor's performance is recorded more than once. The same action is usually shot from two or more different camera setups, and each shot may be repeated to provide multiple takes. Since no two shots are exactly identical, your first task is to select the best shots for your purposes.

The selection task is easy if one entire shot is better than the others, but this is not always the case. Sometimes, the best performance is in separate pieces that are distributed among several setups and multiple takes of each one. When this happens, your task is to cement the best sections together with joints that are invisible to the audience.

To demonstrate, imagine that an actor misreads a line, as in **Figure 21-18**. The problem is that the actor was supposed to say, "I need the *cash*," in order to avoid repeating the word *money*. To correct the error, the director shot a second take, **Figure 21-19**. In Take 2, the actor misread the *other* half of the speech, saying

Figure 20-18 BOB (Take 1): Please lend me that money. I need the money, and I need it now!

Figure 20-17 Bill and Bob at the used car lot.

Figure 20-19 BOB (Take 2): Please lend me the money. I need the cash, and I need it now!

"the" money instead of "that" money. To make the entire speech read correctly, you need to join the first line in Take 1 to the second line in Take 2. If you butted two takes of the same angle together, the result would be an unacceptable jump cut, **Figure 21-20**.

To avoid this problem, you have three options:
- Cut to a different angle of the performer.
- Cut to another actor in the scene.
- Provide an insert or a cutaway buffer shot between the two halves of the line.

Which option you choose depends on the materials available.

Changing Angles

If the director has covered the actor from more then one angle, you can try for a matched cut, as in **Figure 21-21**.

Changing Performers

If you lack another angle of the performer, you can sometimes substitute another actor in the scene and let the performer's off-screen voice carry over the shot, as in **Figure 21-22**.

Using Buffer Shots

Another way to hide the join is by adding a *buffer shot*—an image of something completely

Figure 21-20 Butting two similar shots results in a jump cut.

Figure 21-21 Changing angles.

BOB (Take 1): Please lend me that money. **BOB (Take 2): I need the cash and I need it now!**

BOB (Take 1): Please lend me that money.

BOB (voiceover from Take 2): I need the cash...

BOB (Take 1 cont.): ...and I need it now!

different. There are two types of buffer shots: inserts and cutaways. An insert is a small detail of the action that excludes the actor, **Figure 21-23**. A cutaway is an image outside the action, **Figure 21-24**.

All of these methods of selecting pieces of a performance require cutting from the first shot to another angle of the same performer, a shot of a different performer, or an insert or cutaway. These same techniques are also employed to adjust actors' performances, and to control the pacing of sequences.

BOB (Take 1): Please lend me that money.

BOB (voiceover from Take 2): I need the cash...

BOB (Take 1, cont.): ...and I need it now!

BOB (Take 1): Please lend me that money.

BOB (voiceover from Take 2): I need the cash...

BOB (Take 1, cont.): ...and I need it now!

Controlling Performance Pace

With judicious editing, you can change the pace of an actor's performance, or even the speed of a whole sequence. Usually the goal is to shorten it.

You can do this if the available footage includes individual shots of each performer and/or over-the-shoulder two shots in which one actor's mouth is invisible to the camera.

Some actors tend to pause before and during their lines or business (actions) to make their responses appear more deeply felt, but these pauses slow down the sequence. If you can cut away from the pause, you can often shorten or omit it and speed up the scene. The techniques are the same as the ones demonstrated earlier.

In other situations, you may want to stretch a scene out rather than tighten it up. To do this, you could leave in the performers' pauses, but too much of this device can seem artificial. To lengthen a sequence more naturally, try adding performers' reaction shots or cutaways, or both.

As you can see from **Figure 21-25**, lengthening the sequence heightens the suspense—will Bill lend Bob the money or not?

Enhancing or Adding Meanings

Figure 21-25 uses Bill's reaction shots to build suspense. Notice that the shots also make him seem reluctant to lend the money. We can imply all kinds of performer feelings by the way we time the edits.

To use a different example, we can make Bob seem as reluctant to ask for a loan as Bill is to give it, **Figure 21-26**. By repeating the word "please" and inserting significant pauses, we seem to drag the request out of the borrower.

Now, suppose the director did not shoot Bill's reaction, but we do have that insert of his hand holding the money. By timing that insert, we can make the hand act for us. In **Figure 21-27**, Bill has no problem with giving up his cash. In **Figure 21-28**, however, he is reluctant to part with his cash.

Figure 21-25 Lengthening a sequence.

BOB (Take 1): Please lend me that money.

BOB (voiceover from Take 2): I need the cash…

BOB (Take 1, cont.): …and I need it now!

Figure 21-26 Adding meaning—Bob is reluctant to ask for a loan.

(A long pause as Bob looks at Bill.)

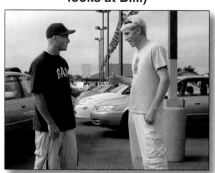

BOB: Please…

BOB (after another pause): Please lend me that money.

Figure 21-27 Adding meaning—Bill willingly gives up his cash.

BOB: Please lend me that money.

(Bill's hand thrusts the cash forward at once.)

Figure 21-28 Adding meaning—Bill is reluctant to give up his cash.

BOB: Please lend me that money.

After a long hesitation…

…Bill's hand thrusts the cash forward.

As a general rule, showing pauses *ahead of* actors' reactions makes them appear to be thinking. Since viewers know what is taking place in the sequence, they will generally supply the thoughts.

Conferring Importance

In fiction programs especially, one performer may be more important, dramatically, than other players in a sequence. You can increase a performer's prominence by selecting image size, assigning screen time, and giving him or her the critical moments in the scene.

Selecting Image Size

In many cases, sequences will be covered by more than one camera angle—for example, a medium (waist) shot and a closeup (head) shot. From the standpoint of viewer psychology, the bigger the image, the more impact it has. You can make key players more important by intercutting closer angles of them with wider shots of others. **Figure 21-29** makes Bill bigger and, therefore, more important.

Apportioning Screen Time

A second way to enhance a character's importance is by giving him or her more time on screen. For example, in **Figure 21-29**, Bob has two shots to Bill's one. To strengthen Bill's status as the more important character, we could reverse the shot ratio, as in **Figure 21-30**. By giving Bill most of the screen time, we make him the focus of the sequence, even though he does not say a word.

Giving the Moment

In addition to image size and screen time, performers also gain importance according to *when* they are on screen. Showing a character at a critical point may be called "giving him or her the moment". **Figure 21-31** focuses our attention on Bob's emotions about getting his loan. On the other hand, if Bill is the more important character, you can give the moment to him instead by putting him on screen as he makes up his mind, **Figure 21-32**.

Figure 21-29 A closer angle of Bill makes him appear more important.

BOB: Please lend me that money. **BOB (off screen): I need the cash.** **BOB: I need it right now!**

Figure 21-30 More screen time makes Bill appear more important.

BOB: Please lend me that money. **BOB (off screen): I need the cash.** **BOB (off screen): I need it right now!**

Figure 21-31   Giving Bob "the moment."

**BOB: I need the cash.
I need it right now!**

**(His expression turns to delight
as he looks down at...**

**...Bill's hand holding
the money out.)**

Figure 21-32   Giving Bill "the moment."

BOB: I need the cash.

**BOB (off screen):
I need it right now!**

**(Bill smiles, giving in,
and looks down at...**

**...his hand holding
out the money.)**

Of the three methods of conferring importance on a performance, giving the screen to an actor at critical moments is perhaps the most powerful.

Emphasis

After continuity and performance, the next editing principle concerns emphasis. We have already seen how editors emphasize one aspect of video programs: actor performance. As a matter of fact, *all* aspects of movies are shaped by editorial emphasis. You emphasize something by making it more prominent, so that the audience pays more attention to it. In the hands of a skilled editor, controlling emphasis can be a sophisticated process, but the basic techniques are quite straightforward. Simply put, you highlight selected material by managing:

- *Content*—What you show.
- *Angle*—How you show it.
- *Timing*—How long you show it.
- *Shot order*—When you show it.
- *Reinforcement*—How you call attention to it with additional material.

In this discussion, we will assume that the director has supplied the editor with complete coverage of the sequence.

To illustrate these techniques for adding editorial emphasis, we will return to the sequence about hitching up a trailer. The essentials of the sequence are reviewed in **Figure 21-33**.

With the footage available, we can create several quite different sequences, simply by controlling the editorial emphasis. We do this by managing *content*, *angle*, *timing*, *shot order*, and *reinforcement*.

Content

To illustrate controlling content, imagine that we decide we want to emphasize the support of the female bystander, while downplaying the skepticism of the male. The result might resemble **Figure 21-34**.

We have emphasized the response of the female bystander by replacing a piece of content: the original two-shot is now a closeup of her reaction.

Figure 21-33 Hitching the trailer.

Figure 21-34 To emphasize her support, the female bystander gets an extra closeup.

Angle

To emphasize the difficulty of the process, we can use our control of camera angles. Up to now, shots of the trailer and truck hitch components mating have been made from neutral angles. By showing this action from the driver's point of view, we show how hard it is to line up the hitch, **Figure 21-35**.

Timing

The sequence still lacks suspense because it all seems too cut-and-dried. By lengthening the first half of the sequence, **Figure 21-36**, we can build suspense before showing the successful hitch.

So, by adjusting content, camera angle, and timing, we have edited a tight, suspenseful sequence.

Figure 21-35 Her worried closeup emphasizes the difficulty.

Now, suppose we decide to change the story slightly by showing that the woman never doubts her ability to hitch up the trailer. Since we have no footage that specifically shows this, we will make the change by simply altering the shot order.

Shot Order

In most programs, events happen in chronological order. That means that the sooner something appears, the earlier it is in story time. Though we have no footage designed specifically

Figure 21-36 Adding shots helps build suspense.

to reveal the woman's inner confidence, we do have a shot of her reacting to the success of her efforts.

To make her seem confident from the start, all we need to do is move her reaction closer to the beginning of the sequence, **Figure 21-37**. By showing her grin *before* she hitches the trailer, we inform viewers that she is confident of success.

Reinforcement

Up to this point we have controlled emphasis by *selecting*, *timing*, and *positioning* shots. To add the final touch to our sequence, we can use editorial *reinforcement*. Reinforcement can take visual form. For example, we could have the hitch components mate in slow motion to lengthen and emphasize the success of the operation. To enhance the effect, we can add further reinforcement with sound, **Figure 21-38**. As the hitch halves mate in slow motion, we lay in the low, reverberating *clang* of a great iron door slamming shut. Though this effect is not

realistic, it will be accepted by viewers because it is subjectively "right" for the feeling of the image.

Pace

After continuity, performance, and emphasis, the last major editing principle involves *pace*. Pace is the sense of progress, of forward movement, that the audience gets from a program. "Pace" is not identical to "speed", although speed is one of its main determiners. In fact, a very fast-moving program can still feel boring if it lacks the other two components of pace: variety and rhythm.

Of course, if the content is boring, it cannot be made interesting by adjusting its pace.

Delivering Content

The easiest way to evaluate the pace of a program is by seeing how well it delivers its content to the audience. If viewers miss important

---

Figure 21-37 Changing shot order.

She grins before the two halves mate.

Figure 21-38 Reinforcing with sound.

details, the pace is too fast. If content points are belabored after the viewers have already absorbed them, the pace is too slow. If viewers feel that they are getting the material while moving briskly forward, then the pace is about right.

Editing pace should not be confused with content *density*. How rapidly a subject is presented also depends on how much detail is included. Suppose, for example, that you finish a how-to video on creating a dinner party with a cherry pie dessert, **Figure 21-39**. Depending on your needs, you could detail every step in baking the pie, or cover only the highlights of the process, or just show the finished pie being served.

Since content density is primarily a function of scripting, it lies outside a discussion of editing. Here, we are concerned with the pacing of the material that the editor is given to work with, whatever its density may be.

Because pacing is judged by the response of the viewers, you will want to consider your target audience in establishing a pace for a program. An elementary school class, a stockholders' meeting, and a classic car club respond to video programs in quite different ways.

The concept of the target audience is covered in Chapter 9, *Project Development*.

Content delivery is also determined by the type of program. Often, fiction programs are paced energetically so audiences can follow the story, but not catch all the enriching details

on the first viewing. (The *Star Wars* episodes and the best Disney™ animated features are good examples.) At the other extreme, training programs are paced quite deliberately to ensure that viewers absorb all the essential information.

In the final analysis, evaluations of pace must be based on personal taste and experience. In effect, if it *feels* right, it *is* right. Though variety and rhythm contribute to the sense of pace, the most important determiner is speed.

Speed

Speed may be expressed as the average length (in seconds) of the shots in a sequence, **Figure 21-40**. Though there are no fixed rules about appropriate speed, here is a rough guide:
- One-half second per shot (120/minute) is quite fast.
- One second per shot (60/minute) is fast.
- Two–three seconds per shot (20–30/minute) is brisk.
- Four–six seconds per shot (10–15/minute) is moderate.
- Ten–fifteen seconds per shot (4–6/minute) is slow.

In nonfiction videos, shots typically average between three and ten seconds apiece. It should be emphasized that these figures are not rules, but only examples.

It is possible to have 30 single-frame shots per second of NTSC video (1,800 per minute). This ultra-high-speed cutting can be used quite effectively, in short bursts. At the other extreme,

Figure 21-39 Good cooking programs deliver their content at a carefully measured pace.

Figure 21-40 Typical shot durations.

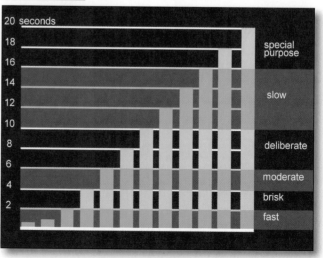

some documentary programs include shots lasting a minute or more.

In the movie *Rope*, Alfred Hitchcock made an entire film out of shots lasting several minutes apiece.

The average speed of the shots in a sequence typically depends on the nature of the subject matter. For instance, a fight or car chase will typically be cut together from very short shots. A business meeting or dinner conversation might consist of moderately long takes, and a languorous love scene might be built from individual pieces that are many seconds long. Alternatively, you can time your shots *against* the subject matter—presenting a chase in long, deliberate pieces to build suspense, or cutting a love scene quickly to mirror its emotional intensity.

Variety

Read these sentences. They are similar. They are the same length. They are short. They are boring.

Shots are like sentences: even short ones grow boring if they are too similar, and long ones can become agonizing if they go on and on without relief. The solution to the problem is variety—mixing shots of varying lengths to avoid falling into repetitive patterns. In most programs, good editors achieve variety instinctively, that is, they do not consciously think, *"The two preceding shots are six and eight seconds long, respectively, so I will follow them with a short shot in order to achieve variety."* Instead, they rely on their feel for the rhythm of the sequence, while keeping the general goal of variety in mind as they work. In doing so, even experienced editors operate by a sort of successive approximation:

- Building a trial sequence.
- Evaluating the effect.
- Fine-tuning the sequence if it seems to be working, or
- Discarding the sequence and starting over if it is not working.

The easiest way to achieve variety in shot lengths is by reviewing your work as you go and adjusting shot timing until you like the effect. When this style of cutting is successful, the result is like a path made of irregular flagstones: though it has no visible pattern, it feels like a coherent design.

Rhythm

But, sometimes, an editor wants a sequence to resemble a patterned brick walk instead of an informal path, **Figure 21-41**. Sometimes you want to time your shots to develop a clear and repetitive rhythm.

Rhythm for Contrast

Instead of varying shot length continually, you may want to group several long, slow shots, punctuate them with a group of very brief shots, then resume the pattern of lengthy shots. This is the classic western gunfight rhythm: slow cutting while the adversaries stalk each other, a hail of gunfire handled in very quick shots, then another slow passage of cat-and-mouse stalking.

Rhythm for Drive

Another classic editing rhythm might be called "marching the prisoner." Imagine that, flanked by Imperial Troopers, our Hero is escorted from the battle cruiser's landing bay to the command center of the Admiral, deep within the giant mother ship. To achieve this sequence, the director has had the actors march through five different sets at exactly the same pace. Using the rhythmic footsteps of the troopers, the editor cuts four out of five shots to exactly the same length:

Figure 21-41 A sequence of rhythmic edits is like a patterned brick walkway.

The Art of Montage

A *montage* is a quick succession of sounds and images designed to condense a lot of material into a brief amount of screen time. For example, a montage dramatizing a winning baseball season might be communicated in a rapid-fire succession of shots showing hitters, scoreboards, umpires signaling "safe!", and so on, intercut with repeated newspaper headlines announcing team wins, pages flying off a desk calendar, and shots of the team bus traveling to one game after another.

Dramatic Effect

Most montages are not absolutely necessary for conveying information. After all, the content of a baseball montage could be fully communicated by a single insert of a newspaper headline, such as this one:

The essential information could be conveyed in a single shot.

The real purpose of a montage is drama: developing a sense of momentum and excitement from the kaleidoscope of images. By showing highlights of a story, a montage preserves the feeling of narrative, in a highly condensed form.

Styles

The simplest montages are linear, with one quick shot after another filling the screen. More complex versions may display multiple images in several forms:

- Superimposed double, or even triple, exposures.
- Screen divided into several separate images.
- Changing background images composited behind a foreground character.

Superimposition and split screen are not usually employed together because the result might be too confusing to follow.

The Role of Sound

Sound is very important in montages for several reasons:

- Because audiences can process audio information very quickly, sound can be several layers deep.
- Compared to video, audio is easy and inexpensive (laying in the sound of a steam train is much easier than shooting a real engine).
- Music, especially, can communicate nostalgia and many other feelings.

As a general strategy, you may wish to keep the visuals relatively simple, while mixing a complex and evocative composite sound track.

An Example

Here is an example of a simple montage. For demonstration purposes, it progresses through only six stages, and contains just six elements. A real montage might run longer and contain many more layers of information. Our montage dramatizes the rise of the brilliant actor Laurence O'Livery from a callow youth to a revered icon of the theater. The montage contains six kinds of information:

1. A succession of theater programs that show O'Livery's rise from extra to superstar.
2. Calendar years changing constantly, to show the passage of time.
3. Titles of the plays our hero appears in.
4. Lines spoken by the great O'Livery.
5. Evocative music of each period.
6. Sound effects of applause.

Notice how inexpensive this montage is to create. There is not a single shot of an actual play in progress.

Montage and Video

Though popular up through the 1950s, montages almost disappeared for a few decades—partly because they went out of fashion, and partly because the film laboratory work required was very expensive. But with digital editing, most software can create complex and effective montages for just the price of the editor's time.

1945

The Faceless Crowd

Solicitor	Craig Sassoon
Landlady	Walla Warfield
First Policeman	Jason Quigley
Second Policeman	Harlan Stander

The Crowd
Angela Aames, William Bassett, Charles Dark, Melissa Evans, George Grimes, Cedric Haddon, Mary Jones, Gwyneth Kalem, **Laurence O'Livery**, Michael Porter, Daryl Ransom, Gale Starr, Herschel Tannen, Gregory Wilson.

Lyceum Theatre Programme / The Faceless Crowd

ACTION: Flipping calendar pages show the years advancing.
VOICEOVER: Crowd noise.
EFFECTS: Scattered applause.
MUSIC: Big band.

1950

Smedley Soldiers On

with	
Lady Landsdowne	Bernice Duprey
Lord Landsdowne	Guy Bainbridge
Rev. Cheadle	Jack Crammell
A. Hobo	Launce Gobbo
Parlor Maid	Fanny Pryce
Butler	**Laurence O'Livery**

Presented by arrangement with his Majesty's Theatre, Hammersmith. Grateful acknowledgement is made to the estate of Lord Wolverhampton for provision of the suits of armour appearing in Act II.

Knightsbridge Theatre Programme

ACTION: Flipping calendar pages show the years advancing.
VOICEOVER: "You rang, Sir?"
EFFECTS: Polite applause.
MUSIC: Pop song.

1955

Weekend at Windemere

The Theatre Company, Ltd.
presents
Edward Maybridge and Vanessa Prawn
in
Weekend at Windemere
with
Cedrec Cubbins, George Brandt
and Laurence O'Livery as Freddie

ACTION: Flipping calendar pages show the years advancing.
VOICEOVER: "Anyone for tennis?"
EFFECTS: Strong applause.
MUSIC: Do-wop pop song.

1960

A Love for the Ages

Selwyn and Selwyn
presents
A Love for the Ages
by
Hamish Korngold
starring
Laurency O'Livery   Angela Bredhan
with
Percy Adline
Maurice Rample

ACTION: Flipping calendar pages show the years advancing.
VOICEOVER: "I love you, my darling!"
EFFECTS: Very loud applause.
MUSIC: Beach Boys song.

1970

Hamlet

The Imperial Shakespeare Company
presents
Laurence O'Livery
in
Hamlet
Prince of Denmark
by
William Shakespeare

ACTION: Flipping calendar pages show the years advancing.
VOICEOVER: "To be or not to be..."
EFFECTS: Thunderous applause.
MUSIC: Paul Simon song.

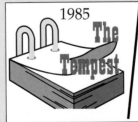

1985

The Tempest

presents the
Laurence O'Livery
production of
The Tempest
starring
Laurence O'Livery
by
William Shakespeare

ACTION: Flipping calendar pages show the years advancing.
VOICEOVER: "We are such stuff as dreams are made on..."
EFFECTS: Distant, filtered applause.
MUSIC: New Age/Celtic instrumental.

Shot 1. Landing bay: TROMP! TROMP! TROMP! TROMP!

Shot 2. Curved corridor: TROMP! TROMP! TROMP! TROMP!

Shot 3. Straight corridor: TROMP! TROMP! TROMP! TROMP!

Shot 4. Antechamber: TROMP! TROMP! TROMP! TROMP!

Shot 5. Command room where the Admiral is waiting: TROMP! TROMP! TROMP! TROMP! TROMP! TROMP! TROMP!

Notice the rhythm of the final shot, which contains three additional steps. The extra paces contribute a sense of closure to the sequence. Also, it consists of an odd number of steps (7), where the other shots contain even numbers (4 each). For psychological reasons that are not perfectly understood, a sequence often feels more powerful when it concludes on an odd number.

The two most common English verse forms are iambic pentameter with five beats per line:

Shall I compare thee to a Summer's day? Thou art more lovely and more temperate.

and ballad measure with seven beats per line:

It looked extremely rocky for the Mudville nine that day.

The score stood four to two with but one inning left to play.

Rhythm for Comedy

Comedy often relies on surprise, or the unexpected. Since rhythmic cutting sets up the expectation that it will be continued, an editor can help a joke by building rhythmically to the payoff, and then suddenly breaking the pattern. For example, suppose that the storm trooper marching sequence was shot by a comic director like Mel Brooks. The edited result might resemble this:

Shot 1. TROMP! TROMP! TROMP! TROMP!—Landing bay

Shot 2. TROMP! TROMP! TROMP! TROMP!—Curved corridor

Shot 3. TROMP! TROMP! TROMP! TROMP!—Straight corridor

Shot 4. TROMP! TROMP! TROMP! Whoa!—Somehow, they are back in the landing bay.

Here, the editor sets up a strong, repetitive rhythm and then deliberately breaks it in order to reinforce the climax of the joke. If you study classic movie comedy, from Charlie Chaplin to the present, you will see how dependent comedy is on rhythmic editing.

Rhythm for Montage

The sidebar *The Art of Montage* explains how multiple images and sounds can be presented to dramatize events over time. Because montages are constructed in repetitive patterns, they are usually edited to strict rhythms set by musical underscoring. Occasionally, however, a montage may be purposely built without any discernible rhythm. For example, imagine a football game. If the story calls for one side to march triumphantly over the other, this might be expressed in a classic rhythmic montage. But, if the game is fought to a messy, brutal scoreless tie, then the montage might be as formless and confusing as the game itself.

Cutting to Music

Rhythmic cutting is widely used in matching visuals to music. This is usually done for music videos and for sequences in other programs in which shots are linked by association, rather than by chronological order. For instance, a succession of views of the Italian countryside might be edited to the lyrical rhythms of Vivaldi's *The Four Seasons*. A montage of New York City going about its complex daily business (**Figure 21-42**) could be cut to the rhythm of a modern jazz composition.

At its best, editing picture to music is like dancing: though you follow the basic beat, you do not match the music exactly, phrase-for-phrase and measure-for-measure. Instead, you maintain an organic collaboration between picture and sound. Handled this way, cutting to music can be the most satisfying kind of rhythmic editing.

Figure 21-42 A New York city street suggests up-tempo jazz music.

(Sue Stinson)

The End Result

In this discussion of editing principles, we have considered continuity, performance, emphasis, and pace. There is more to the craft of editing, especially in subtle matters of tone, mood, and style. We have not addressed these matters directly because as an editor, you cannot do so either. Tone, mood, and style are not guiding principles, but ultimate goals: they are what you will achieve through the way in which you handle *continuity*, *performance*, *emphasis*, and *pace*.

Summary

+ Important editing principles include continuity, performance, emphasis, and pace.
+ Methods of organizing content include narrative, argument, association, and subjective.
+ Sequence continuity may be linear, associative, or subjective.
+ Traditionally, sequences and acts are separated by transitions.
+ The four major tasks involved in editing performance include selecting, adjusting, enhancing, and directing.
+ Create emphasis by managing content, angle, timing, shot order, and reinforcement.
+ Pace is the sense of progress, of forward movement, that the audience gets from a program.
+ Variety involves mixing shots of varying lengths to avoid falling into repetitive patterns.
+ A montage condenses a lot of material into a brief amount of screen time.

Technical Terms

Associative continuity: Organizing a sequence of shots related by similarity of content.
Buffer shot: A shot placed between two others to conceal differences between them.
Linear continuity: Organizing a sequence of shots chronologically and/or logically.
Montage: A brief, multilayer passage of video and audio elements designed to present a large amount of information in a highly condensed form.
Parallel cutting: Presenting two or more sequences at once by showing parts of one, then another, then the first again (or a third) and so on. Also called *cross cutting*.
Subjective continuity: Organizing a sequence according to the feelings of the director/editor, rather than by chronological or other objective criteria.

Review Questions

Answer the following questions on a separate piece of paper. Do not write in this book.

1. Important editing principles include continuity, performance, _____, and pace.
2. *True or False?* Video is usually a linear form of communication.
3. List the four methods of organization for continuity discussed in this chapter and briefly explain each method.
4. Reversing an image to correct screen direction is known as _____.
5. What is *parallel cutting*?
6. Identify some of the typical reasons an act comes to an end.
7. Editing performance involves four major tasks. What are the tasks?
8. What is a *buffer shot*? What are the two types?
9. How is emphasis created in a program through timing?
10. *True or False?* "Pace" is identical to "speed" in program editing.
11. How does variety contribute to the pace of a program?
12. What is a *montage*?

STEM and Academic Activities

STEM

1. **Engineering.** Video is usually a linear form of communication—occurs in a predetermined order from start to finish. Identify other processes that are linear in form. List the typical steps for each process you identify.

2. Language Arts. Record a training or "how-to" program. Review the information in this chapter on scenes and sequences. Watch the program and make note of individual scenes, as well as sequences contained in the program. Summarize the content of each scene, and indicate into which program sequence each scene falls.

A work screen for Apple's Final Cut Pro™ editing software.

(Apple Inc.)

(Avid)

Objectives

After studying this chapter, you will be able to:

- Explain the processes of capturing and importing materials for digital editing.
- Understand the methods involved in building a program.
- Summarize how transitions and effects are used in a program.
- Identify the different tracks of audio in a program and recall some of the options for processing audio.
- Explain the options to consider when publishing a program.
- Recall the common criteria that should be identified when configuring a digital project, and how each affects the completed program.

CHAPTER

22 Digital Editing

About Digital Editing

Now that we have covered editing operations and principles, we are ready to apply both of them to the procedures of digital postproduction. It is here that we first encounter the problem that different editing programs look different and, to some extent, operate in different ways. To address the problem, this chapter focuses less on the menus, screens, and icons, than on the digital operations *behind* them—operations that are common to most editing programs.

The term *digital editing* refers, of course, to postproduction using video and audio stored in digital form and processed on a computer. Traditional "analog" editing used video and audio stored as recorded copies of electrical signals. Once the standard method for amateur and offline professional postproduction, analog editing has now been essentially supplanted by digital.

Postproduction Workflow

The typical workflow for digital editing involves several basic operations:

- Configure the project.
- Import or capture video and audio.
- Assemble, trim, and time the clips.
- Add transitions, titles, and effects.
- Build the multiple audio tracks.
- Archive the completed program.
- Export the program for publishing or distribution.

As we will see, some of these operations may be undertaken concurrently.

The rest of this chapter considers each of these operations in order, beginning with importing video and audio materials.

Though configuring the project is actually the first editing operation, this rather technical process has been saved for later discussion.

Capturing and Importing Material

The first step in building a new program is to bring the raw materials for it into the computer. How this is done depends on whether the video has been recorded continuously on tape, or laid down with a file name assigned to each separate shot.

Capturing Tape Footage

Until recently, all video was initially recorded on tape, and tape is still a common recording medium. When original video is on tape, it must be acquired by the editing system, divided into appropriate segments (usually clips) that are assigned individual file names, and stored on hard drives (**Figure 22-1**). Several considerations apply to capturing source files.

Device Control

In the process of capturing footage from a camera tape, ***device control*** is the ability to operate your source camera from your computer screen, without touching the actual controls, **Figure 22-2**. Device control is a great convenience in capturing footage because you do not have to work with two separate pieces of hardware at one time. Device control is useful for selectively copying clips and assigning them file names for storage.

As the popularity of file-based recording grows, some postproduction software no longer includes on-screen device control.

Individual or Batch

When capturing footage that is not file-based, most editing software allows you to copy

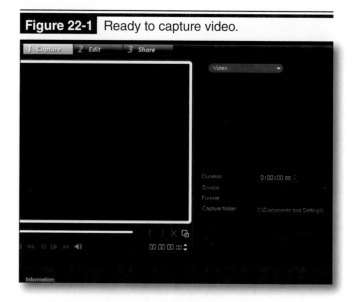

Figure 22-1 Ready to capture video.

The Advantages of Digital Postproduction

The obvious benefits of digital editing are the concentration of all postproduction on a single desktop, and the ability to perform operations like titles and effects that would otherwise be difficult or impossible. But beyond graphics and transitions, computer-based editing offers a whole range of advantages.

- **Random access.** Because your material is stored on a drive or disc, you can jump almost instantly to any point in it.
- **Flexible work order.** You can work on any section of your project at any time. And, you can return to previously edited material and revise it without disturbing the rest of the program.
- **Multiple operations.** You can work on one element at a time, such as adding transitions or laying audio effects.
- **Speed.** Random access, additive editing, and easy revision let you work more quickly.
- **Permanence.** Though digital recordings are not themselves permanent, they can be copied through many generations without appreciable quality loss.
- **Flexibility.** Digital recordings are easy to modify, so effects, composites, filters, titles, and other processes are simpler to work and revise. Digital audio is equally flexible.
- **Economy.** Because desktop computer systems can perform almost all the operations required for sophisticated postproduction, amateur, prosumer, and small professional operations can afford resources that were once available only to large, well-funded organizations.

- **Universality.** Whether standard- or high-definition, video is ultimately displayed in several incompatible formats, including NTSC, PAL, and SECAM. The three major high-definition formats (720p, 1080i, and 1080p) are customized for displays in different regions of the world. Video motion pictures are often recorded in still other, ultra high-quality formats. Internet streaming and computer video displays use several different protocols. With digital video, *transcoding* can translate footage in camera recording systems into the protocols used by the editing software. When postproduction is complete, the finished project can again be transcoded into formats for different systems required by any and all types of publishing and distribution.

Today, some high-definition cameras record in the protocol codec used by specific editing software. ("Codec" is the name coined for the software application that *codes* the video digitally for storage and playback.) Taking the opposite approach, some postproduction software can be set to work in the "native" format used by a particular type of camera. Both methods allow camera-to-editing workflow without the need for transcoding the recordings.

The advantages of "digital post", as it is called, are many, but they can be quickly summed up by these contrasting diagrams.

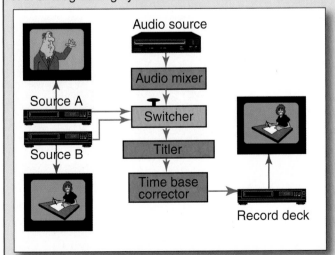

An analog editing system.

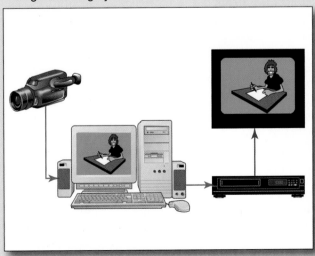

A digital editing system.

Figure 22-2 In some programs, the preview screen controls can be used for device control.

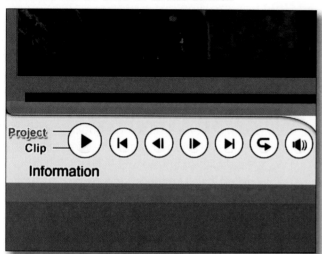

Project
Clip
Information

individual shots one at a time or to identify all the shots you want captured, and have them processed automatically after you have finished selecting them. Both methods have advantages:

- Individual capture allows you to examine each shot, select in- and out-points, capture the shot, and verify that the copy is free of dropped frames and artifacts.
- Batch capture (**Figure 22-3**) allows you to concentrate on picking the footage you want, without waiting for each one to be copied. This can make it easier to keep all the shots in your head as you make decisions about which ones to include.

Capturing versus Importing

The two terms *capturing* and *importing* can be confusing. When all video recording was on analog tape, capturing always meant bringing footage into the editing computer through a special capture card or a port (such as a 1394 port), and converting it to an appropriate digital codec before storing it on a hard drive. Today, however, most video does not require conversion to digital form, and it often arrives via USB port, hard drive, or the Internet. Acquiring footage by these routes is often called "importing" rather than "capturing." But, to avoid confusion, we save this word for the process of copying files into a project's library.

In the Corel® VideoStudio® Pro editing program, the options "Import from digital media" and "Import from mobile device" are included on the Capture screen.

As used in this book:
- *Capturing* means copying video and audio to a computer hard drive or server from an outside source, such as a tape, flash drive, camera hard drive, DVD, music CD, or an Internet site.
- *Importing* means *copying* video, audio, and/or still image files already on a computer drive *to the working directory* ("bin" or "library") of a particular editing project.

Capturing means bringing original footage into the editing computer.

Importing means copying a file into the library of a project.

Figure 22-3 A batch capture command center.

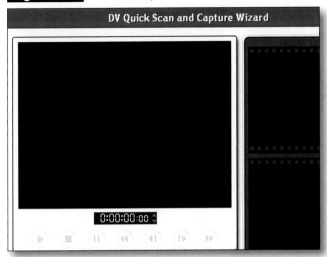

Figure 22-4 A clip is captured, trimmed, and filed as a video-only clip.

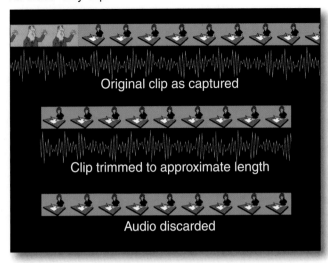

Original clip as captured

Clip trimmed to approximate length

Audio discarded

The *in-point* is the first frame of program to be captured, and the *out-point* is the last frame. A *dropped frame* is a frame that was not copied, usually because some components of your computer could not keep up with the speed of data transfer from the original recording media. An **artifact** is a small blemish on a digital image.

Most editors of tape-based footage employ both methods, often using batch capture for initial copying and individual capture for picking up other shots during the course of editing.

Many editing programs take batch capture a step further. If you select the option, the software will watch for evidence of change from one shot to another and capture the changed image as the first frame of a new shot. When the capture is complete, you can review the shots as transferred and make any needed adjustments.

Saving Tape-Based Material

Shots are saved to disc as part of the capturing process. To conserve space, you can then bring them up, trim unwanted footage, and re-save them under the same file names, **Figure 22-4**. But, it is faster to do some of the selecting and trimming as you capture. You can also save storage space by discarding unwanted components, such as unwanted camera takes or the production audio track when you want only the video.

Capturing File-Based Materials

As you can see, the capturing process was originally set up to separate long, continuous files recorded on tape, into separately identified shots. Today, however, many cameras record video and audio on "swappable" discs or flash memory cards, internal hard drives, external digital video recorders, or other systems. These methods usually treat each shot as a separate file, with a unique file name (**Figure 22-5**). Instead of separating and labeling the footage, as explained above, the editor need only copy the already-separated files to a directory on the editing system.

Figure 22-5 File-based shots.

P1090748.MOV	.mov	77,168,934
P1090749.MOV	.mov	85,570,270
P1090757.MOV	.mov	55,799,154
P1090758.MOV	.mov	48,158,170
P1090759.MOV	.mov	48,451,546
P1090766.MOV	.mov	52,233,450
P1090772.MOV	.mov	55,054,326
P1090773.MOV	.mov	43,997,898
P1090778.MOV	.mov	41,064,822
P1090779.MOV	.mov	48,551,454
P1090780.MOV	.mov	33,787,154
P1090781.MOV	.mov	20,196,494

Capturing Other Components

Most editing programs allow you to incorporate still photographs, graphics, and presentations from PowerPoint® and similar applications. Digital photographs can be uploaded from hard drives, discs, or still camera storage devices. Photographic negatives and prints can be digitized and uploaded from scanners. Graphics from sophisticated design programs can also be incorporated.

Once stored on an internal drive, video, audio, photos, graphics, and other materials can all be handled in much the same way. These materials become program resources that can be imported to a project and manipulated to fit its needs.

Of course, photos, graphics, and presentations are often created on the same computer system. So, they need only be imported to editing projects from the folders where they are already stored.

Managing Program Elements

Some editing software allows you to add descriptive comments about each shot and lets you organize your shots in a searchable database. It is a good idea to annotate your footage while it is still fresh in your mind. This will save time later, when you need to hunt for a particular shot.

Importing Elements into a Project

Once captured, all editing elements exist as files stored at various locations in your computer's filing system. To be used in editing a project, these files must be imported into it. Typically, you locate and copy the files into one or more holding areas called *libraries*, **Figure 22-6**. No matter how you trim, split, or discard elements in your project library, you will not affect the original files because you are working only with copies.

Building the Program

With the raw materials captured and imported into your project, the next step is to assemble the program, piece by piece.

Online or Offline

Historically, video postproduction has been designated "online" or "offline".

- *Online editing* always took place at a full-scale postproduction facility, either in a studio or at an independent operation called a *post house*.
- *Offline editing* was completed elsewhere and on simpler equipment, using lower-quality copies of the original high-quality materials. This process created a frame-accurate database record (called an edit decision list or EDL) of every element in the program for later use in making a high-quality final version online.

As used in editing, the term "online" has nothing to do with being connected to the Internet.

The offline/online editing workflow was typical because high-quality post houses charge steep hourly rates, and the many hours required to complete postproduction in their studios would be ruinously expensive. Instead, a fully edited "work print" of the program would be created inexpensively "offline" (perhaps in the program producer's offices) with professional, but much less expensive equipment. Then, the completed project, its edit decision list, and the original high-quality materials would be taken to the online facility, where a perfectly matched high-quality version could be quickly created through a semiautomated process.

Postproduction houses still play important roles in many professional productions. In recent years, however, the power of personal computers and the sophistication of editing software have grown so much that many highly-professional programs are completed without ever leaving the desktop.

Nevertheless, the concept behind online versus offline editing is still relevant, even on the desktop. Since high-definition files make heavy demands on computer resources and slow down the editing process, some desktop systems can work with "proxy files", which are lower quality copies of the high-definition originals. When a project is completely finished, the software can automatically render an identical full-quality version.

Figure 22-6 In VideoStudio® Pro, video files are organized in folders and displayed in the folder's library.

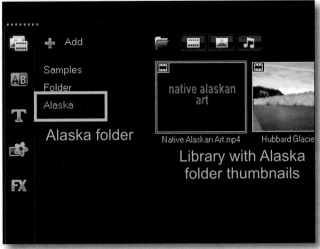

Alaska folder

Library with Alaska folder thumbnails

Native Alaskan Art.mp4   Hubbard Glacie

(Corel)

Assembly Editing

The simplest method to assemble a program is to select shots and place them on the storyboard or timeline in order. This is called *assembly editing*, **Figure 22-7**. The next job is to adjust each shot so that it contains exactly the right content and runs for exactly the desired time.

Shots and Clips

Though the term "shot" is quite clear when used in production, it can be misleading in postproduction. The original camera shot is

never changed. Once in the editing system, it is copied and re-copied as often as necessary, and *only the copies* are used in the video program. To distinguish these copies from the original shots, they are usually called *clips*—any piece of a shot (or shots) imported to the library for use in a program. You can create clips that use or reuse all or parts of shots, you can make clips out of pieces of several shots together, and you can modify clips to change color, speed, direction, length, and other attributes. Even after you have placed clips in your project's clip library, you can continue to divide them into multiple clips and manipulate each one indefinitely.

When you select a shot for the library, the program automatically copies it and imports the working copy as a clip labeled with the original file name. You can then rename the full clip and assign new names to clips you create by manipulating it, **Figure 22-8**.

Trimming Clips

Since you will rarely use all of an original shot, your first task is to trim it. Trimming means deciding exactly where the action in the clip will start and stop, and/or how long the clip will remain on screen.

One way to trim a clip is to display it in a window of its own, complete with timecode readout and movement controls (play/stop/single frame, etc.). By doing this, you can adjust

Figure 22-7 The work screen. (1) Program is in "edit" mode. (2) The "clips" library is selected. (3) The selected clips are displayed in storyboard style.

(Corel)

Figure 22-8 Renaming clips helps identify them quickly.

boat folks 1.mov

(Corel)

the length and content to your liking before placing the clip in the video program. After trimming the clip, you drag and drop it into position on the storyboard or timeline.

Some programs use separate preview windows for clips and projects. Others, display all preview views in one window by selecting either "project" or "clip" (**Figure 22-9**).

Another way to trim a clip is by adding it to the assembly immediately and making adjustments in place, **Figure 22-10**. By trimming on the timeline, you can see the clip in relation to the clip ahead of it. This is useful for establishing cut points and matching action. In practice, many editors use both methods at once, establishing the basic length and content of the clip in the preview, placing it in the assembly, and fine-tuning the trimming there. It is important to understand that the trimming of a clip is not final. You can always change the in/out-points to adjust the length or the content.

When you have every clip trimmed and placed in order, you have assembled the program (or program sequence).

Insert Editing

Once you have selected, trimmed, and placed clips on your timeline, you will generally fine-tune the results, using various forms of insert editing.

If you simply interrupt a program by inserting a new clip, the program's in- and out-points will be the same—the frame division at which you interrupted the flow with the added clip. But, often you will also adjust the existing program material immediately before and/or after the added clip. How you do this determines the type of edit you make. These different edit types include *three-point*, *four-point*, *rolling*, *ripple*, *slip*, and *slide*.

In the following discussion, the differences between edit types may be more easily understood by studying the accompanying graphics.

Figure 22-9 Using the preview window to edit a clip from the library.

(Corel)

Figure 22-10 Using the preview window to edit a clip from the timeline.

(Corel)

Three-Point Edits

As we have seen, every clip has an in-point (where it begins) and an out-point (where it ends). Think of these as points *1* and *2*. When you are inserting a clip into the middle of an existing program, you also have a *program in-point* (where the inserted clip will start) and a program out-point (where it will end). Think of the program in- and out-points as points *3* and *4*.

It often happens that you need to replace an exact range of material in the original program with a new clip in order to precisely maintain the same length. To specify that range, you specify three points:

- Program in-point
- Program out-point
- Replacement clip in-point

When you insert the replacement clip at the program in-point, your software will start it at the clip in-point, run it until it comes to the program out-point, and then set the clip out-point automatically. This is called a ***three-point edit***, **Figure 22-11**. Most frame-for-frame replacements are three-point edits.

Four-Point Edit

In some cases, you may have to fit a new clip that is longer or shorter than the program section that it replaces. In this situation, you specify four edit points:

- Program in-point
- Program out-point
- Insert in-point
- Insert out-point

When you make a ***four-point edit*** (**Figure 22-12**), you may choose either of two ways to match the clip length to the program spot: change the speed or change the length. You can change the speed of the clip to fit. If the speed difference between the old and new material is not great, the resulting slow or fast motion may be undetectable. But if you must stretch, say, a three-second insert to fill a six-second slot, the speed difference will be obvious. Alternatively, you can change the length of the insert clip to fit the target slot. When you do this, you essentially revert to a three-point edit, letting the software insert the clip until it runs out of space and then sets its out-point automatically.

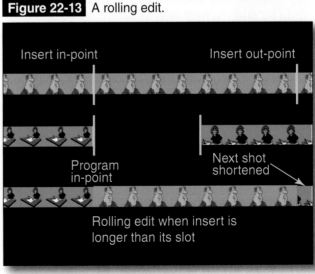

Figure 22-12   A four-point edit.

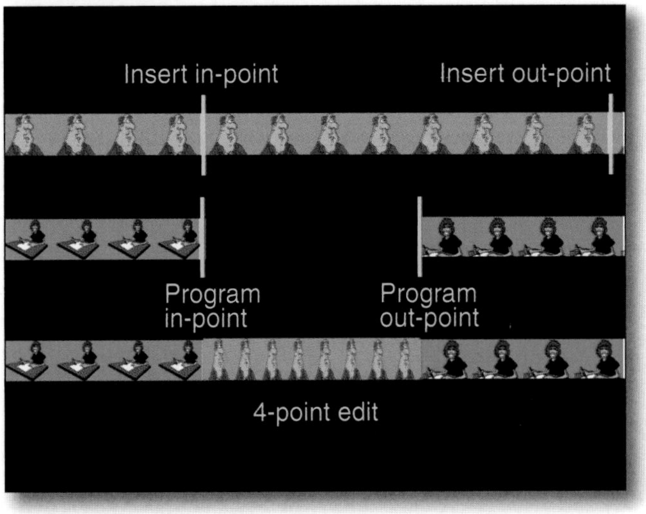

Three- and four-point edits are useful when you cannot change the length of the program or the running time of other clips in it. However, it is more common to insert clips while allowing the rest of the program to adjust to them.

Rolling Edits

One way to do this is with a ***rolling edit***, **Figure 22-13**. When you make a rolling edit, you retain the original program length. However, the clip following the insert adjusts to compensate. If the replacement is *shorter* than the original, the in-point of the following clip moves forward to fill the gap. But if the replacement clip is *longer*, the in-point of the following clip moves backward, to open up the necessary space. To make a rolling edit with a shorter replacement

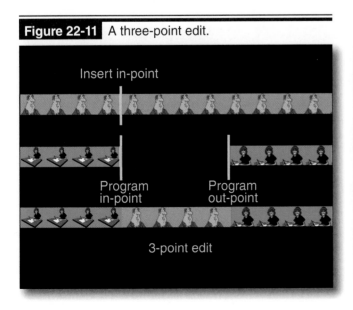

Figure 22-11   A three-point edit.

Figure 22-13   A rolling edit.

clip, the next clip must have previously trimmed material at its head. This is a good reason for digitizing and saving at least 5 extra seconds at the beginning and end of each clip you capture.

Like four-point edits, rolling edits are useful when a program, such as a commercial, must retain a precise overall length. When length is not critical, however, it may be easier to perform a ripple edit instead.

Ripple Edits

The most common type of insert is the ***ripple edit***, **Figure 22-14**. A ripple edit is a three-point edit in which all the rest of the program following the insert adjusts automatically to fit it. If the insert is *shorter*, all the following material closes the gap by moving up to the end of the new clip. But if the insert is *longer*, all the following material makes room for it by moving back.

If you do not edit fixed-length programs (such as commercials), you may use the easy ripple edit technique almost exclusively.

Slip and Slide Edits

If you wish to adjust the program on both sides of the new material, you can select either a slip edit or a slide edit. A ***slip edit*** preserves the in- and out-points of the preceding and following clips by shifting the in- and out-points of the source clip, **Figure 22-15**. This allows you to fine-tune the part of the new clip inserted into the program.

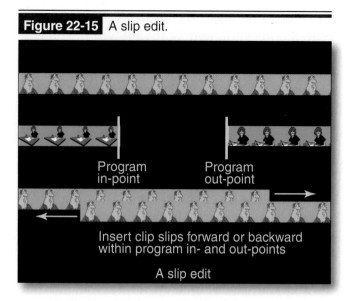

Figure 22-15 A slip edit.

Program in-point Program out-point

Insert clip slips forward or backward within program in- and out-points

A slip edit

A ***slide edit*** preserves the in- and out-points of the new clip by shortening the clips on both sides, **Figure 22-16**. Generally, you can slide the new clip back and forth to decide how much of the previous and/or following clips to shorten. This allows you to change the length of the source clip without affecting the length of the program.

Editing Tools and Options

Three-point, four-point, rolling, ripple, slip, slide—these options can seem confusing at first. And, professional editing software offers many more options as well, including other forms of edits, different tools for making the actual edits, different windows in which

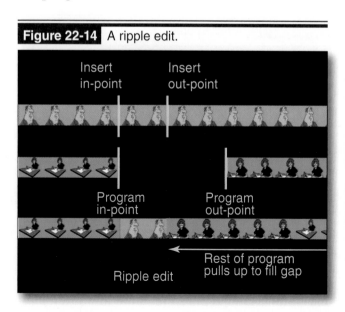

Figure 22-14 A ripple edit.

Insert in-point Insert out-point

Program in-point Program out-point

Rest of program pulls up to fill gap

Ripple edit

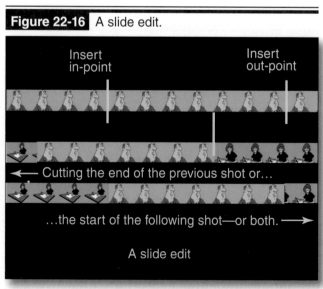

Figure 22-16 A slide edit.

Insert in-point Insert out-point

◄— Cutting the end of the previous shot or…

…the start of the following shot—or both. —►

A slide edit

to adjust clips, and a choice of keyboard or mouse-based commands for many editing functions. In practice, most editors gradually evolve personal working methods to suit their own needs, utilizing the features they prefer and ignoring the others. In other words, you need not master and use every option in a sophisticated postproduction application. Instead, use new features as you need them, and then decide whether to continue with them or not.

Adding Transitions and Other Effects

Program material in digital form is easy to adjust, modify, or change into something quite different. There are hundreds, perhaps thousands, of ways to transform video originals. Each editing application comes with its own proprietary set of effects. Plug-ins can be added to the original package to extend its capabilities even further.

Plug-ins are programs that become part of your editing software. They do not have to be opened separately because their commands and other features are automatically displayed in the menus of your application.

You can also export footage to completely separate software applications, work on it, and then import it back into your project.

With so many options available, it would be impractical to discuss every one. So, this section covers the general types of image effects that you are likely to encounter.

Transitions

The most common effects are *transitions* from one sequence to another, **Figure 22-17**. Traditional transitions, such as fades, dissolves, and wipes, are supplemented by almost numberless two- and three-dimensional transitions made possible by digital video effects (DVEs). See **Figure 22-18**.

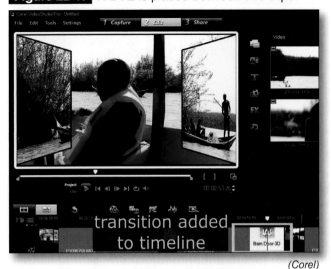

Figure 22-17 A DVE is placed between two clips.

(Corel)

Strictly speaking, fades, dissolves, and wipes are also digital effects. But, the term "DVE" is typically used for the more elaborate transitions and effects developed for digital editing.

Superimpositions

Superimpositions are combinations of two or more images displayed together, **Figure 22-19**. The most common superimposition is the double exposure, in which two full-screen images are visible at once.

Compositing

Compositing is the process of placing part of one image (usually the subject or other foreground object) into a different background. This is commonly done by recording the foreground against a background of a very specific and very uniform solid color. Then the software replaces every pixel of this key color with the corresponding pixel from the background image. See **Figure 22-20**.

"Compositing" is a generic term. Professionals who come from the television industry often call the process "chroma keying," and some video editing programs also use this term. People from the film industry tend to speak of "blue screen" or "green screen."

Figure 22-18 In this "barn door" transition, the first image swings open in two halves to reveal the second.

(Corel)

Figure 22-19 A superimposition.

(Corel)

Titles and Graphics

All editing software includes titling capability, **Figure 22-21**. For more elaborate titles and effects, you can also export material to a separate application, such as a painting and photo program. There, you work one frame at a time, adding and animating elements to your original clip. When you have finished, you import the result back into your project. Since full-motion video requires 30 frames per second, working frame-by-frame can be very time consuming. However, many photo/paint applications can automate parts of the process—copying the changes made in one frame to the next frame, where they can be further modified as the sequence progresses.

Image Processing

You can also change the basic nature of your visuals to suit the requirements of your project.

Speed. You can create slow motion and fast motion clips.

Direction. You can reverse the direction of the action by "flipping" clips left to right. If, for instance, a left-to-right clip violates a general right-to-left screen direction, you can correct it by flipping it digitally.

Composition. You can zoom in to fill the screen with just part of the original composition. This is useful to add emphasis, redesign clip compositions, or eliminate unwanted elements, like microphones.

Figure 22-20 Compositing images.

The subject in front of a green screen.

A background image.

The two elements composited together.

Photographic characteristics. You can change the brightness, contrast, and color balance of your original footage in order to correct errors in shooting, to match surrounding clips, or to create special effects, like a sunset or a moonlit night. You can also turn color footage to black and white, or give it an old-fashioned sepia tone.

Special effects. Most applications include special effects. Though their number is almost limitless, a representative example is "duotone", which changes an image from full-color to a pair of selectable colors, **Figure 22-22**.

Designing the Audio

When assembling a program, you typically place the production sound for each clip with its video. But, that is only the start of audio editing, which includes production sound, background and effects, narration, and music (**Figure 22-23**).

Figure 22-21 A title over the background animation.

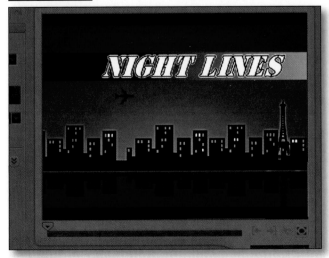

Figure 22-22 Duotone special effect.

Original clip.

Clip processed with a "duotone" video filter.

(Corel)

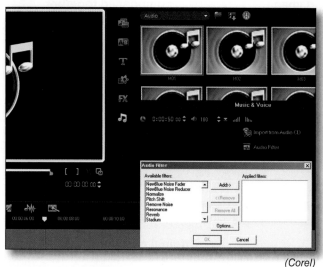

Figure 22-23 An audio editing menu. The speaker icons in the library window above it represent sound clips ready for use.

(Corel)

Digital Audio

Digital audio editing offers so much creative potential, that its major advantages are worth considering in detail.

Multiple Tracks

To begin with, you can have as many audio tracks as you want—up to the limit of your editing software. These tracks can be mono, stereo, or a mixture, depending on the sound sources used. With so many tracks to work with, you can create audio as complex as that of many Hollywood films right on your desktop.

Timeline Control

Working with a timeline, you can see all your audio elements in front of you, where you can select them, audition them, move them, and modify them with mouse or keyboard commands.

"Visible" Sound

You can see each track as a waveform: a graphic representation of the sound's variations in loudness (**Figure 22-24**). With this tool, you can make extremely precise edits and time music passages exactly.

Synchronization

You can unlock a sound track from its video and move it freely. You can synchronize two different video and audio tracks, and then lock them together so that they become a single editing clip. In short, digital audio editing is so powerful that when you are working on a modest desktop computer, you have more creative control than you could find in even the most sophisticated analog sound studio. Editing software applications vary in their audio capabilities, but they all include the basic functions of sound *capture*, *editing*, and *processing*.

Because production sound tracks are typically edited with their video tracks, they do not normally appear separately on the timeline. To edit them, you use the Split Audio command and assign them separate rows on the timeline.

Capturing Audio

Digital sound editing can be so flexible because you work from audio files on your hard drive, **Figure 22-25**. So, the first step is to get those files onto your hard drive in a form that your editing software can use.

Production sound tracks are recorded with video and transferred to your computer as part of the video capture process. If you want only the audio part, you can save it separately and discard the video.

Figure 22-24 A waveform helps make precise edits.

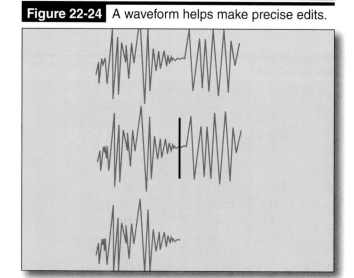

Figure 22-25 Audio files can be imported to the project library. Here, Track02 and Track03 have been imported, and Track03 has been renamed for easier identification. The highlighted musical note icon indicates that the audio library window has been selected.

(Corel)

Audio Editing

With your sound components in the audio library of your project, you are ready to build a sound track.

Production Sound

The first step in sound editing is to assemble all the production audio. This happens automatically as you select, place, and trim video clips. With the basic production tracks in place, you can modify every one to suit your needs: delete it, replace it with another take (usually for better sound quality), or substitute audio from other clips. If there is a separate production track of ambient (background) sounds, you can lay that under the scene, adjusting its volume to compensate for clip-to-clip differences created by different miking positions.

In the example used in previous figures, the sounds of the boat's outboard motor changed drastically from shot to shot, and random off-screen talking marred the audio. To solve the problem, a long take was found with "clean" audio of the motor sound. This audio track was repeatedly cloned and re-used to replace all the other audio in the sequence.

A Simple Audio Production Design

Although editing software may offer close to 100 separate audio tracks, you can achieve very professional results with 12 tracks or less.

To demonstrate, here is a simple audio design for a dramatic program.

Track	Content
1	Production soundtrack A
2	Production soundtrack B
3	Background A
4	Background B
5	Effects A
6	Effects B
7	Music A
8	Music B
9	Misc. A
10	Misc. B

As you study this layout, notice that all components have at least two tracks (A and B) so that sounds can be cross faded between them. A pair of miscellaneous tracks can be used occasionally for dialogue, effects, or music, at places where the regular tracks already have sound in them.

Sound Effects

With the production tracks complete, the next step is to add sound effects. These may have been recorded as part of the production, copied from sound effects discs, or recorded in a Foley studio.

Music

Generally, music is the last set of tracks to be laid down. Blended together, the production, background, and effects tracks will sound like a single "live" sound track. Music, however, is perceived by the audience as a separate set of sounds. For this reason, it is often helpful to create the composite live sound track and then balance it with the music.

Transitions

To make transitions from one sequence to another, or simply from one sound to the next, you can blend audio elements together with fades, cross fades, and split edits (**Figure 22-26**). Most editing software allows you to fade sound volume up or down by dragging a volume bar on the timeline.

Fade

A fade in or out makes a gradual transition. A fast fade in will prevent a sound from seeming sudden and abrupt. A very long, slow fade in can bring up a sound so unobtrusively that the audience is not aware of it. Fade outs are especially useful for dispensing with music that does not have a clean ending. A three- to eight-second fade out will generally deliver a graceful exit.

Cross Fade

A cross fade overlaps the fade out of one sound with the fade in of another sound. Cross fades are often employed to match dissolves on the video track, but you can also use them to blend from one sound to another so that the audience does not notice it.

Figure 22-26 Editing audio. (1) The wave form permits precise edits. (2) The volume level can be adjusted by clicking and dragging. (3) The edit function is selected by the Sound Mixer icon.

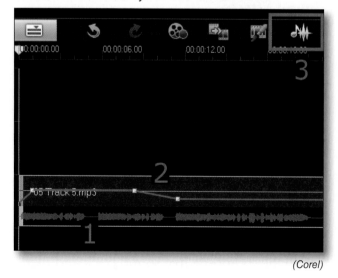

(Corel)

Split Edit

As explained elsewhere, a split edit is a transition in which picture and sound do not change at the same time. When *video leads*, the sound from the old clip continues over the video of the new clip. When *audio leads*, the audio from the new clip begins before the picture, giving the audience a brief foretaste of what is to come. Split edits are especially easy with timeline editing because you can see the overlaps exactly.

Audio Processing

Fading volume is only the simplest of many ways to manipulate the quality of individual audio tracks. Every editing application has its own set of tools and plug-in products are available, as well. In addition, you can process audio with either the software tools provided with computer sound cards or computer-based recording studios that are sold separately. Here are just a few of the many ways to process audio tracks.

Stereo Imaging

If you are working in stereo, you can place a sound anywhere on a line from left to right or have it move from one side of the sound image to the other, **Figure 22-27**. In addition, many programs allow you to create "surround sound" tracks of various kinds.

Figure 22-27 Audio mixing area with track selector icons (1), volume slider control (2), and stereo imaging slider (3).

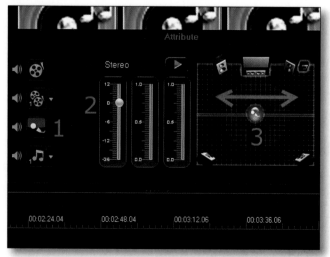

(Corel)

Sweetening

You can clean up audio by adjusting treble and bass, suppressing some sounds, compressing or limiting range of loudness, and eliminating hiss or hum.

Distorting

You can change the character of sound by adding reverberation, changing pitch, or applying one or more of the many sound filters supplied with your editor or others that are available separately.

Storing and Publishing

When you have completely finished your program, you need to preserve it permanently. In order to distribute your program, you also need to transcode it to a format that can be displayed on a theater screen, a computer monitor, a TV set, an Internet site, or any combination of these. In other words, you need to *store* your program and then *publish* it.

Storing your program means archiving it in a high-quality format. This might be the original camera format, such as mini-DV or AVCHD. Or, it might be another codec that is optimized for long-term storage. This stored program is the master from which other copies are made for showing.

Corel® VideoStudio® Pro, discussed in the next chapter, stores finished programs in a proprietary format that allows you to reopen them later for revisions.

Publishing your program involves converting it into the format appropriate for the intended display medium and, often, creating support materials for presenting it. For instance, to show your program on TV or on a computer monitor, you would probably produce a DVD version of it. If you will publish your program on the Internet, you will convert it to the codec preferred by the video website(s) you have chosen, and then upload your program.

Some editing software offers built-in functions for converting and publishing programs on widely-viewed websites and personal electronic devices.

Although storing and publishing may not seem like glamorous activities, they are so important that they deserve closer attention, **Figure 22-28**.

Storing Finished Projects

Whenever you save a word processing or spreadsheet document, the file you store is a full version of the work you are creating. It is important to understand, however, that this is not always true of video programs. When you save your work during the editing process, you are saving not a **program**, but a **project**.

A *project* is a record of everything you have done to create the program. A project is like a complete set of blueprints for a complex building, plus a detailed scale model; but it is not the building itself.

A *program* is the actual video, which can be played back at full quality on some form of standard display, or be transcoded for publishing. It is the completed building described in the blueprints.

Project Files

Project files usually do not contain full-quality versions of the clips, graphics, and transitions themselves, but only lower-quality versions. It is a good idea to save your projects frequently for protection, **Figure 22-29**.

Figure 22-28 Ready for storing and publishing.

(Corel)

Figure 22-29 VideoStudio® Pro projects are stored as ".vsp" files.

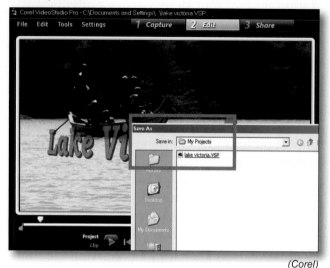

(Corel)

Because of their complexity and their demands on computer resources, video editing programs can crash more often than other kinds of software. This inconvenience makes frequent project saving even more important.

In addition, each project includes a detailed "blueprint" for creating a high-quality program identical to the working version.

Project files are not easily transported from one computer to another because all the camera shots and other original materials for the finished program are in other files that remain behind. Without these originals, the project blueprint cannot construct a full-quality version of the program.

Program Files

Program files are finished videos. They consume larger amounts of disc space because, unlike project files, they do contain every frame of video and audio in full-quality versions.

Rendering Program Files

Your editing software creates a finished program file by automatically assembling it at full quality, following the frame-by-frame blueprint in the project file. This assembly process is called *rendering*. Powerful editing programs running on high-speed computers may render in real time. That is, they create and recreate the program at full-quality as you work

on the project. This means that you can see full-quality results at any time, you can catch any defects in the rendering process and correct them on the spot, and you do not need to wait while the software builds your final program.

More modest setups may require you to render your program by batch processing, when you have finished editing. With this approach, you instruct the system to assemble the program in its finished form and then wait a considerable time while it does the job.

Most editing programs include trial versions of transitions, effects, titles, etc., in the working version, so that you can review them as you work.

Aside from the inconvenience of long rendering times, the other drawback of batch rendering is quality control. If the final program contains dropped frames, artifacts, or other imperfections, you will have to repeat the whole process.

Rendering versus Archiving

Rendering gives you a final program file stored on a hard drive. But in most cases, you do not want to leave it there, tying up many gigabytes of hard drive space. Instead, you want to *archive* the final program file by storing it permanently in a place where you can preserve it, **Figure 22-30**. Typically, you archive a program by exporting it to a disc recorder, or to tape in a camera or other recorder. In professional setups, you may archive to large servers or special-purpose, high-quality tape.

Figure 22-30 Rendering and archiving.

Archiving

Rendering

Both discs and tapes will decay with time, and digital servers used for long-term storage can fail. For this reason, all programs should have at least one backup copy. (Some high-level professional productions are archived on special types of film, because, to date, film has the best track record for longevity.)

Formats for Archiving

If the program may be distributed in multiple formats (or if it might be revised or re-edited later), it is best to keep it in its original digital format or in the special archiving format provided by your software.

With high-capacity, external hard drives now so affordable, some editors of professional productions archive a project by copying its original video and audio sources, its project files, and its fully rendered final version to a dedicated drive, which is then unplugged and stored for safety.

Publication and Display

With the growth of broadband Internet connections, it has become possible to view video programs on personal computers and personal electronic devices. This means that video programs must be converted into digital formats suitable for streaming or downloading.

For TV display, DVDs and Blu-ray Discs™ are the standard media for disc recording in regular and high-definition versions. Increasingly, standalone (set-top box) and computer-based digital video recorders (DVRs) are used for storage and TV playback.

With the variety of formats and display systems now in use, preparing programs for distribution has become a whole separate operation.

Varying the Task Order

For clarity, we have discussed each editing operation in order—as if you always capture and import all of your raw materials, then assemble all of your clips, add all of your effects, etc. However, one great advantage of digital postproduction lies in your freedom to work on any part of your program at any time,

and to perform operations on clips, sequences, or the entire program—or in any combination that you find convenient.

Sequence-Based Editing

Many editors prefer to build one sequence at a time—usually, action that happens at one point in the program continuity. Using this approach, you would:

- Capture the footage for this sequence.
- Catalog, organize, and import shots for use.
- Assemble the clips in the sequence, trimming each to approximate length.
- Fine-tune the trimming of all the clips in the sequence together.
- Replace and/or add clips by inserting them into the sequence.
- Process the production audio.
- Lay background and sound effects audio.
- Correct clip-to-clip luminance, contrast, and color balance.
- Apply special overlays.

In effect, you would complete almost all the editing work on each sequence before going on to the next.

Multiple Pass Operations

Some operations are better performed on entire programs, or at least on segments that are several sequences long. Scene-to-scene transitions, for example, are easier to keep consistent when you add them all at once, **Figure 22-31**.

Music may be easier to evaluate when heard under several sequences together. Musical choices that seem fine individually, may not work as a group or may feel excessive when heard consecutively.

Successive Approximation

The biggest single advance in digital postproduction is the ability to revisit any part of a program and improve it. In old-style film editing, even changing in- and out-points was tedious and time-consuming, and you could never see the result of adjusting scene transitions because they had to be made later in a laboratory.

Figure 22-31 Three dissolves were added after the scenes were completed.

(Corel)

In 16mm film editing, removing two frames to adjust a cut point required the editor to identify and keep track of a piece of film less than one inch long!

In digital postproduction, you are not working with your video and audio originals, but only with copies of them. And even when you instruct your software to render the finished program, the original materials remain untouched. To take full advantage of this major improvement in post-production procedures, be prepared to re-work parts of your program until you are fully satisfied.

Configuring Digital Projects

Before you can begin digital editing, you must tell the computer the precise electronic characteristics of the program you will work on. That is, you must *configure* the project.

As noted previously, though configuring is the first editing operation, it has been saved for last because it is a rather technical and because it is not always necessary. If all your projects use the same configuration, you can set it once and then leave it as the default setting, **Figure 22-32**.

Most editing software is supplied with default values that match the most common program settings. So, it is possible to begin working with these applications without configuring projects. To take full charge of your programs, however, you need to understand the variables in these parameters and how to set

them. The following sections provide general descriptions and explanations of the many factors involved in configuring a project, starting with the program's basic format(s).

Video Formats

The format of a video is the set of electronic protocols (usually called a *codec*) used to record and play it. As we have seen, you may start with source material in one format, process it in your computer in a second format, and export it for storage or viewing in yet a third format. In many setups, the format is the same for both source and destination. For instance, you may capture .AVI footage from your camera, process it in that format on your computer, and export it back to your camera to store your edited program on a blank tape.

In other cases, you may be working with as many as three different formats. For example, if you are editing footage for streaming to the World Wide Web, your source may be analog footage, your editor will work in its native digital format, and your destination may require transcoding to yet another digital protocol.

Even within basic formats, there are choices to be made: Full-screen image or partial? Smooth motion or economical 15 frame per second movement? The list goes on through many more such decisions. For this reason, you need to understand the major project parameters and set them systematically. The rest of this section discusses each one.

Figure 22-32 Project configuration menu under "Settings."

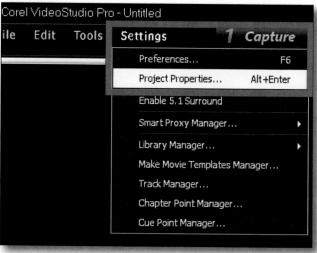

(Corel)

Time and Frame Measurement

Video (and film) systems do not record continuously. Instead, they divide each second of time into several equal parts and record a picture of each part. Displayed in real time, these separate pictures give the illusion of a continuous, moving image. There are several common ways to record video images, and in every one, the first recording parameter is *frame rate*.

Frame Rate

The *frame rate* is the number of individual images (frames) recorded each and every second. It is important to know the frame rate of both your source material and your eventual storage medium, so that your software can process the source material and prepare a compatible final project.

Here are some of the principal frame rates:

- NTSC and ATSC television: 30 fps (the North American standard). The actual frame rate of NTSC video is 29.97 fps, but it is informally referred to as "30 fps."
- PAL and SECAM television: 25 fps
- Computer full motion: 30 fps
- Computer limited motion: 15 fps (10 fps is a less-common alternative)
- Sound film: 24 fps

Since the conversion to digital broadcasting and cable, the NTSC system has been replaced by the ATSC system (A for "Advanced"). To further complicate matters, the "NTSC" designation is still used on North American DVDs (**Figure 22-33**). Standard-definition European PAL and SECAM material can also be displayed by ATSC systems, although Europe and much of the rest of the world now use their own DVB-T digital standard.

Time Base

Time base refers to the way in which your system calculates time. In most formats, the time base is identical to the frame rate. But, the NTSC format used in North America is an exception. Though we speak of 30 fps for convenience, the true NTSC frame rate is actually 29.97 fps.

Figure 22-33 Box label of a DVD in NTSC format, for sale in region 1 (North America).

The change from a true 30 fps to 29.97 was made for technical reasons, when color television broadcasting was developed.

If your computer uses a true 30 fps as the time base, even the tiny difference of 0.03 frames per second will gradually make the frame count increasingly inaccurate. For this reason, it is preferable to set your editing software to a 29.97 fps time base when you are editing projects intended for TV display.

Timecode

The way in which frames are labeled is called *timecode*. Timecode provides a unique address for every frame of video in your program. The most common timecode system is SMPTE, which labels each frame of picture and/or sound with a unique identifier. This address consists of the number of hours, minutes, seconds, and frames elapsed since the start of the timecode (**Figure 22-34**). So, the timecode address 00:37:22:09 labels a frame that is 37 minutes, 22 seconds, and nine frames from zero. The leading "00" indicates that no hours have elapsed.

SMPTE stands for the Society of Motion Picture and Television Engineers, and is pronounced "SIMP-tee."

Timecode is essential for accurate editing because it allows you to specify the exact frame on which you want every operation to start or end.

Understanding Timecode

The statement that "each frame is assigned a permanent timecode address" applies only to a single recording of that frame (and the other frames around it). This means that when you copy a clip, you can sometimes copy its timecode address with it. But in other cases, your system may assign the material a new set of timecodes.

In working with a project, you may be dealing with three separate sets of timecode:

- *Source timecode:* Each clip you import will have timecode addresses created in the camera.
- *Clip timecode:* In some editing systems, each separate piece of captured material has its own timecode.
- *Project timecode:* Your software will timecode your project with a single set of continuous addresses.

Drop-Frame and Nondrop-Frame Timecode

Timecode cannot label an interval shorter than a whole frame. So, NTSC timecode counts frames as if they were displayed at 30 fps, rather than the actual rate of 29.97 fps. If timecode assigned an address to every frame, the count would very slowly grow inaccurate. After ten minutes, the difference between the timecode readout and the actual elapsed time would be over half a second—a huge difference where synchronizing picture and sound is concerned.

To solve the problem, "drop-frame" timecode skips (drops) two frame numbers at the beginning of each new minute. (Since this method over-compensates just slightly, the two frame numbers are not dropped after every tenth minute.)

It is important to understand that "drop-frame" timecode does not actually drop any frames, it simply omits numbers. In other words, between timecode addresses ending in "00:29" and "01:02" there are numbers missing, but not pictures.

To continue the address analogy, if two houses side-by-side on Elm Street are 305 Elm and 309 Elm, it does not mean that a house between them has been torn down, but only that the number 307 has been skipped in assigning house addresses.

"Drop-frame" timecode does not really drop any frames, but only address numbers.

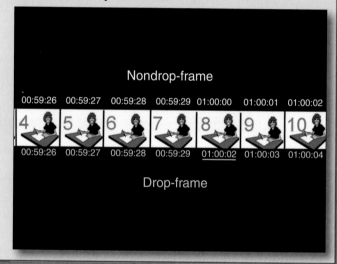

Keeping Them Straight

To understand the differences between frame rate, time base, and timecode, remember the following:

- *Frame rate* is the actual number of frames per second recorded by a timeline or created in the editing process.
- *Time base* is number of units per second with which editing software calculates all its operations. It may or may not be the same as frame rate.

- *Timecode* is the numbering system used to assign a unique address label to every frame. ("Drop-frame" timecode is used to prevent errors in applying 30 fps timecode to a 29.97 fps frame rate.)

Video Scanning System

Each video image (frame) is divided into several hundred horizontal lines, which are "painted" on a digital TV set or computer display. This process is called *scanning*. There are two approaches to video scanning: interlaced and progressive.

Figure 22-34 The parts of a timecode address.

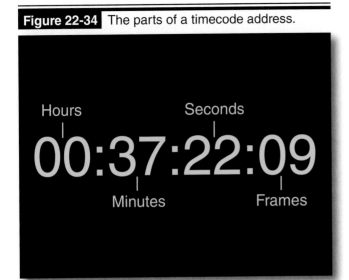

Interlaced Video

With interlaced video, every frame of video is made up of two *fields*, one showing only the odd-numbered lines and the other only the even-numbered lines. Since each of the two fields is scanned in 1/60 second, the complete frame is displayed for 1/30 second (in the NTSC standard), with the two fields interlaced (mixed together—a/b/a/b/a/b), **Figure 22-35**.

Interlaced video was developed in the early days of television to work around technical limitations that have since been overcome. But, it remains the system used for all standard-definition and some high-definition TVs.

Figure 22-35 Interlaced (2-field) video. (Scan lines exaggerated for clarity.)

First field

Second field

Progressive Scan Video

In a progressive scan video system, every frame line is shown in one pass—there are no separate fields (groups of "a" and "b" lines), **Figure 22-36**. Many digital video systems use progressive scanning.

In digital postproduction, you may need to convert between interlaced and progressive scan video. An interlaced frame must be "de-interlaced" into a single scan frame in order to create a full-quality still image. A progressive scan frame must be broken down into two interlaced fields for export to a conventional video display standard, such as NTSC.

Although most North American high-definition screens now show 1080p images (progressively scanned lines), broadcasters transmit in 720p or 1080i formats because they require less bandwidth.

Picture Quality

The apparent quality of a displayed image depends on three factors:
- *Resolution*: The perceived sharpness and amount of detail.
- *Color fidelity*: The accuracy and subtlety of the colors.
- *Integrity*: The freedom from "artifacts", such as spots or areas with a blocky, mosaic appearance.

Of these factors, *resolution* is perhaps the most important.

Figure 22-36 Progressive scan video.

Resolution

Resolution is the ability of a video system to resolve (distinguish) fine detail. In digital editing, you can set the resolution of each new project. If you are working with digital source material, it is often best to set the project resolution to match. So, if you are using standard-definition footage with a 640 × 480 resolution, the project resolution will also be 640 × 480. High-definition original material should usually be edited at its native resolution.

To compensate for small losses during the editing process, it may be advisable to set the digital project resolution one level higher than the minimum needed to capture detail in the originals (unless, of course, your originals are 1080p).

Color Fidelity

Color fidelity means the accuracy of the colors in the picture; this quality depends heavily on bit depth. **Bit depth** refers to the length of the string of code numbers representing each picture pixel. The more bits recorded per pixel, the greater the amount of information about it. Typical color depths are 8-bit, 16-bit, and 24-bit. Consumer cameras can record at least 16-bit color, and most editing software works in at least 24-bit.

Image Integrity

Because of the tremendous amounts of data involved, digital video is usually compressed.

One popular camera for feature movie production does not compress data for recording.

Digital compression can be complicated because there are at least three variables to consider:

- *Compression type.* Some systems, like the common **M-JPEG**, reduce the data for each individual frame. Others, such as **MPEG**, record all the information for a "base frame", but on the frames that follow it, only the pixels that have changed since the base frame are recorded.

Some standards have grown into whole families. There are now several varieties of MPEG codecs.

- *Compression algorithm.* Each type of compression is available in more than one variety, with designations like "4-2-2" and "4-4-4".
- *Compression level.* Compression systems allow you to set the severity of the compression, from slight to vigorous. The milder the compression, the better the picture quality—and the bigger the file size, as well.

Usually, the compression type and particular algorithm are determined by the codecs in your camera and computer. However, you can often set the level of compression for each project.

Aspect Ratio

After frame rate, scanning system, and image quality, the last picture parameter to set is **aspect ratio**—the relationship of the frame's width to its height. The traditional aspect ratio of video is 4 to 3: four units of width to three units of height, **Figure 22-37**.

The original 4 to 3 ratio matches the aspect ratio of all movies at the time broadcast TV was developed. Since then, films have been shot in many different aspect ratios, as wide as 2.39 to 1 or (for comparison purposes) 7.17 to 3.

Figure 22-37 A traditional 4 to 3 monitor.

(JVC)

Nowadays, however, TV is also displayed in an aspect ratio of 16 to 9, **Figure 22-38**. This proportion is close to the shape of most theatrical movies. Many digital cameras can be set to shoot 16 to 9, and some editing software can modify traditional 4 to 3 video to the wider 16 to 9 aspect ratio. As high-definition cameras and TV sets have become standard, the 16 to 9 aspect ratio has replaced the older screen shape. However, many Internet sites still use the traditional 4 to 3 aspect ratio display.

Audio Quality

In addition to setting project parameters for video, you can also select parameters for your sound tracks.

Sample Rate

All digital recording is created by sampling the signal and recording the result thousands of times per second. The rate at which this process occurs is the *sample rate*. CD audio is sampled at a rate of 44.1 kHz (kilohertz). This means that the signal is sampled over 44,000 times per second. Audio for the Internet or CD ROM may use a 22 KHz sample rate, and there are other sample rates between the two. (A sample rate lower than 22 KHz is not generally recommended.)

Figure 22-38 A 16 x 9 monitor.

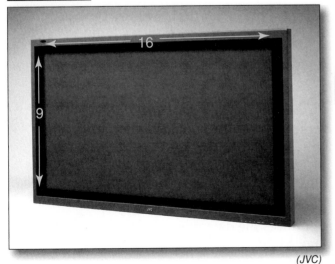

(JVC)

Bit Depth

As with video, bit depth for audio this is the number of digits (bits) used to record each sample of the audio. 16-bit audio is CD quality. Selecting a lower sample rate and bit depth will reduce the size of your audio files, but the results will be somewhat lower quality.

Matching Video

If you set your video at 30 fps for editing, but then play it back at 29.97 frames for NTSC recording, your audio may grow slowly out of synchronization because it was captured to match the slightly faster 30 fps. For this reason, it is best to standardize all operations at 29.97 fps throughout the capture/edit/export process. That way, your audio will remain in sync with the picture.

It is possible to re-sync 30 fps audio with 29.97 fps video by slowing the audio playback sample rate from 44.1 kHz to 44.056 kHz.

Working for NTSC storage and playback, you can, as a general rule, safely set a video frame rate of 29.97 and an audio sample rate of 44.1 kHz at 16 bits.

Unlike video, which is recorded in separate frames, audio is recorded continuously. For convenience, however, timecode addresses are used for audio, as well.

Though sizeable, audio files are smaller than video files. So, you may not significantly reduce overall project size by skimping on audio quality.

A Lot of Standards

At this point, you may feel that all these technical standards are a confusing nuisance. Keep in mind, however, that they must be set only when changing program specifications. If you work entirely in one format—say, making programs from mini DV footage for DVD distribution—you can establish your standards once and leave them that way.

Summary

- Among the many benefits of digital postproduction are speed, flexibility, and universality.
- Capturing original footage is the first step in building a new program.
- Two basic methods used in building a program are assembly editing and insert editing.
- The most common effects are transitions from one sequence to another.
- Audio editing includes production sound, background and effects, narration, and music.
- Publishing a program involves converting it into the format appropriate for the intended display medium (Internet, TV set, theater screen, etc.).
- A *project* is a record of everything you have done to create a *program*.
- Archiving the final program file stores it permanently in a place where it can be preserved.
- To take full charge of editing programs, you need to understand the variables available in program settings and how to set them. Program configuration settings are available for both video and audio.

Technical Terms

Archiving: The process of assembling all the materials used to create a finished program and placing them together in permanent storage.

Artifact: A spot or other blemish on a digital image.

Aspect ratio: The relationship of an image's width to its height; traditional aspect ratio of video is 4 to 3, high-definition video formats are typically 16 to 9.

Assembly editing: The process of building a program by selecting shots and placing them in order on the timeline.

Bit depth: The number of digits (bits) used to encode each piece of digital data. 8-bit, 16-bit, and 24-bit are common bit depths.

Capturing: To bring program material that has been recorded without file names into a computer from an external source.

Clip: A unit of program material in a digital editing project. A clip may contain anything from part of one clip to several clips together.

Device control: The ability of software to operate external hardware (such as a video camera) by remote control.

Four-point edit: An insert edit that specifies the in- and out-points of both the source clip and the place in the program at which the clip will be inserted.

Frame rate: The actual speed with which video frames are displayed. The frame rate of NTSC video is 29.97 frames per second.

Importing: Adding a file stored on the editing computer to the library of a project.

Library: In editing software programs, a holding area for video, audio, and other elements to be used in a project.

M-JPEG: A video compression codec that reduces the data required to record each frame. Common formats include .AVI and .MOV.

MPEG: A video compression system that records all the data of a "base frame". For subsequent frames, the system records only data that differs from that of the base frame.

Plug-in: A special-purpose application that, when installed on an editing computer, is accessed directly from within the editing software.

Post house: Short for "postproduction house"; a professional facility that rents editing setups and personnel for high-quality video editing.

Render: To execute any digital processing of an image at full quality. A dissolve between clips, for example, is ultimately rendered, frame-by-frame, by the computer.

Resolution: The ability of a video system to display fine image detail.

Ripple edit: An insert edit in which all the material following the inserted clip is moved up or back to accommodate it.

Rolling edit: An insert edit in which the clip following the inserted material is shortened or lengthened to accommodate it.

Sample rate: The number of times per second at which an analog signal level is inspected and digitally recorded; expressed in kilohertz (kHz).

Slide edit: An insert edit in which the out-point of the preceding clip and the in-point of the following clip are adjusted to accommodate the inserted material.

Slip edit: An insert edit in which the in- and out-points of the inserted material are adjusted, so that the preceding and following clips remain unchanged.

Superimposition: A video effect in which two or more images are displayed together in layers, such as the double exposure effect.

Three-point edit: An insert edit in which both the in- and out-points of the new material are specified, but only the in-point of the program into which it is inserted is specified.

Time base: The basis on which timecode is assigned to frames. The time base of NTSC video is 30 frames per second.

Review Questions

Answer the following questions on a separate sheet of paper. Do not write in this book.

1. What is *digital editing*?
2. Identify the basic operations in a typical workflow for digital editing.
3. Explain the difference between *capturing* and *importing*.
4. A(n) _____ is a small blemish on a digital image.
5. *True or False?* Insert editing usually follows the assembly editing process.
6. What are the three points that need to be specified to perform a three-point edit?
7. What is a *rolling edit*?
8. A(n) _____ edit preserves the in- and out-points of a new clip by shortening the clips on both sides.
9. Full-motion video requires how many frames per second?
10. What is a *cross fade*? How is it used in audio editing?
11. Storing a program means _____ it in a high-quality format.

12. Explain the difference between *program* files and *project* files.
13. What is *frame rate*? What is the NTSC frame rate for video?
14. What do the numbers in a timecode address mean?
15. Explain the purpose of "drop-frame" timecode.
16. How does *interlaced* video scanning operate?
17. What is the relationship between color fidelity and bit depth?
18. What is *sample rate*? How is this rate expressed?

STEM and Academic Activities

1. **Science.** As many programs on videotape are being converted for display and distribution on DVD or other tapeless media, old videotape cassettes are being discarded. What are the raw materials used to manufacture videotape cassettes? What, if any, of the components of a videotape cassette is biodegradable or recyclable?

2. **Technology.** Research the technology and equipment involved in video recording over the last 25 years. How have advancements in technology affected video equipment? How have the advances in video equipment and media affected the video editing process?

23 Mastering Digital Software

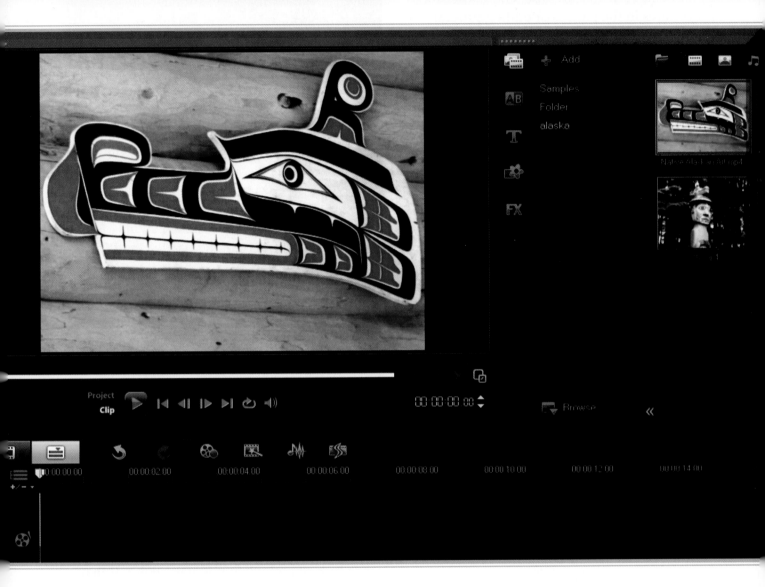

Objectives

After studying this chapter, you will be able to:

- Recall how to optimize your computer for digital editing.
- Identify sources of help for your editing program.
- Explain the use of common program operations.
- Understand the benefit of exploring more advanced functions and programs.

About Mastering Digital Software

The previous chapter (*Digital Editing*) covers the basic procedures common to all desktop postproduction systems. This chapter continues by showing how to learn, use, and master those procedures in the particular editing program that you have chosen. Although there are many such programs available, the discussion is illustrated with Corel® VideoStudio® Pro. If you will be using different software, you may wonder how this chapter can benefit you. The answer is that this is not primarily a chapter on how to use VideoStudio® Pro, but on how to approach *any* new editing software, using VideoStudio® Pro as a typical example. The goal is to communicate strategies for mastering editing software in general.

The discussion does not require access to VideoStudio® Pro. As you read, however, you may wish to explore suggested procedures on whichever editing program you are using.

Digital Editing Software

Learning any powerful software demands an investment in time, effort, and patience. Video editing software is particularly powerful because of the volume of data it handles and the variety of tasks it performs. No matter what the product package promises, you cannot expect to install the program and start editing immediately. So, no matter how eager you are to begin a project, you will avoid frustration if you budget the time and effort to learn your program first.

Some programs include simple options for beginners. In VideoStudio® Pro, selecting the Instant Project button on the main screen brings up a window in which to create quick and easy projects, **Figure 23-1**.

Different Learning Styles

How you actually learn your software depends on your personal style, and no single approach is superior to the others. You may read the whole manual first, and complete all

Figure 23-1 Selecting the Instant Project button on the VideoStudio® Pro main screen brings up the Instant Project work screen.

Instant Project

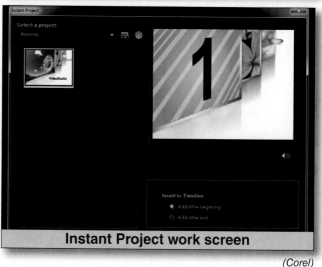

Instant Project work screen

(Corel)

the tutorials (if provided). You may start with simple projects and work toward more ambitious programs. Or, you may just fool around until you get the basic idea and then figure out procedures if and when you need them. If that works for you, fine. But, if you want to grow comfortable with your editing software quickly, you will probably benefit from a more systematic approach to learning it.

Preparing Your System

Today's computers have powerful processors, large memory and storage capacities, and fast system speeds, communication ports, and hard drives.

When desktop digital editing became available in the 1990s, a fast computer ran at 166 megahertz and a big hard drive held five whole gigabytes of data.

Even with all this power, video editing places severe demands on computers. When the computer cannot keep pace with the flow of data, video frames may be dropped and artifacts (visual blemishes) may mar the images. To prevent, or at least minimize these problems, you will want to optimize your computer for editing.

Recommended Hardware

First, you need two fast hard drives. Having two drives lets you install the editing software on the system drive, and use the other drive for storing the files needed for editing. That way, one drive can handle editing operations while the other manages data input and output. Dividing the tasks speeds up the process and reduces problems, **Figure 23-2**.

If you are capturing analog footage, you will need extra hardware to convert the *analog video* to a digital format for editing. Video capture cards can be installed in your computer, but they must be coordinated with the computer's sound card (or sound circuits on the motherboard)—often a tricky and time-consuming process. External capture hardware plugs into a computer port and is sometimes easier to use.

Even the best systems may drop occasional frames during capture. So, it is prudent to screen captured footage immediately and re-capture any shots with problems.

Video editing screens are packed with data. So, the bigger monitor you have the better. If your video card allows it, a second monitor can let you spread out your editing tools (**Figure 23-3**).

This works only with software that features "floating" windows, toolbars, preview screens, etc. You can also display other programs, like an asset management database (which is discussed later in this chapter).

System Tweaks

Today's do-everything operating systems can be set up in many different ways, and different editing programs can benefit from different adjustments. For example, VideoStudio® Pro works better if you make (or verify) these settings:

- *DMA enabled.* If you have older IDE-type hard drives, they will work faster when Direct Memory Access is enabled. (Today, most drives are SATA-type, rather than IDE.)
- *Write caching disabled.* Though this feature improves the performance of some operations, it is generally not suitable for capturing video.
- *Paging file set at twice RAM size.* If you have, say, 4 gigabytes of main memory, set the paging ("swap") file size to 8 gigabytes.

Figure 23-2 Using two hard drives improves capturing by dividing the workload.

Figure 23-3 Many graphics programs allow you to "float" elements to different positions, and even place them on a different monitor.

(Corel)

The Corel® VideoStudio® Pro user's guide recommends additional settings for *Windows®* systems running *Windows 7®* or later versions.

Other Programs

File *capture* and program rendering place heavy requirements on a system. So, all other programs should be closed during these operations.

If you are comfortable with computer management operations, you can also temporarily disable many of the programs that run automatically at startup (**Figure 23-4**).

However, the editing process itself is not quite as demanding. So, you may want to keep at least one support program open as you work. An up-to-date desktop computer will let you run, say, a graphics asset management program at the same time to help you track your video and photo files.

Reviewing the Basics

No matter what your learning style, you will be well-rewarded by an initial review of the materials supplied with your software. Even the best editing programs are not perfectly intuitive, and settings changed during hunt-and-peck explorations can have disagreeable effects on later projects.

Getting Started

If there is an installation guide, use it by all means. "Quick Start" guides, on the other hand, are compromises addressed to users who otherwise would not read anything at all. In adjusting your system, installing software, and configuring your program, it is better to follow detailed, step-by-step instructions.

Software Manuals

Program user's guides are useful for reference, but they offer only limited help in getting started. Their drawbacks are often inherent in the form itself. For example, the VideoStudio® Pro user's guide is well-organized and presented, but it omits many details and its index is incomplete.

VideoStudio® Pro software includes a user's manual as a digital PDF document. Because these virtual manuals can be searched using the PDF display software, information may be easier to find electronically than in the printed manual distributed with "boxed set" versions of the VideoStudio® Pro (**Figure 23-5**).

Most programs include direct links to their Internet sites. In VideoStudio® Pro, access is provided via arrow icons (**Figure 23-6**).

Many well-established software applications are supported by a wide variety of training and reference materials. Most large bookstores

Figure 23-4 Shareware programs can help you manage startup applications.

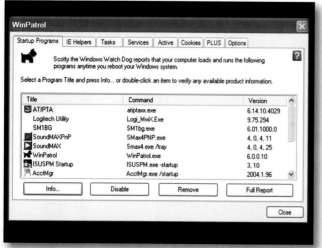

(WinPatrol)

Figure 23-5 Use the PDF search field instead of the user manual index.

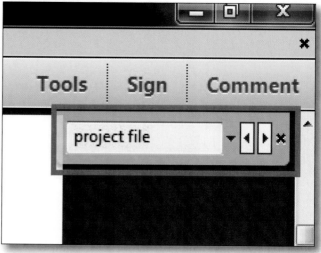

(Adobe® Systems, Inc.)

Figure 23-6 VideoStudio® Pro offers a window of additional resources on the Internet, accessed by the red dot (moved into this window from the far-right corner of the screen, for clarity).

(Corel)

carry third-party manuals for popular software programs. Review them carefully before selecting one, as they vary widely in quality and relevance to your needs. Third-party tutorials are sometimes available (usually on DVDs), but their quality may be doubtful and they are difficult to evaluate before purchasing.

The Internet offers a large selection of video editing tutorials. Corel® VideoStudio® Pro is the subject of many tutorials on individual topics, **Figure 23-7**.

You can also use a search service, such as google.com, to find independent websites and

user groups devoted to your software. These sites are often rich sources of information, advice, and support.

Learning the Interface

Like all desktop software, editing programs present you with a *graphical user interface*—a set of pictorial elements that you manipulate to execute commands. Because of the variety of postproduction tasks, editing program interfaces can look dismayingly complicated compared to the simpler screens of, say, a word processing program.

To get past this initial barrier, you need to learn your program's **interface**. If it has been well thought out, you will soon see the logic of each component, and moving around the screen will become intuitive.

Since examples using VideoStudio® Pro appear throughout the previous chapter, expect some content overlap in this chapter.

VideoStudio® Pro is a good representative example. Its interface has seven main parts (**Figure 23-8**):

- The step panel
- The menu bar
- The options panel
- The preview window
- The navigation panel
- The library
- The timeline

Figure 23-7 Many online tutorials are available for the various functions and features of Corel® VideoStudio® Pro.

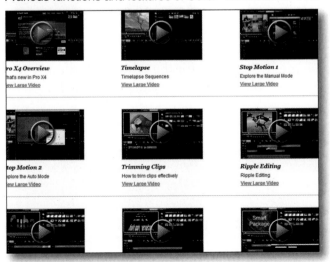

Figure 23-8 The VideoStudio® Pro interface. 1—The step panel. 2—The menu bar. 3—The options panel. 4—The preview window. 5—The navigation panel. 6—The library. 7—The timeline.

(Corel)

The Step Panel

The program organizes postproduction into three basic steps, which you perform more or less sequentially. Each tab on the step panel changes the interface to present the tools needed for the operations performed there, **Figure 23-9**.

In **Figure 23-10**, the Options tab (moved here from its actual position at the far right side of the screen) is explained below.

The Menu Bar

The standard *Windows®* drop-down menu bar provides access to commands and operations that you may need at different times during the editing process (**Figure 23-11**). As usual, the choices (here, "save trimmed video") that are not available for a certain operation are grayed out.

The contents of the step panel and menu bar remain unchanged, no matter which screen you are working on.

The Options Panel

The options panel includes information, settings, and controls appropriate to the operations you are performing. To allow more space for the clip library, the options panel is closed until opened by selecting its tab (**Figure 23-12**).

The Preview Window

The *preview window* always shows the graphic element you are working with, whether it is a clip, an effect, a filter, a title, or the assembled clips on the timeline (**Figure 23-13**). Here, the window shows a selected clip. The preview window can also show you the entire project. You *toggle* back and forth between

Figure 23-10 The options panel with the different sources from which footage can be captured.

(Corel)

Figure 23-11 The menu bar with the "File" menu opened.

(Corel)

individual graphic elements and the whole program by clicking a button to the left of the navigation bar—"Clip" for the individual element, "Project" for the whole program. The selected button turns white, **Figure 23-14**.

Figure 23-9 The step panel with "Capture" selected.

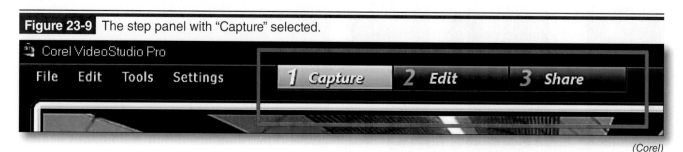

(Corel)

Figure 23-12 The options window open in capture mode. The small blue tab opens and closes it.

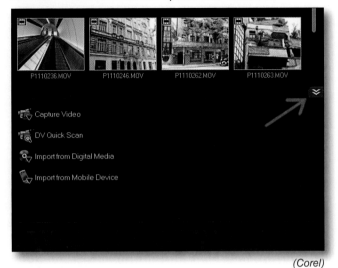

(Corel)

Some programs show individual elements and assembled clips in separate windows. The VideoStudio® Pro's single preview window requires switching back and forth, but simplifies the overall screen display.

The Navigation Panel

The navigation panel allows you to move (navigate) through the clip or program in the preview window, **Figure 23-14**.

When you are working in "capture" mode, the navigation panel is used to control the functions of a connected camcorder.

Figure 23-13 The preview window, with a clip displayed.

(Corel)

Figure 23-14 The navigation panel.

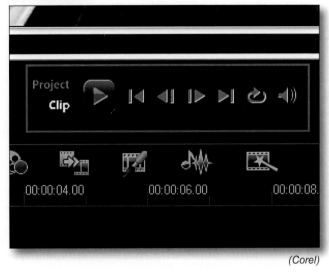

(Corel)

The Library

The library is the collection of resources (video clips, still graphics, titles, effects, audio, overlays) that you access from time to time as you work on your project (**Figure 23-15**). Like a physical library, this virtual one is organized into sections by subject, and each section contains many individual "books." These sections contain different types of program elements. Some items, such as video clips, still graphics, and audio segments, are imported into the library by the editor. Others, like *effects* and title *templates*, are part of the program.

The Timeline

The timeline provides a graphic representation of your program, with the first element on the extreme left and each succeeding element to the right, in order. As you play or *scrub* (scroll) forward through your project, timeline elements move off-screen to the left and new ones appear on the right(**Figure 23-16**).

The timeline extends the full width of the editing screen. **Figure 23-16** shows only the far left- and right-side portions.

Referring to **Figure 23-16**, the video, audio, title, and overlay tracks are identified by icons **(1)**. By dragging the white triangle **(2)**, you can move through your program, which displays on the preview screen. Numbers above the timeline **(3)** display timecode locations at every

Figure 23-15 The library. A—The icons on the left side of the library represent the library sections. B—The drop-down menu **(1)** selects the elements to be displayed in the library **(2)**.

A

B

(Corel)

point in the program, in hours/minutes/seconds/frames. Controls at the right side **(4)** let you change the scale of your timeline, and a timecode window **(5)** displays the total length of your program.

When first placing and arranging shots on a program, it is sometimes easier to work in **storyboard** mode (**Figure 23-17**). Select storyboard mode by clicking its icon and then dragging shots into a preferred order, like slides on a light table. Once you have selected and ordered your shots, you will usually return to timeline mode to adjust them.

The names and appearances of work areas in other editing programs may be somewhat different from those in VideoStudio® Pro, but their functions will be the same or similar.

Mastering Your Program

Once you know the layout of your program, you are ready to learn how to use it. And, there lies one difficulty with postproduction software—no two programs work quite alike. In word processing or spreadsheet analysis, you can move fairly easily from one program to another, because most tasks have the same labels and are performed in the same general way. In video editing, that is often not the case.

To overcome this problem, you need to know the basic operations *behind* the windows, labels, and buttons of the particular program

Figure 23-16 The left and right sides of the timeline.

left side   right side

1 2

3 4 5

(Corel)

Figure 23-17 Working in storyboard mode.

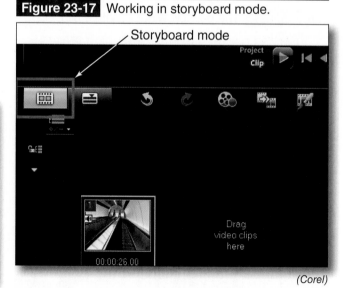

Storyboard mode

Drag video clips here

00:00:26.00

(Corel)

that you are using. If you understand the concepts underlying the actions, you can, later on, transfer your understanding to more complex software or programs offered by other vendors.

What follows, then, is the way in which the basic operations that are common to all editing software are organized and presented in one popular program, Corel® VideoStudio® Pro. Again, if you are working in a different program, you will want to translate the actions of each basic operation into the procedures that your software uses.

Since revised versions of software appear frequently, the illustrations in this chapter may differ slightly from the current version of VideoStudio® Pro.

Figure 23-18   From the Settings menu, select "Project Properties".

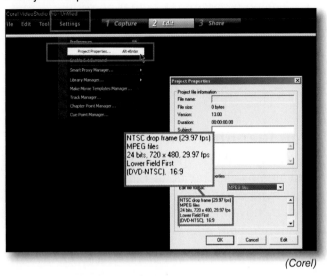

(Corel)

Starting a New Project

VideoStudio® Pro refers to postproduction on a video program as a *project*, because editing uses so many different kinds of content files and software operations, and because the end result of your work is not just a program. Instead, it is:

- The completed program *master* in the form of a "video file."
- The *program* in one or more digital formats.
- The *project* file, containing all the instructions for automatically creating the completed, full-quality program.
- The video, still photo, graphic, and audio *files* from which the program was created.

As we will see, the video file and the program are sometimes the same file.

The major steps in starting a new project include configuring it and capturing raw materials to use in it.

Configuring the Project

As noted in the previous chapter, each new video project is started by configuring it. With the capture step selected, (**Figure 23-9**) VideoStudio® Pro begins a new project at the main menu, where one choice under Settings is "Project Properties", **Figure 23-18**. Selecting this item opens a series of screens on which you set the technical parameters of your project.

Capturing Materials

Capturing raw footage is the next editing step in any project. So, VideoStudio® Pro devotes a screen to it. The Capture screen includes a field for entering information about each shot, **Figure 23-19**.

If you check the pull-down menu, you will notice that many selections are unavailable (grayed-out) because they do not apply to the capture process.

Once your program's raw materials are on your hard drive, you will switch from the Capture screen to the Edit screen for the bulk of the postproduction process.

Figure 23-19   The Capture screen information field. It extends the full width of the screen.

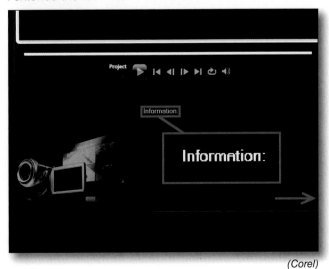

(Corel)

Small but Important Features

Editing programs, like other powerful applications, are well-supplied with unobtrusive controls that add versatility and/or convenience. A good editing program offers many of these small features, such as these typical examples from Corel® VideoStudio® Pro.

Icon Identification

Many settings, operations, and windows are activated by clicking icons. Since most of these small symbols are not labeled, you can identify them by hovering the cursor over them—a label will then appear briefly.

This label identifies the integrated paint program.

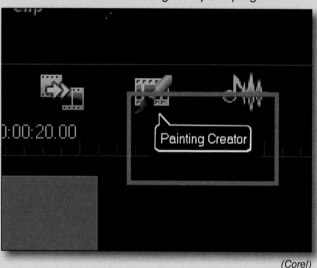

(Corel)

Full-Screen View

Many times, you want a closer view of a clip than the preview window can provide. To enlarge the view, click the big screen/small screen icon below the lower-right corner of the preview window.

The full screen option toggles between regular and large-size preview.

(Corel)

Audio Track Selection

The volume and stereo imaging controls operate on one audio track at a time. Clicking on the down-arrow to the right of the music symbol opens a list of available audio tracks.

Figure 23-9 includes the Edit tab, next to the highlighted Capture tab.

Editing Your Project

The previous chapter reviewed the basic operations in video postproduction. Now, we will see how those operations are carried out in VideoStudio® Pro. (Once again, if you are using different software, you can use what follows as typical examples.)

Assembling Your Clips

The first task is to select your video and/ or still clips, and place them in order on the timeline. After selecting and sequencing shots, you can *trim* them to exact length. On short projects, you will probably assemble all your clips. But with programs longer than, say, five minutes, it is generally easier to work one sequence at a time.

The arrow beside the music icon opens a track-selection window.

(Corel)

Clip Trim Window

By double-clicking on a clip in the library, you can open it in a special window that makes it easier to fine-tune the clip. Virtual jog-shuttle controls help you move frame by frame.

Keyboard Shortcuts

When you become proficient in using your software, you may find that keystrokes are faster than mouse operations. Every editing program includes many keyboard shortcuts.

You can obtain special keyboards for some of the more popular editing programs with the keystrokes already labeled and color-coded. Most include jog-shuttle controls similar to game controllers.

The clip trim window.

(Corel)

Keyboard shortcuts are listed in the VideoStudio® Pro's user's guide.

Navigation Panel shortcuts	
F3	Set mark-in
F4	Set mark-out
Ctrl + 1	Switch to Project mode
Ctrl + 2	Switch to Clip mode
Ctrl + P	Play/Pause
Space	Play/Pause
Shift + Play button	Play the currently selected clip
Home	Home
Ctrl + H	Home
End	End
	End

(Corel)

As you work, keep playing your program (selecting project in the navigation panel) to see how everything fits together. In most cases, you will find several places where edit points should be changed, or where a different shot might replace one previously selected. It is wise to spend as much time as you need on the assembly phase. Once you add transitions, overlays, titles, and several audio tracks, program changes become much more complex.

When you have placed and timed the shots in a program or sequence, you can make additions and adjustments. Video Studio® Pro calls the icons for these *transition*, *title*, *graphic*, and *filter* (**Figure 23-20**).

Inserting Transitions

Selecting the A|B icon displays the library of *transitions*—dozens of ways to signal a move from one sequence to another (**Figure 23-21**).

Figure 23-20 Edit screen icons for modifying shots.

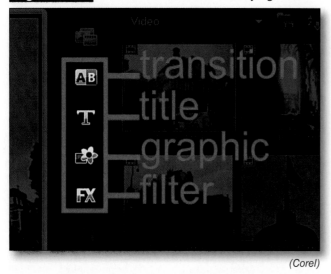

(Corel)

Figure 23-21 The AIB icon **(1)** opens the transitions library. The menu **(2)** allows you to select all transitions, or individual transition categories.

(Corel)

As noted in Chapter 21, *Editing Principles*, you will probably settle upon just a small number of transition types for any one program. To simplify access to them, you can add your selections to a "My Favorites" category (**Figure 23-22**).

Designing Titles

Titles, too, have their own work area. As you can see from **Figure 23-23**, you can choose one of several types of movement, or simply type your text in the preview window.

Some examples in **Figure 23-23A** look incomplete because the ***thumbnails*** are constantly moving.

Figure 23-22 You can copy the commands for selected transitions into the My Favorites category.

(Corel)

They were in transition when this screen capture was made.

When selected, all moving titles say "Lorem ipsum". You double-click these words and then replace them with your own title. It is important to understand that the title thumbnails control only the *style* of the movement. Once a style is selected, you can alter the movement itself, as well as the type (**Figure 23-24**).

Lorem ipsum is the start of a nonsense text widely used to show type fonts. It looks like Latin because it is based (very roughly) on a passage by Cicero.

When you have entered your copy, you can then design it just as you would in a graphic design program (**Figure 23-25**). In addition to the usual type design steps, you can set the title's duration **(1)** and screen position/alignment **(2)**. You can also choose many special color and border effects through the type effects drop-down menu **(3)**.

Using Graphics

Graphics is a catch-all term for non-photographic visual materials (**Figure 23-26**). The graphics library includes several types, including:
- solid colors for title backgrounds
- objects
- frames
- flash animations (**Figure 23-27**).

Figure 23-23 The title work area. A—Title design tools are accessed by opening the Options panel by its tab, off-screen to the right here. B—If you do not want moving titles, simply type them in the Preview window.

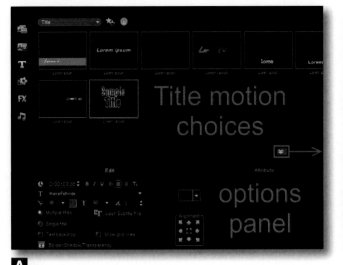

A

B

(Corel)

Figure 23-24 This moving title grows until it fills the screen.

(Corel)

In **Figure 23-27** (flash animation), the "heads-up display" overlay wavers slightly, as if the virtual "scanning device" were hand-held. The graduated background shown in **Figure 23-28** looks more professional than a plain color, and is easily created in a paint program.

Like the media and music libraries, the graphics categories can be added to by importing files to them.

Employing Filters

Filters are effects that can be laid over shots to change their character, **Figure 23-29**.

It would be impractical to discuss the dozens of different filters included in VideoStudio® Pro, but the FX Sketch filter is a typical example. This filter turns a video or still clip into a graphic,

Figure 23-25 The type effects drop-down menu **(3)** is one of those tools that are unlabeled.

(Corel)

Figure 23-26 Solid color graphics accessed by the graphics icon **(1)** and the selection menu **(2)**.

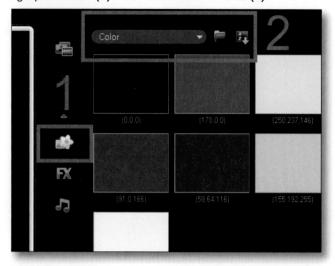

Figure 23-27 A flash animation **(1)** placed on the overlay track **(2)**.

(Corel)

Figure 23-28 A fountain fill background created in a paint program.

Figure 23-29 The filter library icon **(1)** opens a menu of 14 types of filter **(2)**.

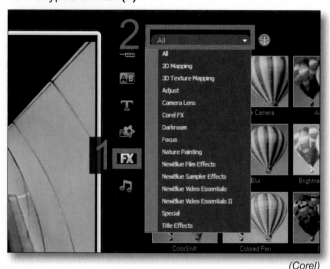

(Corel)

Figure 23-30. To create the effect, the live action clip was first cut into two halves. Next, the sketch filter was laid over the first half only, converting it into a graphic. Finally, a cross fade transition was inserted between the two halves of the shot.

In VideoStudio® Pro, the filter categories and names are not always self-explanatory (even the icon says "FX" rather than "filter"), and many of the filters are quite specialized. It is worth the effort to try each one out to discover how it modifies a clip.

Panning and Zooming

Panning and zooming around still photos and graphics is a popular technique that is available in VideoStudio® Pro.

This "digital rostrum" effect is discussed in Chapter 10, *Program Creation*. Note that pan-and-zoom works only with still images, not video clips.

This procedure is accessed from the Options tools in the photo library. You can select a pre-programmed zoom, or click "Customize" to create your own (**Figure 23-31**).

The Customize button opens the pan and zoom work window, **Figure 23-32**. The first cross (+) marks the center of the entire image **(1)**—the starting point for the pan and zoom movement. You drag the second cross **(2)** to the center of the end point of the movement. To zoom, you drag the handles on the marquee

Figure 23-30 The escalator shot is turned into a graphic.

A cross fade transition...   **...turns it back into a video clip.**

(Corel)

Figure 23-31 A—The pan and zoom function is accessed from the photo library. B—The Pan & Zoom tool **(1)** appears in the options panel for the still library. The Customize button **(2)** opens the work window.

A

B

(Corel)

Figure 23-32 The pan and zoom work window.

(Corel)

(3) to frame your new composition **(4)**. When you are satisfied with the effect, click the Okay button **(5)** to add the effect to the graphic.

Although the pan and zoom information remains with the graphic in the library, it does not automatically apply if the graphic is reused. To repeat the movement, the Pan & Zoom button must be selected each time the graphic is reused.

The versatility of the pan and zoom feature will make it worthwhile to explore thoroughly. For example, you can:

- Move the starting point cross and adjust the starting point marquee to begin with a different framing.
- Change the duration of the movement, which will change its speed.
- Add more crosses/marquees to continue the movement around the image.

If you are creating a slide video, frequent use of the pan and zoom effect can greatly add to the liveliness and interest of your program.

Compositing

As we have noted previously, *compositing* is an increasingly important part of video postproduction. Briefly, in compositing, computer software takes a subject recorded against a monochrome backing (usually a special green or blue) and replaces the monochrome backing color with another image (**Figure 23-33**).

VideoStudio® Pro calls this procedure *chroma key*, but *compositing*, *green screen*, and *blue screen* are now more common terms.

Here is the compositing procedure in VideoStudio® Pro (**Figure 23-34**):

1. The background clip is placed on the video track of the timeline **(1)**.
2. The foreground subject is placed below it on the first overlay track **(2)**.
3. The overlay track is selected by clicking, so it appears on the preview screen **(3)**.
4. The two clips are trimmed to the same length.

With the overlay track still selected, the Options panel is then opened to reveal the overlay options (**Figure 23-35**)—the Mask & Chroma Key tool is selected. The Options panel now displays the overlay tools. Select the Apply Overlay Options tool **(1)**. Then set the range of colors that will become transparent **(2)**.

Since monochrome screens can never be lit with perfect uniformity, some color variations always occur. This tool lets you widen the "definition" of the background color to include the variations. Selecting the eyedropper icon allows you to sample different shades on the background image.

When the settings are complete, the background image has replaced the overlay background color (**Figure 23-36**).

When you play the section with the Project button selected on the preview screen, the work marquees are invisible, as in the last image of **Figure 23-33**.

The range of effects you can achieve through compositing is limited only by your ingenuity.

Figure 23-33 The process of compositing.

The subject in front of a monochrome screen.

A background "plate" of full motion video.

The resulting composite.

(Corel)

Figure 23-34 The background and foreground clips in position on the timeline.

(Corel)

Figure 23-36 The completed composite image on the preview screen.

(Corel)

Designing the Audio

Corel® VideoStudio® Pro allows up to four stereo audio tracks: one on the timeline voice track and three on the music tracks. You can also have audio as part of the video track, and six video overlay tracks.

The seven tracks for visuals accept audio only as part of a video recording.

You will often want to "uncouple" an audio track from its video track (for example, when you are making split edits). To do this, you right-click on the track in the timeline and select "Split Audio" (**Figure 23-37**). You can

also use the "Split Audio" command in the Options panel.

VideoStudio® Pro includes many tools for shaping audio tracks, which are accessed in a number of different ways, as the sidebar explains.

Audio Mixing

If you need more than the seven audio tracks on the timeline, you can get around this limitation by pre-mixing two or more tracks. To do this, create a copy of your project (renaming it, of course), then remove all the audio tracks you do not want to mix, leaving only the ones to be combined. Then, select "Create Sound

Figure 23-35 The options window tools that appear when an overlay is selected on the timeline.

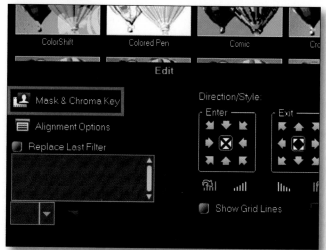

(Corel)

Partial Background Compositing

Unless you have a complete compositing studio, your background color will often be too small to fill the screen. Without a screen-filling backing, you cannot replace the whole background.

Subject in front of a painted background screen made from a 4′ × 8′ sheet of board.

Only the green area is replaced by the background clip.

(Corel)

To solve this problem, VideoStudio® Pro lets you *digitally* expand the undersized background as needed. In the Overlay Options tools, you adjust the "Width" and/or "Height" controls to "grow" the background area.

The background control tools in the overlay options window.

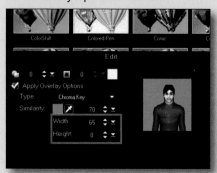

As the width control is moved, more of the background appears…

…until the entire background fills the screen.

(Corel)

Of course, it is important to keep the actual subject within the borders of the real green screen. In this example, the 4′ × 8′ vertical screen could be used for a full shot, as long as the subject kept her arms and legs inside the green area.

File" from the main menu in the Share window (**Figure 23-38**) and make a composite track. You can then reopen your original project, remove the audio tracks that you mixed, and substitute the new sound file or files. Depending on how many tracks you mixed, this can free up to six audio timeline rows for additional tracks.

Completing Your Project

When you have finished creating all the picture, titles, graphics, transitions, and audio tracks, you are ready to complete your project. So far, everything on the various tracks and in the preview window has been only a rough approximation of the finished program,

(Corel)

Figure 23-38 The Options panel for the Share screen.

(Corel)

constructed on the fly according to the blueprint called the "project file".

Different editing software packages call project files by different names. In VideoStudio® Pro, project files are easy to find on your drives because their names have a unique ".vsp" extension (for **V**ideo**S**tudio® **P**ro).

You can now play your program only on your editing computer, because doing so requires all the original files used to make it. To store your program and to show it elsewhere, you need to render a full-quality version of it as a single file. There are basically two ways to do this:

- Render a master program file in a form that can be viewed on a computer, stored, or transferred to other computers and file media, and converted to any format you might choose at a later time.
- Render the project immediately in a selected output format, such as *DVD*.

The master file option has the advantage of finishing your program in the highest-quality format that you will need. The disadvantage is that when exporting it later in DVD or another display format, you will have the extra first step of "transcoding" the master into the desired file type. On the other hand, if you render directly to DVD, for instance, you will not be able to later output a higher quality, such as *Blu-ray*™.

You can, indeed, make a Blu-ray Disc™, but the result will still be only DVD quality. The universal rule applies here: what you get out can be only as good as what you put in.

Either way, VideoStudio® Pro performs this task from the Share screen. As usual, the relevant commands are accessed by opening the Options panel.

Creating a Master Program File

A good choice for the master program file is what VideoStudio® Pro calls a "video file." Selecting the Create Video File option opens a menu of available program formats (**Figure 23-39**).

Of the many different file formats offered, MPEG-4 is a versatile choice. Selecting "MPEG-4" opens several different sub-formats. MPEG-4 HD is especially useful for preserving high quality. Selecting "MPEG-4 HD" opens a save menu with its own Options button for customizing your file, **Figure 23-40**. Name the file and press Save; the program creates and saves the file for you in 1080p format.

If your program is more than a few minutes long, this step will be time-consuming, even on a fast computer. Since it is completely automatic, you may want to do something else while you are waiting.

From the completed master file, you can then transcode your program into any of the available formats.

Audio Work Areas

The way in which VideoStudio® Pro distributes audio controls and commands is a good example of the individuality of different editing programs. In addition to commands accessible from the timeline and the main menu pull-downs, the program uses two different audio works areas.

The Audio Options Panel

The audio Options panel is opened, in the usual way, by clicking the options button. It contains two tabs, "Music & Voice" and "Auto Music".

The Music & Voice tab under audio options.

The Auto Music tab. (The Options button is actually at the far-right side of the screen.)

(Corel)

The Sound Mixer Panel

The Sound Mixer panel is opened by clicking the gold-colored icon underneath the preview screen. Its two tabs are "Attribute" and "Surround Sound".

You will notice that this arrangement has a few quirks:

- The tabs accessed through the Options panel appear only when the audio library is selected. However, the tabs accessed through the Sound Mixer icon can appear under any of the libraries.
- With Auto Music and Surround Sound, the active tools seem to appear under the inactive tab label.
- The purpose and even the name of the "Attribute" tab are not intuitively clear.

Since some of the more specialized audio commands may be unfamiliar to you, it would be useful to invest some time in learning where they appear and what they do.

The Attribute tab under Sound Mixer.

The Surround Sound tab. (The Sound Mixer button is actually to the left, below the preview window.)

(Corel)

In some cases, the tab tools are not below the active label. Unless you look carefully, these commands seem to be under Music & Voice.

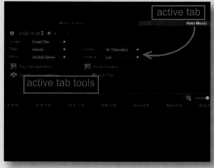

(Corel)

Figure 23-39 Selecting "Create Video File" opens its menu.

(Corel)

Figure 23-41 Options for uploading videos to the web.

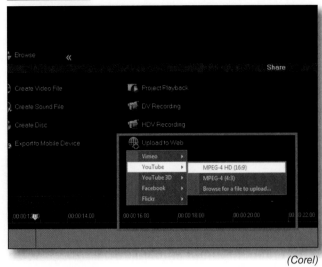

(Corel)

Figure 23-40 The MPEG-4 HD file save screen.

(Corel)

Figure 23-42 The finished YouTube version appears as a highlighted clip.

(Corel)

Creating an Output File

If you are sure you will need only a single file format for your project, you can skip creating a master file and finish your program directly in the format you choose, such as YouTube, for example (**Figure 23-41**). As usual, you select the format on the Options screen, customize the file format with the sub-options, and save the project, ready for uploading to the YouTube website. Also as usual, the finished program appears as a video clip (**Figure 23-42**).

Blu-ray Discs™ and DVDs

Blu-ray and DVD discs are perhaps the easiest and most popular ways to store programs and play them on TV sets, as well as computers.

For simplicity, both types will be referred to here as DVDs.

However, they cannot be made by simply copying files from a hard drive to a disc. The media files must first be transcoded into a special format that is readable by a DVD player. In addition, most DVDs are equipped with specialized menus that allow viewers to select various programs (or parts of programs) for playback.

Storing

Creating permanent storage for your completed program (also called "archiving") will protect your work if your hard drive crashes or if you wish to move to a different computer. Storing means copying every single file associated with the program (video, audio, still image, etc.) onto a portable storage medium, so that every element is in one place. This includes the project (.vsp) file and the master computer file and/or other output formats.

To save time later, you may want to copy each file to the portable medium *before using it in the project*. Then, use the copy when adding the elements to your program. When you have finished postproduction, these elements will already be located and assembled.

If you want to do more work on the program later, simply access the files on the portable media and open the project. The program will tell you that the source files are not where the project file "remembers" them, and will ask you to identify their current location.

Portable storage media evolve and change rapidly, so the following are only suggestions:

- **DVD or Blu-ray Discs™.** For small projects, these media work perfectly. Keep in mind though, that file storage methods typically become obsolete (for example, the floppy disks common some years ago are not playable on modern computers).
- **Thumb drives.** USB-connected thumb drives can hold all the files for longer programs.
- **Flash media.** The solid-state "data chips" used in still and video cameras also work well for storage.
- **Portable hard drives.** The simplest require no external power supply. Though more expensive, the diskless types with no moving parts may be expected to last longer.

Professional production companies typically use redundant storage for safety, backing up projects onto large servers, and also archiving them onto external media. If you have the resources to do this (and programs valuable enough to justify the expense) this is a prudent approach to long-term storage.

Many modern DVD players can play back *some* computer file formats and fetch some types of content from the Internet.

Creating a DVD in Corel® VideoStudio® Pro involves several steps:

1. Selecting and placing one or more programs (Corel calls them "clips," as if they were single shots).
2. Selecting and customizing a "top menu" that shows each program.
3. Creating chapter points for each program. Chapter points appear on the program's sub-menu. By selecting one, you can jump to its section of the video. (This step is optional.)
4. Rendering the final file in DVD playback format and *burning* the DVD playback file onto a disc.

Selecting Programs

You begin creating a DVD by opening the output format menu from the Share screen (**Figure 23-43A**). Selecting the "Create Disc" option opens a sub-menu with the various types of discs. Selecting "DVD" opens the DVD creation work screen, **Figure 23-43B**. Like the main program screen, the three tabs at the top show the sub-screens you will use. The Add Media tab **(1)** is automatically selected. Selecting the Add Video icon **(2)** opens a file menu containing the video files to be used **(3)**. Selecting the video files, in turn, places them on the DVD timeline, **Figure 23-43C**.

In this example, the two program files have been gathered in their own sub-directory.

Figure 23-43 Creating a DVD. A—Disc type sub-menus. B—Disc creation work screen. C—Programs placed on the timeline.

A

B

C

(Corel)

Creating Menus

Pressing Next at the bottom of the window (hidden by the drop-down menu in **Figure 23-43B**) then moves you to the Menu & Preview screen, **Figure 23-44A**. The Gallery menu of title templates is now open. Selecting a template (framed in yellow) adds your programs to the template.

In this example, "Native Alaskan Art" is the title of the first program on the disc. The characters "PRJ_20120310" will be replaced by the name of the DVD as a whole.

When the menu template has been selected, it can be extensively edited using the Edit menu (**Figure 23-44B**). You can change the background and the way it does (or does not) move, as

well as the size, position, type face, and color of each item on the menu you are building, and the music (if any) heard. When you have customized your DVD *menu*, you can preview it (**Figure 23-44C**) and then go back and adjust the customization as needed.

Unlike the tabs at the top of the VideoStudio® Pro main screen, the three tabs over the DVD creation work screen are for information only. You move back and forth through the process by selecting the Back, Next, or Close buttons on the bottom-right side of the work screen (**Figure 23-45**).

Adding Chapter Points

With the top menu designed, you can create sub-menus with chapter points for any

Figure 23-44 Creating menus. A—Gallery menu with title templates. B—Title Edit menu, before customization. C—Customized DVD title, as shown in the preview.

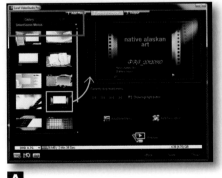
A

B

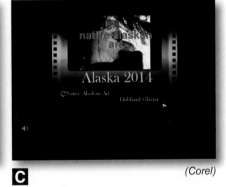
C

(Corel)

Figure 23-45 The navigation buttons.

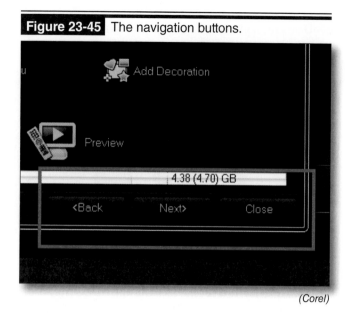

(Corel)

or all the programs on the disc. To begin, you select the Back button on the work screen.

Because this step is performed at the Add Media screen, this may seem like going backward. But since the sub-menus automatically take the same layout as the top menu, it is often easier to create the master layout first.

Chapters are segments of a single program. Chapter points mark the start of each segment. The Add/Edit Chapter command (**Figure 23-46**) on the Add Media screen opens the chapter work

Figure 23-46 Use the Add/Edit Chapter command to open the chapter work screen.

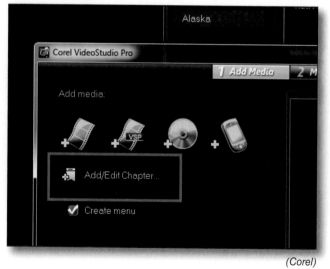

(Corel)

screen. Use the play commands (**Figure 23-47**) to set chapter points in a program:

- Use the slider arrow and jog wheel below the screen to move to each chapter point (**1**).
- Select the Add Chapter command (**2**).

A red vertical line marks the start of each new chapter.

For a faster (though less precise) method, just press the forward arrow (**1**) and click the Add Chapter command as the program reaches each point in turn.

To see the chapters you have chosen, press the OK button to return to the Add Media screen. Then, press Preview to open the preview screen so you can check your work (**Figure 23-48**). To check the results:

- Go to the video screen and select a program (here, it is "Native Alaskan Art").
- On the virtual remote control, select the Menu button (below Top Menu).

The preview screen will display the sub-menu with the chapters marked, **Figure 23-49**.

When you are satisfied, press the Back button to return to the Menu & Preview screen.

Figure 23-47 Use the controls set chapter points and add chapters to a program.

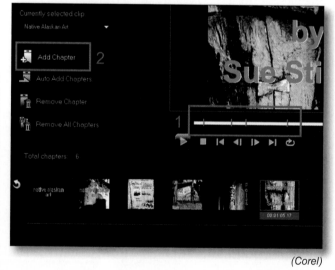

(Corel)

Figure 23-48 Preview the chapters chosen for your program.

Preview button

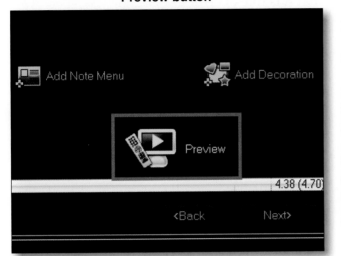

Preview screen

(Corel)

Figure 23-49 Program chapters listed.

(Corel)

Disc Burning

The last step in creating a DVD is to render the program in DVD file format and copy ("burn") it onto a DVD disc.

Some disc creation software treats rendering as a separate step.

When you press the Burn command (**Figure 23-50**), the program:

- Renders a final DVD-quality version of the programs, menus, and submenus from the elements in the video projects and the elements created in preparing the DVD. This process can be time-consuming, especially with longer programs.
- Copies the resulting DVD files to a disc.

Figure 23-50 Burn button.

(Corel)

Clicking Next on the Menu & Preview screen opens the disc burning work screen. For simple discs, pressing the Burn button begins the process. For more complex programs, you may wish to open the Burning Options sub-menu (**Figure 23-51**) by selecting the disc and "I" logo in the lower-left corner of the work screen (**1**). This allows you to set additional burning parameters (**2**).

Either way, when the rendering is complete, it is a good idea to play the disc in order to verify that it is free of errors.

The Preview button on the work screen does not display the rendered disc, but only the same rough version that you have been using.

Figure 23-51 Disc burning options.

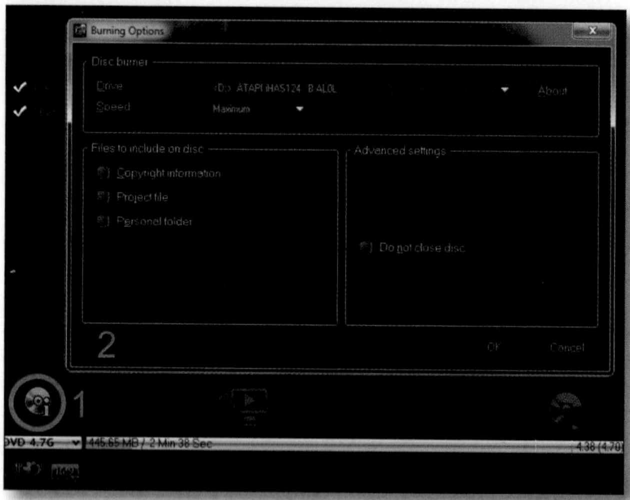

(Corel)

Wrapping Up

As you can see, VideoStudio® Pro packs a great deal of power into relatively few screens and menus. Though it takes some time to explore and learn all the features, this will be well worth your time and effort. But, you do not need to use this particular software in order to benefit from this chapter. We have covered universal editing procedures here by demonstrating them with a single, typical program.

Summary

- To operate editing software efficiently and effectively, optimize your computer for editing—both hardware and system settings.
- Learning resources for editing programs include software manuals, software publisher's online guides and tutorials, and many third-party tutorials available on the Internet.
- Familiarize yourself with the components and features of your editing software's interface, and moving around the screen will become intuitive.
- Starting a new project involves configuring the project and capturing the raw footage.
- The steps involved in editing a project include assembling clips, inserting transition, designing titles and graphics, applying effects, and designing the program audio.
- Creating a master file allows you to later transcode your program into any of the video file formats available in your editing program.
- Blu-ray and DVD discs are perhaps the easiest and most popular ways to store programs and play them on TV sets, as well as computers.
- Menus and chapters allow viewers to navigate the content of a Blu-ray or DVD disc.

Technical Terms

Analog video: Video that is recorded as a continuous electrical signal, rather than a series of codes.

Blu-ray Disc™: A very high-capacity disc for recording high-definition video and other large files.

Burning: Recording onto disc.

Capture: To copy footage from an external source to a computer drive; digitizing the footage if the original is analog video.

Compositing: Digitally replacing a solid color background with a different image.

DVD disc: A high-density recording medium capable of holding two hours or more of video.

Effect: Technically, any digital manipulation of the video, but commonly used to mean a scene-to-scene transition. Short for digital video effect (DVE).

Filter: A digital effect that changes the character (such as color) of the clip(s) it is applied to.

Graphic: Any non live-action visual element.

Interface: The method by which a user communicates with a computer.

Menu: In DVD authoring, a screen displaying buttons for navigating to other menus, programs, or functions.

Preview window: A window that displays what a piece of visual material looks (or will look) like.

Scrub: To move through a program by dragging a handle along a track below a preview window.

Storyboard: In editing, a graphic metaphor that represents the project clips as a succession of slides laid out in order.

Templates: Pre-designed menus that are customized by substituting appropriate buttons and text.

Thumbnail: A small image representing a clip.

Title: Any lettering that appears in a video.

Toggle: To switch back and forth between two settings, such as "off" and "on".

Transition: A visual effect signaling the end of one program part and the beginning of the next.

Trim: To indicate the in- and out-points of a clip in order to specify its content and set its exact length.

Review Questions

Answer the following questions on a separate sheet of paper. Do not write in this book.

1. Identify the recommended hardware to optimize your computer for editing.
2. *True or False?* In addition to the materials that accompany software, resources for many software applications can be found online and in independent bookstores.
3. A graphical user _____ is a set of pictorial elements that you manipulate to execute commands.
4. What is the *preview window*?
5. How are templates used in creating a program?
6. A(n) _____ signals a move from one sequence to another.
7. *True or False?* "Title" refers only to the name assigned to a program.
8. What are *filters*?
9. The compositing process in editing is also commonly called what other names?
10. If you require more audio tracks than your editing program allows on the timeline, how can you accommodate more tracks?
11. *True or False?* Both Blu-ray and DVD versions of a program can be made by simply copying files from a hard drive to the disc.
12. What are *chapters*?

STEM and Academic Activities

STEM

1. **Engineering.** Choose two (2) popular video editing software products that would be appropriate for an at-home, non-professional video maker. For each product, perform a cost-benefit analysis that considers the following factors:
   - The purchase price of the software.
   - The features and functions the software offers (including effects packages, file compatibility, and multiple audio and video tracks).
   - The cost of any hardware required to effectively operate the software.
2. Language Arts. The interface of a software application is often described as "intuitive". Research how software developers define an intuitive interface and explain what makes one interface more intuitive than another.

Glossary of Technical Terms

A

Act: A major section (usually between 10–45 minutes) of a longer program. (7)

Action line: An imaginary line separating camera and subject. Keeping the camera on its side of the line maintains screen direction. (19)

Additive editing: Creating a program from raw footage by starting with nothing and adding selected components. (20)

Amp (ampere): In lighting, the amount of electrical current drawn by a lighting instrument. (16)

Analog video: Video that is recorded as a continuous electrical signal, rather than a series of codes. (23)

Angle of view: The breadth of a lens' field of coverage; expressed as an arc of a circle, such as 10°. (13)

Aperture: The opening in the lens that admits light. The iris diaphragm can vary the aperture from fully open to almost or completely closed. (13)

Archiving: The process of assembling all the materials used to create a finished program and placing them together in permanent storage. (22)

Artifact: A spot or other blemish on a digital image. (22)

Aspect ratio: The relationship of an image's width to its height; traditional aspect ratio of video is 4 to 3, high-definition video formats are typically 16 to 9. (22)

Assembly editing: The process of building a program by selecting shots and placing them in order on the timeline. (22)

Associative continuity: Organizing a sequence of shots related by similarity of content. (21)

Asymmetrical balance: A composition in which dissimilar elements have equal "visual weight." (6)

Audio: 1—Collectively, the sound components of an audio-visual program. (8) 2—Sound as electronically recorded and reproduced. (17)

Auto exposure: An automatic camera control that regulates the amount of light admitted through the camera lens; may be disabled for manual operation. (2)

Auto focus: An automatic camera control that focuses the incoming light to keep the picture sharp and clear; may be disabled for manual operation. (2)

AV: Abbreviation for "audiovisual," a catch-all term for all nonfiction video genres. Pronounced "a-vee." (9)

Available light: The natural and/or artificial light that already exists at a location. (14)

B

Background light: A light splashed on a wall or other backing to lighten it and add visual interest. (15)

Background track: An audio track of the characteristic sounds of an environment, such as ocean, city traffic, or restaurant noises. (3)

Balanced line: A three-wire microphone cable designed to minimize electrical interference. (17)

Barn doors: Metal flaps in sets of two or four, attached to the front of a spotlight to control the edges of the light beam. (14)

Beat: A short unit of action in a program; often, though not always, corresponding to a scene. (18)

Bird's eye angle: An extremely high, vertical camera angle that simulates the view from a plane or high building. (7)

Bit depth: The number of digits (bits) used to encode each piece of digital data. 8-bit, 16-bit, and 24-bit are common bit depths. (22)

Blocking: The predetermined movements that a subject makes during a shot. Camera movement is also blocked. (13)

Blu-ray Disc™: A very high-capacity disc for recording high-definition video and other large files. (23)

Boom: A studio microphone support consisting of a rolling pedestal and a horizontal arm. (17)

Booming: Moving the entire camera up or down through a vertical arc. Also called *craning*. (19)

Breaking news: News that is covered as it is happening, or soon after. (10)

Brightness: The position of a pictorial element on a scale from black to white. (6)

Broad: Small, portable floodlight in the form of a shallow rectangular pan. (14)

Budgeting: Predicting the costs of every aspect of a production, and allocating funds to cover it. (11)

Buffer shot: A shot placed between two others to conceal differences between them. (21)

Buildup: A sequence of title cards, each one adding a new line of information. (10)

Bulb: See *incandescent lamp*.

Burning: Recording onto disc. (23)

Business: The activities a subject performs during a shot, such as writing a letter or filling a vase with flowers. (18)

C

Camcorder: An appliance intended solely for capturing sound and motion pictures (camera), and stores them on tape, disc, or other media (recorder). (1)

Camera angle: The position from which a shot is taken, described by horizontal angle, vertical angle, and subject distance. (7)

Camera dolly: A rolling camera support. A *cart dolly* resembles a wagon; a *pedestal dolly* mounts the camera on a central post. (12)

Camera stabilizer: A mechanical camera support system carried or worn by the operator that permits perfectly steady hand-held shooting. (12)

Capture: To copy footage/program material from an external source to a computer drive; digitizing the footage if the original is analog video. (22)

Cardioid: A spatial pattern of microphone sensitivity, named for its resemblance to a Valentine heart. (17)

Cast: Collective term that refers to all the on-camera performers in a production. (11)

CCD (charge-coupled device): The camera imaging chip that converts optical images into electronic signals. (12)

Century stand: A telescoping floor stand fitted with a clamp and usually an adjustable arm for supporting lights and accessories. Commonly referred to as a *C-stand*. (14)

CG: Short for computer graphics, which are visuals created (in whole or in part) with a computer rather than recorded by a camera. (3)

Character arc: The growth, development, or simply change in a character during the course of a script. (10)

Cheat: To move a subject from its original place (to facilitate another shot) in a way that is undetectable to the viewer. (4)

Cinema verité: A style of documentary that presents material with as little intervention by the program makers as possible. (3)

Clip: A unit of program material in a digital editing project. A clip may contain anything from part of one clip to several clips together. (22)

Codec: A particular computer protocol for encoding audio-visual data to record it and decoding it to display it. (11)

Color shot: A view of the scene, or a detail of it, not directly part of the action. (19)

Color temperature meter: A light meter that measures the relative blueness or redness of nominally "white" light. (14)

Color temperature: The overall color cast of nominally white light; expressed in degrees on the Kelvin scale—sunlight color temperature (5,200K) is cooler (more bluish) than halogen light color temperature (3,200K). (12)

Commercial: A very short program intended to sell a product, a person, or an idea. (3)

Composite: 1—A multilayer image that digitally combines foreground subjects with different backgrounds. Creating these images is called "compositing." (3) 2—Digitally replacing a solid color background with a different image. (23)

Composition: The purposeful arrangement of visual elements in a frame. (3)

Concept: The organizing principle behind an effective program. Often called an *angle*, *perspective*, or *slant*. (9)

Conflict: The struggle between opposing sides that creates dramatic action. (10)

Continuity: The organization of video material into a coherent presentation with continuous flow. (7)

Continuity script: The master document that shows, at a glance, what has been recorded, how it has been covered, and which takes are preferred. (20)

Contrast: 1—A technique of emphasizing one element in a picture by making it look different from the other elements. 2—The ratio of the brightest part of an image to the darkest. (6)

Cookie: A sheet cut into a specific pattern and placed in the beam of a light to throw distinctive shadows, such as leaves or blinds. (14)

Correcting: Making small continuous framing adjustments to maintain a good composition. (19)

Cover: Additional angles of the main subject or shots, like inserts and cutaways, recorded to provide the material needed for smooth editing. (18)

Craning: See *booming*.

Crew: Production staff members who work behind the camera. In larger professional productions, the producer, director, and management staff are not considered "crew." (11)

Cross-cutting: An editing technique that allows two actions to be shown at once by alternating back and forth between them—presenting part of one, then part of the other, then back to the first, and so-on. Can be used to show three or even more parallel actions. (5) Also called *intercutting* or *parallel cutting*.

Cut together: To follow one shot with another. Also, two shots are said to "cut together" when the edit is not apparent to the viewer. (7)

Cutaway: A shot other than, but related to, the main action. (18)

D

Day-for-night lighting: A method of shooting daylight footage so that it appears to have been taken at night. (16)

Decoding: The process of identifying and understanding the elements in a composition. (6)

Default: An action or condition that is selected automatically by the equipment, but the user may change it manually. (2)

Delivery system: The method by which a program will be presented (such as website, TV monitor, or kiosk), as well as the situation in which it will be watched (alone at a desk, in a training room, in a crowded store, etc.). (9)

Depth of field: The distance range, near-to-far, within which subjects appear sharp in the image. (13)

Device control: The ability of software to operate external hardware (such as a video camera) by remote control. (22)

Dialogue: Speech by performers on-screen; often as conversation between two or more people. (8)

Diffusion: White spun glass or plastic sheeting placed in the light path to soften and disperse it. (14)

Digital: To record images and sounds as numerical data, either directly in a camera or during the process of importing them to a computer. (1)

Digital assets: Electronic files of still and moving images. (20)

Digital intermediate (DI): Original camera film that is converted into ultra high-definition video for use in postproduction. (1)

Digital Video Effect (DVE): Any digitally created transitional device, other than a fade or dissolve. However, fades and dissolves are usually digital, too. (7)

Digital zooming: Increasing the subject size by filling the frame with only the central part of the image. (13)

Diopter correction: A lens in front of an internal viewfinder that can be adjusted to compensate for the eyesight of the operator. (12)

Disjointed time flow: The way in which video sequences are usually recorded in production; most programs are shot out of chronological order and reorganized during the editing process to follow script order in the finished video. (5)

Dissolve: A fade in that coincides with a fade out; the incoming shot gradually replaces the outgoing shot. Typically used as a transition between sequences that are fairly closely related. (7)

Documentary: A type of nonfiction program intended to communicate information about a real-world topic. (3)

Dollying: Moving the entire camera horizontally. Also called *trucking* and *tracking*. (19)

Drag: The resistance of a tripod head to panning or tilting. On professional tripods, the drag is adjustable. (12)

Dramatic action: The essential story line of a script; the plot. (10)

Dramatic structure: The organization of a story to build interest and excitement. (10)

Dress: To add decorative items to a set or location. Such items are called *set dressing*. (11)

Dutch: Referring to an off-level camera. The phrase, "Put dutch on a shot" means to purposely tilt the composition. (18)

DVD disc: A high-density recording medium capable of holding two hours or more of video. (23)

E

Editing: Creating a video program from production footage and other raw materials. (20)

Editorial documentary: A documentary with a specific position or point of view that attempts to persuade viewers, as well as inform them. (3)

Effect: Technically, any digital manipulation of the video, but commonly used to mean a scene-to-scene transition. Short for digital video effect (DVE). (23)

Emphasis: Calling attention to a pictorial element in a composition. (6)

Equalization: The adjustment of the volume levels of various sound frequencies to balance the overall mixture of sounds. (8)

Equalizer: A device for adjusting the relative strengths of different audio frequencies. (17)

Expressionism: A lighting style that adds a heightened emotional effect, without regard for lighting motivation. (15)

F

Fade in: A transition in which the image begins as pure black and gradually lightens to full brightness. Used to signal the start of a major section, such as an act or an entire program. (7)

Fade out: A transition in which the image begins at full brightness and gradually darkens to pure black. Used to signal the end of a major section, such as an act or an entire program. (7)

Fast motion: A video effect that contracts the screen time during which an action happens; created by recording a shot at a slower rate or by changing the frame rate digitally in postproduction. A typical fast motion recording speed would be 15 fps. Played back at the normal 30 fps cuts the apparent time in half. (5)

Field: One-half of a single frame of video, consisting either of the odd lines or the even lines. Two fields are interlaced to make a complete frame. (12)

File name: The identification of an editing element as it is stored in a computer. File names may or may not be identical to timecode addresses or slate numbers. (20)

Fill light: The light that lightens shadows created by the main (key) light. (15)

Film: An audiovisual medium that records images on transparent plastic strips by means of photosensitive chemicals. (1)

Filter: 1—An optical glass placed (usually) in front of the lens to modify the image being recorded. (12) 2—A digital effect that changes the character (such as color) of the clip(s) it is applied to. (23)

Filtering gel: See *sheet filter*.

Fishpole: A location microphone support consisting of a handheld telescoping arm. (17)

Flag: A flat piece of opaque metal, wood, or foam board that is placed to mask off part of a light beam. (14)

Flash media: Data storage on small, thin cards without moving parts. (12)

Flashback: A sequence that takes place earlier in time in the story than the sequence that precedes it. (5)

Floodlight: A large-source instrument that lacks a lens, but is still moderately directional; used for lighting wide areas. (14)

Fluorescent lamp: A lamp that emits light from the electrically charged gases it contains. (14)

Focal length: Technically, one design parameter of a lens, expressed in millimeters (4mm, 40mm). Informally, the name of a particular lens, such as a 40mm lens. (13)

Focus: 1—The part of an image (measured from near-to-far) that appears sharp and clear. 2—The object of a viewer's attention. (6)

Foley studio: An area set up for recording real-time sound effects synchronously with video playback. (17)

Following focus: See *pulling focus.*

Four-point edit: An insert edit that specifies the in- and out-points of both the source clip and the place in the program at which the clip will be inserted. (22)

Frame: 1—The border around the image. (3) 2—To *frame* something is to include it in the image by placing it inside the frame. (3) 3—A single film or video image. (4)

Frame off: To exclude something from an image by placing it outside the frame. (3)

Frame rate: The number of frames (images) per second at which video is recorded or displayed. Expressed as frames per second (fps). The frame rate of NTSC video is 29.97 frames per second. (5)

f-stop: A particular aperture setting. Most lenses are designed with preset f-stops of f/1.4, f/2, f/2.8, f/4, f/5.6, f/8, f/11, f/16, and f/22. (13)

G

Gaffer: The chief lighting technician on a shoot. (15)

Gag: Any effect, trick, or stunt in a movie. (4)

Gain: The electronic amplification of the signal made from an image, in order to increase its brightness. (13)

Genre: A specific type of program, such as story, documentary, or training. (9)

Glamorous lighting: Lighting that emphasizes a subject's attractive aspects and de-emphasizes defects. (16)

Graphic: Any non live-action visual element. (23)

Grip: Production staff member who practices many of the technical crafts associated with program production. (1)

H

Halogen lamp: A lamp with a filament and halogen gas enclosed in an envelope of transparent quartz. (14)

Head room: The distance between the top of a subject's head and the upper edge of the frame. (2)

High angle: Vertical camera angle in which the camera is positioned noticeably higher than human eye level. (7)

High-key: Lighting in which much of the image is light, with darker accents. (15)

I

Image stabilization: Compensation to minimize the effects of camera shake. *Electronic stabilization* shifts the image on the chip to counter movement; *optical stabilization* shifts parts of the lens instead. (12)

Image: A single set of visual information. An image may last for many frames, until the subject, the camera, or both, create a new image by moving. Most shots contain several identifiable images. (7)

Importing: Adding a file stored on the editing computer to the library of a project. (22)

Incandescent lamp: A lamp with a filament enclosed, in a near-vacuum, by a glass envelope. Also called a *bulb.* (14)

Incident meter: A light meter that measures illumination as it comes from the light sources. (14)

Infomercial: A program-length commercial designed to resemble a regular TV program. (3)

Infotainment: A form of documentary whose primary purpose is to entertain viewers. (3)

Insert: A close shot of a detail of the action; often shot after the wider angles, for later insertion by the editor. (19)

Instructional design: The craft of organizing an effective education or training script. (10)

Instrument: A unit of lighting hardware, such as a spotlight or floodlight. (14)

Intercutting: See *cross-cutting*.

Interface: The method by which a user communicates with a computer. (23)

Interlace: In video, the process of combining two fields to create one frame of video image. (12)

Iris (iris diaphragm): A mechanism inside a lens (usually a ring of overlapping blades) that varies the size of the lens opening (aperture). (13)

J

Jump cut: An edit in which the incoming shot is visually too similar to the outgoing shot. (7)

K

Key light: The principal light on a subject. (15)

L

Lamp: The actual bulb in a lighting instrument. (14)

Large-source light: A lighting instrument, such as a scoop or other floodlight, with a big front area from which light is emitted. Light from large-source instruments is relatively soft and diffused. (15)

Lavaliere: A very small omnidirectional microphone clipped to the subject's clothing, close to the mouth. (17)

Lead room: The distance between the subject and the edge of the frame toward which the subject is moving. (2)

Leading lines: Lines on the picture plane that emphasize an element by pointing to it. (6)

LED (light-emitting diode): A cool, low-power light source arranged in panels that are used like flood or soft lights. (14)

Letterboxed image: A widescreen image displayed in the center of a regular TV screen, with black bands above and below it filling the frame. (6)

Library: In editing software programs, a holding area for video, audio, and other elements to be used in a project. (22)

Library footage: Film or video collected, organized, and maintained to be rented for use in documentary programs. Also called *stock footage*. (10)

Line reading: A vocal interpretation of a scripted line that includes its speed, emphasis, and intonations. In simple terms, "You came back!" is one reading of a line, and "You came back?" is a different reading. (18)

Linear continuity: Organizing a sequence of shots chronologically and/or logically. (21)

Lines: Scripted speech to be spoken by performers. (18)

Live: A program that is recorded and, sometimes, transmitted for display continuously, in real-time. (1)

Look room: The distance between the subject and the edge of the frame toward which the subject is looking. (2)

Looping: Replacing dialogue in real-time by recording it synchronously with video playback. (17)

Low angle: A vertical camera angle in which the camera is noticeably positioned below human eye level. (7)

Low-key: Lighting in which much of the image is dark, with lighter accents. (15)

M

M-JPEG: A video compression codec that reduces the data required to record each frame. Common formats include .AVI and .MOV. (22)

Magic hour: The period of time, up to two hours before sunset, characterized by long shadows, clear air, and warm light. (16)

Magic realism: A lighting style that creates a dreamy or unearthly effect, often enhanced digitally in postproduction. (15)

Magnification: The apparent increase or decrease of subject size in an image, compared to the same subject as seen by the human eye. Telephoto lenses magnify subjects; wide angle lenses reduce them. (13)

Marks: Places within the shot where the performer is to pause, stop, turn, etc. These spots are identified by marks on the floor made of tape or chalk lines. (18)

Match point: The specific places in two shots with matching action, where the shots can be cut together to make the action appear continuous. (7)

Media opportunity: An event created specifically for the purpose of being covered by news organizations. (3)

Medium key: Lighting in which neither light nor dark tones dominate the image (a term used only in this book). (15)

Menu: In DVD authoring, a screen displaying buttons for navigating to other menus, programs, or functions. (23)

Metadata: Supplementary information stored in a still or video image file that is not visible in the actual image. (20)

Microphone (mike): A device that converts sound waves into electrical modulations for recording. (17)

Mixing: The blending of separate audio tracks together, either in a computer or through a sound mixing board. (8)

Montage: A brief, multilayer passage of video and audio elements designed to present a large amount of information in a highly condensed form. (21)

Motivated lighting: Lighting that imitates real-world light sources at a location. (15)

MPEG: A video compression system that records all the data of a "base frame". For subsequent frames, the system records only data that differs from that of the base frame. (22)

N

Narration: Spoken commentary on the sound track provided by a person not seen in the frame. (3)

Naturalism: A lighting style that imitates real-world lighting so closely that it is invisible to most viewers. (15)

Neutral angle: Vertical camera angle in which the camera is positioned more or less at human eye level. (7)

Neutral density filter: A gray filter that reduces the incoming light without otherwise changing it. Typically supplied in one- to four-stop densities (ND1 through ND4). (12)

O

Objective insert: A detail of the action presented from a neutral point of view. (18)

On-camera light: A small light mounted on the camera to provide foreground fill. (14)

Optical zoom: Changing a lens' angle of view (wide angle to telephoto) continuously by moving internal parts of the lens. (13)

P

Pan: 1—To pivot the camera horizontally (from side to side) on its support. (2) 2—In lighting, a large, flat instrument fitted with fluorescent or other long tube lamps. (15)

Parallel cutting: Presenting two or more sequences at once by showing parts of one, then another, then the first again (or a third) and so on. Also called *cross-cutting*. (21)

Parallel time flow: Form of video time progression in which two or more sequences of action happen at the same time. (5)

Parallel time streams: Two or more lines of action presented together, either by alternating pieces of the various streams or by splitting the screen for simultaneous presentation. (3)

Pedestaling: Moving the camera up or down on its central support. (19)

Perspective: The simulation of depth in two-dimensional visual media. (3)

Pickup: Video and/or audio material recorded later than the principal production, to add to or replace parts of material already recorded. (10)

Pickup pattern: The directions (in three dimensions) in which a microphone is most sensitive to sounds. (17)

Pictorial realism: A lighting style in which lighting, though motivated, is exaggerated for a somewhat theatrical effect. (15)

Picture plane: The actual two-dimensional image. (6)

Playback: 1—Previously recorded video and/or audio reproduced so that actors or technicians can add to or replace parts of it synchronously, in real-time. 2—Studio-quality music recording reproduced so that performers can synchronize lip movements with it while video recording. (17)

Plug-in: A special-purpose application that, when installed on an editing computer, is accessed directly from within the editing software. (22)

Point-of-view: A vantage point from which the camera records a shot. Usually abbreviated "POV." (7)

Polarizing filter: A rotatable filter that can suppress reflections and specular (very small) highlights, as well as darken blue skies. Also called a *polarizer*. (12)

Post house: Short for "postproduction house"; a professional facility that rents editing setups and personnel for high-quality video editing. (22)

Power switch: A camera control that turns the camera on and off. (2)

Practical: A lighting instrument that is included in shots and may be operated by the actors. (14)

Presence track: See *room tone*.

Presenter: An on-camera narrator who speaks directly to the viewer. (10)

Preview window: A window that displays what a piece of visual material looks (or will look) like. (23)

Pre-visualizing: The process of creating manual or computer images to plan shots and shot sequences prior to actually recording them. Often abbreviated as "previs." (9)

Production mixer (mixer): A device that balances the input strengths of signals from two or more sources, especially microphones. (17)

Production sound: The "live" sound recorded with the video. (8)

Profile angle: Camera angle in which the camera is placed at a right angle to the front angle shot. (7)

Program: Any complete video presentation, from a five-second commercial to a movie two or more hours long. (7)

Progressive scan: The process of displaying 30 complete frames of video per second. (12)

Properties: All the items that are called for in a script; anything that appears in a program. Usually shortened to *props*. (11)

Protection shot: A shot taken to help fix potential problems with other shots. (19)

Publish: To distribute a video program publicly by uploading it to a website. (9)

Pulling focus: Changing the lens focus during a shot to keep a moving subject sharp. Also called *following focus* and *racking focus*. (13)

R

Racking focus: See *pulling focus*.

Realism: A lighting style that looks like real-world lighting, but it is slightly enhanced for pictorial effect. (15)

Rear angle: Camera angle in which the camera is placed directly opposite its front position and fully behind the subject. (7)

Reconstruction: A re-enactment of past events that is recorded and used in a program. (10)

Record switch: A camera control that starts and stops the actual camera recording. (2)

Reflective meter: A light meter that measures illumination as it bounces off the subjects and into the camera lens. (14)

Reflector: A large silver, white, or colored flat surface used to bounce light onto a subject or scene. (14)

Releases: Legal documents granting permission to include people, places, objects, and music in a program. (11)

Render: To execute any digital processing of an image at full quality. A dissolve between clips, for example, is ultimately rendered, frame-by-frame, by the computer. (22)

Rendering: Creating a full-quality version of material that was previously edited in a lower quality form. (20)

Resolution: The ability of a video system to display fine image detail. (22)

Rim light: A light placed high and behind a subject to create a rim of light on the head and shoulders to help separate subject and background. (15)

Ripple edit: An insert edit in which all the material following the inserted clip is moved up or back to accommodate it. (22)

Roll to raw stock: To advance a videotape through previously recorded sections to a portion of blank tape in preparation for additional recording. (2)

Rolling edit: An insert edit in which the clip following the inserted material is shortened or lengthened to accommodate it. (22)

Room tone: A recording of the ambient background sounds in a studio or at a location that is used to fill gaps in the production sound track. Also called a *presence track*. (17)

Rostrum camera: A camera rigged for moving around still images to record different details of them. (Now largely replaced by manipulating images scanned into computers.) (10)

Rugged lighting: Lighting technique that emphasizes three-dimensional qualities and surface characteristics of a subject. (16)

Rule of thirds: An aid to pictorial composition in the form of an imaginary tic-tac-toe grid superimposed on the image. Important picture components should be aligned with the lines and intersections of the grid. (2)

S

Sample rate: The number of times per second at which an analog signal level is inspected and digitally recorded; expressed in kilohertz (kHz). (22)

Scale: The perception of the size of something by comparison to another object. (3)

Scene: A short segment of program content usually made up of several related shots that occur in a single place at a single time. (7)

Scoop: A large bowl-shaped floodlight used mainly in TV studios. (14)

Screen direction: The subjects' orientation (usually left or right) with respect to the borders of the screen. (19)

Screen time: The length of real-world time during which a sequence is displayed on the screen (in contrast to the length of video-world time that apparently passes during that sequence). (5)

Screen: A mesh material that reduces light intensity without markedly changing its character. (14)

Script: Full-written documentation of a program, including scenes, dialogue, narration, stage directions, and effects, that is formatted like a play. (9)

Scrub: To move through a program by dragging a handle along a track below a preview window. (23)

Sell: To add details in order to increase the believability of a screen illusion. (4)

Sequence: A segment of a program, usually a few minutes long, consisting of related, organized material or scenes. (3)

Sequencing: Determining the order in which individual shots are placed. (3)

Serial time flow: Form of video time progression in which a single series of events moves forward in a single stream. (5)

Setting focus: Adjusting the lens to make the subject appear clear and sharp. (13)

Setup: A single camera position, usually including lights and microphone placements, as well. (4)

Sheet filter: In lighting, a sheet of colored or gray-tinted plastic placed over lights or windows to modify their light. Also called a *filtering gel*. (14)

Shoot: To record film or video. Also, "a shoot" is an informal term for the production phase of a film or video project. (1)

Shot: An uninterrupted recording by a video camera. (3)

Shutter: The electronic circuitry in a camera that determines how long each frame of picture will accumulate on the imaging chip before processing. The standard shutter speed is 1/60 per second for NTSC format video. (13)

Signpost: In a training or documentary program, a reminder to viewers of what has been covered and what will come next. (10)

Silk: A fabric material that reduces both light intensity and directionality, producing a soft, directionless illumination. (14)

Slate: A written (video) and/or spoken (audio) identification of a recorded program component, such as a shot, a line of narration, or a sound effect. (17)

Slide edit: An insert edit in which the out-point of the preceding clip and the in-point of the following clip are adjusted to accommodate the inserted material. (22)

Slip edit: An insert edit in which the in- and out-points of the inserted material are adjusted, so that the preceding and following clips remain unchanged. (22)

Slow motion: A video effect that expands the screen time during which an action happens; created by recording a shot at a faster rate or by changing the frame rate digitally in postproduction. (5)

Small-source light: A lighting instrument with a small front area from which light is emitted, such as a spotlight. Light from small-source instruments is generally hard-edged and tightly focused. (15)

Softlight: A lamp or small light enclosed in a large fabric box, which greatly diffuses the light. (14)

Sound effects: Specific noises added to a sound track. (8)

Sound: The noises recorded as audio. (8)

Specular reflections: Hard, bright reflections from surfaces, such as water, glass, metal, and automobile paint, that create points of light on the image. Often controllable by a polarizer. (13)

Speed: The light-gathering ability of a lens; expressed as its maximum aperture. Thus, an f/1.4 lens is "two stops faster" (more light sensitive) than an f/2.8 lens. (13)

Spin: The management of information for somebody's benefit. (3)

Split edit: An edit in which the audio and video of a new shot do not begin simultaneously. When "video leads," the sound from the preceding shot continues over the visual of the new shot. When "audio leads" the sound from the new shot begins over the end of the preceding visual. (5)

Spotlight: A small-source lighting instrument that produces a narrow, hard-edged light pattern. Also called a *spot*. (14)

Staging in depth: Positioning subjects and camera to exploit perspective in the image. (6)

Standup: A report presented on camera, usually by a reporter. (10)

Stock shot: A shot purchased from a library of pre-recorded footage for use in a program; collectively called *stock footage*. (3) See *library footage*.

Storyboard: 1—Program documentation in graphic panels, like a comic book, with or without dialogue, narration, stage directions, and effects. (9) 2—In editing, a graphic metaphor that represents the project clips as a succession of slides laid out in order. (23)

Straight cut: 1—An edit in which audio and video change simultaneously. 2—An edit that does not include an effect, such as a fade or dissolve. (8)

Subjective continuity: Organizing a sequence according to the feelings of the director/editor, rather than by chronological or other objective criteria. (21)

Subjective insert: A detail of the action presented from a character's point of view. (18)

Subtractive editing: Creating a program by removing redundant or poor-quality material from the original footage, and leaving the remainder essentially as it was shot. (20)

Superimposition: A video effect in which two or more images are displayed together in layers, such as the double exposure effect. (22)

Symmetrical balance: A composition in which visual elements are evenly placed and opposed. (6)

T

Tabletop: Cinematography of small subjects and activities on a table or counter. (16)

Take: A single attempt to record a shot. (3)

Talent: Every production member who performs for the camera. (11)

Tally light: A small light on a camera that glows to indicate that the unit is recording. (18)

Target audience: A specific group of viewers for whom a program is designed. (9)

Telephoto: A lens setting that magnifies distant subjects and reduces apparent depth. (3)

Telephoto lens: A lens (or a setting on a zoom lens) that magnifies subjects and minimizes apparent depth by filling the frame with a narrow angle of view. (7)

Teleprompter: A machine that displays text progressively as a performer reads it on-camera. (18)

Television: Studio-based, multi-camera video that is often produced and transmitted "live." (1)

Templates: Pre-designed menus that are customized by substituting appropriate buttons and text. (23)

Tent: A lighting arrangement in which white fabric is draped all around a subject to diffuse lighting completely for a completely shadowless effect. (16)

The Three Ts: Informal name for the basic organization of training programs—**Tell** them what you will tell them; **Tell** them; **Tell** them what you have told them. (10)

Three-point edit: An insert edit in which both the in- and out-points of the new material are specified, but only the in-point of the program into which it is inserted is specified. (22)

Three-point lighting: Classic subject lighting technique that consists of key, fill, rim, and background lights. (16)

Three-quarter angle: Camera angle in which the camera is placed roughly between 15°–45° around to one side in front of the subject. (7)

Three-quarter rear angle: Camera angle in which the camera is placed about 45° around the back side of the subject, so the subject is facing away from the camera. (7)

Throw: A lighting instrument's maximum effective light-to-subject distance. Also, that distance in any particular setup. (The term "throw" is also used to describe camera-to-subject distance.) (15)

Thumbnail: A small image representing a clip. (23)

Tilting: Vertical camera movement to point the camera lens up or down. (6)

Time base: The basis on which timecode is assigned to frames. The time base of NTSC video is 30 frames per second. (22)

Timecode address: The unique identifying code number assigned to each frame (image) of video. Timecode is expressed in hours, minutes, seconds, and frames, counted from the point at which timecode recording is started. (20)

Timeline: The graphic representation of a program as a matrix of video, audio, and computer-generated elements. (20)

Title: Any lettering that appears in a video. (23)

Toggle: To switch back and forth between two settings, such as "off" and "on". (23)

Tracking: See *dollying*.

Training program: An informational program designed to teach specific subjects. (10)

Transcoding: The process of translating material from one codec into another. (11)

Transducer: The component of a microphone that converts changing air pressure ("sound") into an electrical signal ("audio"). (17)

Transition: A visual effect signaling the end of one program part and the beginning of the next. (23)

Treatment: A written summary of a program that is formatted as narrative prose; may be as short as one paragraph or as long as a scene-by-scene description. (9)

Trim: To indicate the in- and out-points of a clip in order to specify its content and set its exact length. (23)

Trimming: Removing unwanted material from the beginning and/or end of a shot. (3)

Tripod: A three-legged camera support that permits leveling and turning the camera. (12)

Trucking: See *dollying*.

U

Umbrella: A fabric-covered umbrella frame. Metallic cloth umbrellas are used to reflect light onto subjects; white cloth models can also filter lights placed in back of them. (14)

Unbalanced line: A two-wire microphone cable that is subject to electrical interference, but is less bulky and expensive than balanced line; for use in amateur applications. (17)

V

Video: An audiovisual medium that records on a magnetic tape or digital storage media by electronic means. Also, single-camera program creation in the manner of film production, rather than studio television. (1)

Visual literacy: The ability to evaluate the content of visual media through an understanding of the way in which it is recorded and presented. (1)

Voice-over: 1— Spoken commentary on the sound track from someone who is not in the image. (3) 2—Narration or dialogue that is recorded independently and then paired with related video. (8)

Voltage: The electrical potential or "pressure" in a system—typically 110 or 220 volts in North America. (16)

W

Wallpaper: Footage intended to take up screen time while the narration presents material that cannot be shown. (10)

Wattage: In lighting, the power rating of a lighting instrument. 500, 750, and 1,000 watt lamps are common. (16)

White balance: A camera control system that neutralizes the color tints of different light sources, such as sunshine and halogen lamps, and matches the camera to the overall color quality of light in the shooting environment. (2)

Wide angle: A lens setting that reduces subjects in size and exaggerates apparent depth. (3)

Wide angle lens: A lens (or a setting on a zoom lens) that minimizes subjects and magnifies apparent depth by filling the frame with a wide angle of view. (7)

Widescreen video: A video image or video display screen proportioned 16 to 9, in contrast to the traditional TV screen's 4 to 3. (6)

Wipe: A transition between sequences in which a line moves across the screen, progressively covering the outgoing shot before it with the incoming shot behind it. (7)

Worm's eye angle: An extremely low camera angle in which the camera "looks" dramatically upward at the subject. (7)

Z

Zoom control: A camera control that zooms the camera lens in to fill the screen with a narrow portion of a scene, or zooms the camera lens out to fill the screen with a wider portion of the scene. (2)

Zoom: Change the focal length of the camera lens while recording to magnify or reduce the size of an element in or a portion of a shot. (2)

Index